W9-BIZ-287

Names and Formulas of Some Common Ions

Positive Ions		Negative Ions	
Ammonium ion	NH_4^+	Acetate ion	$C_2H_3O_2^-$
Copper(I) ion (cuprous ion)	Cu^+	Bromide ion	Br^-
Hydronium ion	H_3O^+	Chlorate ion	ClO_3^-
Potassium ion	K^+	Chloride ion	Cl^-
Silver ion	Ag^+	Cyanide ion	CN^-
Sodium ion	Na^+	Fluoride ion	F^-
Barium ion	Ba^{2+}	Hydrogen carbonate ion (bicarbonate ion)	HCO_3^-
Calcium ion	Ca^{2+}		
Cobalt(II) ion	Co^{2+}	Hydrogen sulfate ion	HSO_4^-
Copper(II) ion (cupric ion)	Cu^{2+}	Hydroxide ion	OH^-
Iron(II) ion (ferrous ion)	Fe^{2+}	Iodide ion	I^-
Lead(II) ion	Pb^{2+}	Nitrate ion	NO_3^-
Magnesium ion	Mg^{2+}	Nitrite ion	NO_2^-
Manganese(II) ion	Mn^{2+}	Permanganate ion	MnO_4^-
Mercury(I) ion (mercurous ion)	Hg_2^{2+}	Carbonate ion	CO_3^{2-}
Mercury(II) ion (mercuric ion)	Hg^{2+}	Chromate ion	CrO_4^{2-}
Nickel(II) ion	Ni^{2+}	Dichromate ion	$Cr_2O_7^{2-}$
Tin(II) ion	Sn^{2+}	Oxalate ion	$C_2O_4^{2-}$
Zinc ion	Zn^{2+}	Oxide ion	O^{2-}
Aluminum ion	Al^{3+}	Sulfate ion	SO_4^{2-}
Chromium(III) ion	Cr^{3+}	Sulfide ion	S^{2-}
Iron(III) ion (ferric ion)	Fe^{3+}	Sulfite ion	SO_3^{2-}
		Phosphate ion	PO_4^{3-}

Introduction to Chemistry

Introduction to Chemistry

Second Edition

by

T. R. Dickson

JOHN WILEY & SONS, INC.

New York | London | Sydney | Toronto

This book was set in Century Schoolbook by
Progressive Typographers. It was printed and
bound by Halliday Lithograph Corporation.
The cover was designed by Eileen Thaxton.
The drawings were designed and executed by
John Balbalis with the assistance of the
Wiley Illustration Department.

Cover Photograph: Courtesy of Bausch & Lomb

Copyright © 1971, 1975, by John Wiley & Sons, Inc.

All rights reserved. Published simultaneously in Canada.

No part of this book may be reproduced by any means, nor
transmitted, nor translated into a machine language without the written permission of the publisher.

Library of Congress Cataloging in Publication Data:

Dickson, Thomas R
  Introduction to chemistry.

  1. Chemistry.   I. Title.
QD31.2.D53   1975   540   74-19024
ISBN 0-471-21289-X

Printed in the United States of America

10-9 8 7 6 5 4 3 2 1

To Students

Preface

This book is intended for a one-semester course in chemistry. It presents the foundations of chemistry to the student who needs preparation for further study and to the student who wishes to take only an introductory course.

Important basic concepts are developed carefully at an introductory level so that the student will have a clear understanding of some of the foundations of chemistry. The major objectives are presented at the beginning of each chapter. The chapters form a logical sequence which can, however, be altered in accordance with the individual preference of the instructor.

The concepts of measurement and units are introduced early. The discussion of atomic theory leads to a definition of the mole. Since the mole concept is fundamental to any meaningful study of chemistry, it is carefully developed and illustrated. Atomic structure is introduced from an orbital but nonmathematical point of view. Chemical bonding is developed to provide a view of molecules and ions. The significance of chemical formulas and nomenclature is explained. A dynamic model for the behavior of matter is continuously developed from the description of the dynamic structure of the atom to the dynamic phenomena involved in the gas, liquid, solid, and solution phases. The idea of dynamic equilibrium is introduced from both a physical and a chemical viewpoint. Approaches to solving chemical problems by reasoning and the unit equation or factor-label method are explained in detail, and numerous illustrative examples are given. Examples of problem solving are presented and explained as often as possible.

This edition is based on five years of teaching from the first edition. The changes were stimulated by numerous comments by students who used the first edition and by instructors who taught from it. The fundamental approach of the first edition has been retained; that is, this book is intended to be a first introduction to chemistry to prepare the students for further study in this field. A previous background in chemistry is not assumed. My purpose is to teach chemistry by providing students with a dynamic and conceptual view of matter and the environment and by introducing them to the logical methods used in chemistry.

The fundamental chemical topics discussed here are essentially the same as the topics of the first edition. In many instances, more emphasis has been placed on those important topics, and new examples and analogies have been included for improved clarity. For example, the discussions of chemical equilibrium and Le Chatelier's principle have been expanded and illustrated with examples. The concepts of pH and buffers are covered more thoroughly. Some topics have been placed in the appendixes for use at the discretion of the instructor. As often as possible, topics of interest (such as toxic metals, air pollution, energy problems, and water pollution) have been added to relate chemistry to current environmental dilemmas. Finally, the chapter objectives have been revised, and new end-of-chapter problems have been developed to emphasize familiar chemicals and topics.

I thank all of the students, instructors, and individuals who gave me the justification for this second edition and the energy to write it. I especially thank Philip Meyer of Skyline College, Jack Powell of Iowa State University, Alan Cunningham of Monterey Peninsula College, James Larsen of Santa Barbara City College, John Ingangi of the University of Maryland, Helen McAferty of Orange Coast College and Ardas Ozsogomonyan for their critical comments and reviews.

Aptos, California, 1974   *T. R. Dickson*

Note to Students

Chemistry includes a vast number of observed facts and many theories. Physical chemistry deals with the physical phenomena associated with matter. Analytical chemistry is the branch of chemistry that deals with the determination of the composition of types of matter. Organic chemistry involves the study of the compounds of the chemical element carbon. Inorganic chemistry involves the study of the chemical elements other than carbon. Biochemistry is the branch of chemistry that deals with the chemistry of biological processes. Chemistry is important in many fields of study. Physics and chemistry are closely related sciences, and chemistry is fundamental to modern biology. Agricultural science, medical science, oceanographic science, engineering, space science, and environmental science all utilize chemistry.

This book is designed as a first introduction to chemistry. From it, you can learn the basic fundamentals of chemical science and an understanding of your environment. Each chapter begins with a list of objectives. These will describe the content of the chapter. Read the objectives and keep them in mind as you read. After completing the chapter, go back and check to see if you have satisfied each objective. Your instructor may choose to indicate which objectives he or she thinks are most important. A topic like chemistry involves many facts and concepts. It is important to read the text carefully and to underline or take notes on the material. Furthermore, if you are looking for a specific topic, look in the table of contents or the index.

T. R. D.

Contents

1

Chapter 1: Chemical Concepts and Measurements

Objectives

The student should be able to:
1 List the basic metric units and the common metric prefixes.
2 Convert a metric measurement made in one unit to another unit.
3 Convert a measurement made in English units to metric units and vice versa.
4 Convert a measurement of volume or area from one multiple unit to another.
5 Calculate the density of a substance from given data.
6 Use the density of a substance as a property conversion factor.
7 List the common temperature scales.
8 Convert a temperature measurement from one scale to another.

Terms to Know

Science
Technology
Chemistry
Liter
Density
Matter
Mass
Energy
Calorie

1

Chemical Concepts and Measurements

1-1   Science and Technology

Science is the attempt to describe physical reality in a systematic and logical manner. Physical reality refers to the physical objects that exist and phenomena that occur within our space-time environment. The practice of science seems to spring from the desire of humans to know about and describe physical reality and to accumulate knowledge. Scientific thought has evolved within the realm of human thought to the highly organized, extensively communicated, and somewhat domineering discipline that includes all fields of science today.

Pure science and scientific research are, in a sense, concerned only with the development of new knowledge. However, the evolution of scientific knowledge provides a better understanding of physical reality and reveals ways in which the environment can be manipulated to serve humans. **Technology** is the application of science to the manipulation of the physical environment especially for industrial or commercial purposes. It can be argued that technological advances have improved the lives of humans, but it also can be argued that technology has been used to destructive ends.

It is important to distinguish between pure science and technology, which is applied science. These two aspects of science are closely allied and practically inseparable. That is, many scientific advances have stimulated technological applications of the advances. Similarly, many technological advances have stimulated scientific thought. For instance, scientific knowledge of the chemical function of the body has stimulated the technology of the production of synthetic chemicals to fight disease.

An example of the influence of technology on science is the development of the steam engine, which stimulated the creation of a field of scientific thought called thermodynamics.

Throughout the ages, the application of science has been a two-edged sword. The discovery of the technology of the refining of iron ore to make iron established the iron age, and iron serves as a backbone of the industrial societies that exist today. However, iron brought new dimensions to the production of weapons of war. The technological production of ammonia is necessary to the manufacture of fertilizers needed by a hungry world, but the technique was developed for purposes of making explosives. The development of the knowledge of nuclear energy has led to many promising uses of this energy, but the development was stimulated by the quest for nuclear bombs as the ultimate weapons of war. The list of useful and reprehensible applications of technology could go on and on. It is true that technology can be used for constructive and destructive purposes, but certainly the decisions concerning such uses are made by humans.

It is somewhat nonproductive to argue whether or not science and technology have been responsible for the many environmental dilemmas that exist today. It is important to realize that scientific advances and the proper use of technology will play an important part in the resolution of environmental problems.

1-2   Chemistry

Chemistry as a science is based on observations of events that occur and materials or matter that exist within our space-time environment. **Chemistry** is a dynamic science that deals with the structure and behavior of matter. Chemistry involves finding out what things are made of and how they undergo changes. We are surrounded by material objects, and human beings as living organisms have a material existence. Chemistry is concerned with investigating the nature of all matter ranging from substances such as water to complex biological material such as deoxyribonucleic acid or DNA. The science of chemistry has developed from the desire of humans to describe what objects are made of and how the structure of these objects causes them to have certain properties. From a practical point of view, desirable properties of matter are sought. Substances that are useful — to cure disease, explode, intoxicate, and smell or taste good — and capable of being fabricated into things — clothing, utensils, and tools — are isolated from nature or manufactured from other substances. Chemists explore nature and experiment with substances to develop and refine theories concerning the structure and behavior of matter.

We all benefit from the practical applications of chemistry. Synthetic and natural chemicals are used as drugs and medicines. Agricultural productivity is greatly enhanced by the use of chemical fertilizers and pesticides. Many important consumer products such as foods, gasoline, and plastics are derived from chemical processing. Actually, modern living would not be as convenient as it is withoug chemical technology.

Moreover, we are learning that when we utilize substances in our environment for convenience, the environment is altered and can become polluted. An understanding of environmental problems requires knowledge of the chemical processes involved. Furthermore, solutions to problems will require the development of appropriate chemical technology. It is important to learn some chemistry to help understand the nature of our environment and the forces that threaten it.

1-3   Matter and Mass

We use our senses—sight, hearing, taste, smell, and touch—to observe those objects that surround us and the changes that continuously occur in and around us. We see the sizes and colors of objects and see changes in the positions of objects (i.e., a moving automobile). We hear the sounds of certain phenomena (i.e., the sound of a musical instrument or a jet airplane). We taste and smell pleasant and obnoxious things (i.e., good food and foul air). We detect heavy or light objects using the sense of touch. We also use our sense of touch to discern hot and cold objects.

A simple description of what we see in our environment is that it consists of objects having substance. We call such objects matter. **Matter,** of course, makes up the material things around us that we conceive of as occupying space and having mass. **Mass** is a property of matter that determines its resistance to being set in motion or resistance to any change in motion. A baseball is much easier to throw than a more massive shotput ball and, obviously, more desirable to catch. It is not too difficult to develop feeling for the idea of mass since we are familiar with objects in our environment. The useful aspect of mass is that it can be used to compare different objects. For example, when you pick up two objects, you can often say that one feels heavier and therefore has a mass that is greater than the other.

As we observe, the environment is not static but instead involves dynamic changes. Day and night come and go, sunlight replaces rain and snow, and plants and animals grow. Another aspect of our environment that is related to the dynamic phenomena is energy. To understand energy we need the concept of force. A **force** is generally an action that may cause some effect. When you push on a boulder you are applying a force in an attempt to set it in motion. There are various types of forces. Water flowing from a higher level to a lower level can apply a force that results in the motion of a water wheel or a turbine. Expanding hot gases such as steam can exert forces that make engine pistons move. We are familiar with the force of gravity—the force that arises when one mass is near another mass—for example, the attraction between you and the earth. In fact, the weight of an object is the force that results from the gravitational attraction of the earth for the mass of the object. When we jump we have to exert a force great enough to overcome the force of gravity. Note that there is a difference between the mass of an object and the weight of an object. The mass is an inherent property of the object, and its weight is a result of the mass and the gravitational attraction. In other words the mass is an unchanging property, and the weight varies as the gravitational attraction varies. In fact your mass on the earth or on the moon would be the same, but your weight on the moon would be about one-sixth of your weight on the earth. This is because the gravitational attraction of the moon is less than that of the earth. On occasion the terms mass and weight are used interchangeably, but, strictly speaking, they are not the same.

Magnetic forces are familiar to most of us. We know that in the immediate vicinity of a typical magnet there is a magnetic force field. This force field results in certain types of objects being attracted toward the magnet when they are introduced into the field. Electrical forces are also important. Electrical forces involve the existence of electrical charges. There are two types of electrical charge: positive and negative. Objects of like charge tend to repel one another, and objects of opposite charge tend to attract one another. These are called electrostatic forces of repulsion and attraction. Electrical charges are involved in certain chemical aspects of matter.

1-4   Energy

Energy is involved in dynamic changes that are used to perform work. However, the basis for dynamic changes is unbalanced forces. When a force is exerted and overcomes any opposing force, motion or some related phenomena occurs, and it is said that work has been done or an energy flow has taken place. For instance, when we pick up an object we must exert a force to overcome the force of gravity. When we pick it up we have done work or expended energy. The object we have picked up possesses stored energy or the potential for doing work. If we release the object, the force of gravity sets it in motion, and it expends the stored energy as it falls. If the object were dropped so that it struck a paddle wheel, the stored energy could do work by setting the wheel in motion. **Energy** is sometimes simply considered to be the capacity for performing work.

Energy appears in many interrelated forms. Observations of energy changes have shown that energy can be stored and transformed from one form to another but cannot be created or destroyed. Let us consider an example of how energy is involved in our environment. Certain processes occurring on the sun produce energy. Some of this energy takes the form of **radiant energy** or sunlight that travels through space and strikes the earth. Some of this radiant energy is absorbed by water in the ocean and converted to **heat energy.** This heat energy causes some water to evaporate and ultimately to form clouds that drift inland. Some of the water is deposited in mountain areas as rain. This water now possesses **potential energy** or **energy of position** since it is attracted back to sea level by gravity. As the water flows through the rivers toward the sea, some of it is trapped behind dams. The **energy of motion** or **kinetic energy** of the water as it is released from the dams is used to turn a mechanical device called a turbine. This results in the conversion of kinetic energy to **mechanical energy.** The turbine is designed to convert the mechanical energy to electricity or **electrical energy.** The electricity is transported to populated areas and is used to perform various types of work. For example, electricity is converted to mechanical energy in an electric motor.

Other portions of the original solar energy are absorbed by plants. Through a chemical process called photosynthesis, chemicals are produced by the plants in which some of the energy is captured in the form of chemical energy. As we shall learn, **chemical energy** is energy that is stored as certain kinds of chemical substances. Humans and other animals eat parts of the plants and, through chemical processes, gain some of the energy that is used in life processes called metabolism and for performing useful work.

The point of the above example is to demonstrate the significance of energy and energy transfers. In chemistry we are concerned with energy transfers involved in chemical processes. Energy use is fundamental to an industrial society. Thus, the concept of energy is of great importance. Energy is discussed in more detail in Sections 8-8 through 8-13.

1-5   Scientific Laws and Theories

To obtain a better view of the environment, scientists have developed instruments and techniques to extend the realm of our normal senses. Distant objects can be viewed with telescopes while small objects can be viewed with microscopes. Masses of objects can be determined with balances. The temperature of an object can be measured with a thermometer.

The increased sophistication of these instruments allows for more precise observations and, thus, more precise descriptions of environmental objects and events.

Through the use of such methods scientists are able to make consistent and reliable observations. In this manner any patterns and consistencies within the environment can be established. By repeatedly observing phenomena, it is sometimes possible to make a general statement about the reliability of certain phenomena. For example, throughout centuries scientists have been observing energy transformations and have noted that no energy is ever created or destroyed but is instead transformed from one form to another. Such a generalization based on years of observation is called a physical principle or law, or simply a **law.** The observation mentioned above is called the **law of conservation of energy:**

Energy cannot be created or destroyed but only transformed from one form to another.

Physical laws are expressions of an apparently consistent pattern in nature that never seems to have an exception. Another such physical law is the **law of conservation of matter:**

Matter cannot be created or destroyed by normal processes occurring on earth.

Physical laws provide a way to describe phenomena that occur around us. Furthermore, such physical laws have significance in our interaction with our environment. For example, the law of conservation of energy tells us that if we try to convert energy in one form to electrical energy, and if the conversion is not completely efficient, then the leftover energy must take on some other form. Many electrical generating plants utilize the burning of coal or oil to obtain heat energy that is ultimately con-

(a)

Observations of the external operations of a watch: Hour hand moves more slowly than minute hand and watch needs periodic winding.

(b)

Theory one: Small animal is running a wheel which turns gears which control the hands.

(c)

Theory two: Falling sand turns paddle wheel which turns gears which control the hands.

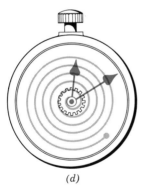

(d)

Theory three: Force of unwinding spring turns gears which control the hands.

(e)

Theory four: Small battery runs a small motor which controls the hands.

Figure 1-1   An Example of a model or theory. A theory or model is a mental picture or explanation of an aspect of reality, which is unobservable with normal senses. The example here shows that we can observe the external result of the watch functioning. To explain how the watch functions without looking inside, we can develop a model or theory. Many theories are possible, but often most can be refuted since they are not substantiated by the external observations. From *Understanding Chemistry: From Atoms to Attitudes* by T. R. Dickson, Wiley, N.Y., 1974.

verted to electrical energy. Another physical law reveals that it is not possible to have 100 percent conversion of heat to work to provide electricity. Much of the heat energy produced in an electrical generating plant is lost as waste heat. This is unavoidable and is expected. However, such waste heat changes the temperature of the immediate environment of the plant. When this waste heat disrupts the animal and plant life around the generating plant, it is called thermal pollution.

As another example, consider that the law of conservation of matter tells us that we cannot destroy matter. When we use some material and then throw it away it does not disappear but instead can accumulate in inconvenient ways. In other words, matter will always exist on earth, and our using it merely moves it from one place to another. It has been said that "it is impossible to throw anything away, because there is no away."

In addition to the physical laws, we have sought a satisfactory explanation of why objects behave as they do and why certain processes occur or can be made to occur. Such explanations involve the development of a description of reality that provides a view of our environment and will allow the prediction of what may happen under specific circumstances. Such a view of reality involves the establishment of concepts or mental pictures of aspects of reality that are unobservable with normal senses. In fact, modern science is based mainly upon such conceptualizations of reality.

A conceptualization that serves as a description of why material objects behave as they do and/or why certain processes occur is called a **theory** or a **model.** As we shall see, there are many important theories and models in chemistry. To illustrate the nature of a theory or model by analogy, consider a case in which you are able to observe a watch from the outside only. By observing you see what the watch does. To explain how it works you have to develop a mental picture of what might be inside the watch that would produce the external effects. (See Figure 1-1.) A mental picture serves as an explanation of something that is not directly observable.

1-6   The International System of Units or Metric System

The qualities of our environment that we can perceive with our senses can be used to describe objects and phenomena. We recall that matter is defined as any object that has mass and occupies space. Thus, one of the apparent qualities is that objects have **mass.** The concept of space results from our ability to observe distances between points. This provides a quality of our spatial environment that we call **length.** Our sense of touch conveys to us that some objects are hotter or colder than we are. This suggests another quality of matter called temperature. Still another quality arises from our ability to observe durations between events and the irreversibility of a sequence of events. This quality is called **time.** To illustrate, consider that if you drop a glass, a certain duration of time passes before it shatters on the floor, and there is no

Table 1-1   Definitions of the Metric Units

Quality	Unit Name and Abbreviation	Definition
Length	Meter (m)	*Original:* One 1/10,000,000 of the quadrant of the earth's meridian passing through Barcelona and Dunkirk. *Current:* 1,650,763.73 times the wavelength corresponding to a specified transition occurring in excited atoms of krypton-86 under specific conditions.
Mass	Kilogram (kg) (The gram, g, which is 1/1,000th of a kg is often used as the basic reference unit, even though the kilogram is the basic defined unit of mass.)	*Original:* Mass of 1000 cm^3 of water at 4 °C. *Current:* Mass of the prototype kilogram number one kept at the International Bureau of Weights and Measures at Sèvres, France.
Time	Second (sec or s)	*Original:* One 1/86,400 of the mean solar day. *Current:* The duration of 9,192,631,770 cycles of the radiation corresponding to a specified transition in the excited atoms of cesium 133.
Temperature	Kelvin (K) or Degree Celsius (°C) (K = °C + 273)	*Original:* One 1/100 of the interval between the freezing point of water and the boiling point of water. *Current:* One 1/273.16 of the thermodynamic temperature of the thermal state corresponding to the triple point of water. This is the thermal state at which solid, liquid, and gaseous water exist in equilibrium.

From Standards of Measurement by A.V. Astin.
Copyright © June 1968 by Scientific American, Inc. All rights reserved.

way to reverse the time sequence so that the pieces restructure the glass. With the added dimension of time, we refer to our space-time environment.

Using these four qualities of mass, length, time and temperature, we can describe what exists around us and the events that are occurring. To give precise descriptions it is necessary to establish a reference unit for each of these qualities. For instance, if we want to measure the length of an object we need a reference for comparison. In other words, we need to give some definite length a name (i.e., the inch) so that other distances can be measured in terms of this defined unit of length. Throughout the centuries many different kinds of reference units have been defined for the four qualities of length, mass, time, and temperature. The cubit, a biblical measurement, corresponded to the length from the elbow to the end of the middle finger. The foot originated from the length of the human foot and the inch from the width of a human thumb. A mile was first defined by the Romans as 1000 double paces and later defined as 5280 feet by the English. The grain originated from the weight of a grain of wheat. The fact that our usual system of units is not conveniently logical can be seen by considering that there are 12 inches per foot, 3 feet per yard and 1760 yards per mile or that there are 16 ounces per pint, 2 pints per quart, and 4 quarts per gallon. A more logical system of units is one that is based on a decimal relation or multiple of 10 relation between various sizes of units (e.g., 10 mills per cent, 10 cents per dime, 10 dimes per dollar).

The most common system of reference units used today is called the **International System of Units (SI)** or the **metric system.** This set of units is used internationally and, thus, provides universal agreement in measurements. The defined units of the International System are listed below, and the official definitions are given in Table 1-1.

Quality	Unit	Abbreviation
Mass	**Kilogram** or **gram** (the gram is not the basic defined unit of mass in the SI, but it is the unit of mass most often used; 1 g is 1/1000 of a kilogram)	**kg, g**
Length	**Meter**	**m**
Time	**Second**	sec or s
Temperature	**Kelvin** or **degree Celsius** (the Celsius scale is not the basic temperature scale in the SI, but it is often used to express temperature measurements)	**K, °C**

1-7   Measurements

We can describe a measurement of a quality of an object by stating that it is a certain multiple or fraction of the reference unit. Of course, the unit used will indicate the quality described by the measurement. Remember that a measurement must always consist of a number (fraction or multiple) and a unit. For example, if we measure the length of an object and state that it has a length of 2.25, we have failed to convey any meaningful information. We must include the unit involved in the measurement. So, we could say that an object has a length of 2.25 m. (Read m as meter.) This would not be ambiguous but would indicate that the length of the object is 2.25 times the standard unit of length called the meter.

When expressing measurements, it is permissible to use appropriate abbreviations for the units. You must become familiar with the ones that are most often used (e.g., meter = m, gram = g, degree Celsius = °C).

Significant Digits. When numbers are obtained from measurements, they are not exact since the number of digits expressed in a measurement depends on the limits of the measuring instrument used. For instance, if the length of an object is found to be 2.25 m, using a meter stick, we say that the measurement is known to the nearest one hundredth of a meter, which is the smallest unit of measure on the ruler. All of the digits in a measurement including the last digit are called significant digits. The 2.25 m measurement has three significant digits. When dealing with numbers obtained from measurements, the following rules apply.

1. All nonzero digits are significant: 278 sec (three significant digits).
2. Zeros between nonzero digits are significant: 1003 cm (four significant digits).
3. Zeros to the left of nonzero digits are not significant but are used to indicate the position of the decimal point: 0.023 g (two significant digits).
4. When a number ends in zeros, the zeros are not significant unless otherwise indicated: 2200 g (unless otherwise indicated, there are only two significant digits).

Most of the numbers used to represent measurements in this book can be considered to have the same number of significant digits that are written even when they end in zero.

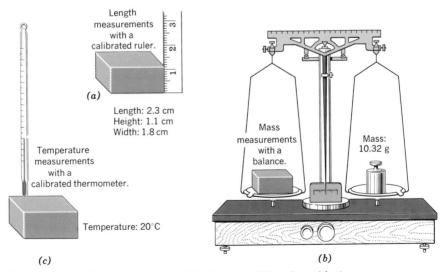

Length measurements with a calibrated ruler.

(a)

Length: 2.3 cm
Height: 1.1 cm
Width: 1.8 cm

Temperature measurements with a calibrated thermometer.

Temperature: 20°C

Mass measurements with a balance.

Mass: 10.32 g

(c)   *(b)*

Figure 1-2   The measurement of the basic qualities of an object.

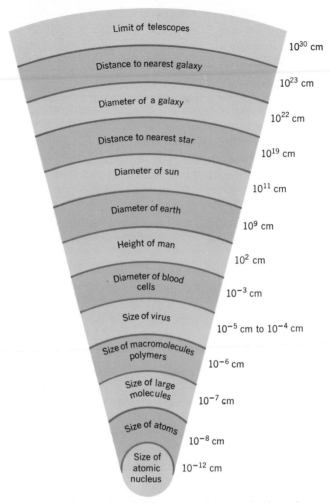

Figure 1-3 shows the following labels and measurements:

Limit of telescopes

10^{30} cm

Distance to nearest galaxy

10^{23} cm

Diameter of a galaxy

10^{22} cm

Distance to nearest star

10^{19} cm

Diameter of sun

10^{11} cm

Diameter of earth

10^{9} cm

Height of man

10^{2} cm

Diameter of blood cells

10^{-3} cm

Size of virus

10^{-5} cm to 10^{-4} cm

Size of macromolecules polymers

10^{-6} cm

Size of large molecules

10^{-7} cm

Size of atoms

10^{-8} cm

Size of atomic nucleus

10^{-12} cm

Figure 1-3   Ranges in the size of objects and distances in the universe.

Using the proper measuring instrument, we can measure the properties of an object and, thus, describe it. An example of such a description is given in Figure 1-2. Sometimes we use a combination of measurements to express a property. For instance, the volume of a rectangular solid is the product of its length, height, and width, as in Figure 1-2. When an object is moving with a certain speed, we can express this by stating the distance or length traveled in a certain period of time. For instance, we express a speed in terms of miles per hour. As we shall see, many other properties and phenomena are expressed in terms of some combination of basic units.

Occasionally, it is necessary to express measurements of very small or large entities. Figure 1-3 illustrates the sizes of objects in the universe ranging from submicroscopic to far reaches of the universe. The distance between two cities may be 200,000 m, or the mass of an object may be 0.0023 g. To conveniently express such measurements, the metric system uses a set of prefixes that denote large or small multiples of the basic units. Three of the most commonly used prefixes are:

Prefix	Abbreviation	Meaning
Kilo	k	1000 times unit
Centi	c	1/100 times unit
Milli	m	1/1000 times unit

(See Appendix VI for other prefixes.)

These prefixes are used in front of the unit and represent the indicated multiple or fraction. The distance 200,000 m can be expressed as 200 km (read km as kilometers) and the mass 0.0023 g can be expressed as 2.3 mg (read mg as milligrams). Note that the prefixes centi- and milli- have the same meaning as when they are used in our monetary system. One cent is 1/100 of a dollar and one mill is 1/1000 of a dollar. To illustrate the relation between these prefixes and the base unit, consider that if we divide a distance of 1 m into 100 equal parts, each part corresponds to 1 centimeter (cm). One hundred centimeters equals 1 m. If the meter distance is divided into 1000 equal parts, each part is a millimeter (mm). One thousand millimeters equals a meter.

The prefixes can be used with any metric unit whenever desired and are often used in the abbreviated form. A few examples of measurements illustrating the use of prefixes are given in Table 1-2. Some measuring instruments allow for measurements directly in terms of a specific metric unit. For instance, we may use a special ruler to measure distance in terms of centimeters or a special clock to measure time in terms of milliseconds.

Table 1-2   Some Examples of Metric and English Measure

	SI	English
Dimensions of postage stamp	22.4 mm × 25.4 mm	7/8 in. × 1 in.
Diameter of United States quarter	2.4 cm	15/16 in.
Distance between New York City and San Francisco	4,868 km	3,025 mi
Mass of 200 lb man	90.7 kg	200 lb
Mass of 5 lb of apples	2.27 kg	5 lb
Mass of a 5-grain aspirin tablet	32.4 cg	0.000714 lb
Mass of a fly's wing	2 mg	0.0000044 lb

From *Understanding Chemistry: From Atoms to Attitudes* by T.R. Dickson, Wiley, N.Y., 1974.

1-8   Unit Conversions

When measurements are made in the metric system it is sometimes necessary to change from the base unit to a prefixed unit or from a prefixed unit to the base unit. This is accomplished by a conversion process in which the meaning of the prefix is used. To illustrate, consider the relation between the meter and the common prefixes used with the meter. (Similar relations can be expressed for any unit.)

$$10^3 \text{ mm} = 1 \text{ m} \qquad 10^2 \text{ cm} = 1 \text{ m} \qquad 10^3 \text{ m} = 1 \text{ km}$$

These relations can be expressed as ratios in the form:

$$\left(\frac{10^3 \text{ mm}}{1 \text{ m}}\right) , \left(\frac{10^2 \text{ cm}}{1 \text{ m}}\right) \text{ and } \left(\frac{10^3 \text{ m}}{1 \text{ km}}\right)$$
$$\left(\frac{1 \text{ m}}{10^3 \text{ mm}}\right) \left(\frac{1 \text{ m}}{10^2 \text{ cm}}\right) \left(\frac{1 \text{ km}}{10^3 \text{ m}}\right)$$

Multiples and fractions of 10 can be represented by expressing 10 raised to a power or exponent.
Positive exponents are used to represent multiples of 10.

$$100 = (10)(10) = 10^2$$
$$1000 = (10)(10)(10) = 10^3$$
$$10000 = (10)(10)(10)(10) = 10^4$$

Negative exponents are used to represent fractions of 10.

$$\frac{1}{10} = 0.1 = 10^{-1}$$

$$\frac{1}{100} = 0.01 = 10^{-2}$$

$$\frac{1}{1000} = 0.001 = 10^{-3}$$

Exponential Notation. Large and small numbers can be represented in a form called exponential notation. To write a number in standard exponential notation form, the decimal point is moved so that one digit is on the left:

3.0200.   0.004.21

Then all significant digits are written with the decimal point in this position, and these digits are multiplied by 10 raised to an exponent corresponding to the number of positions the decimal has been moved:

$$3.02 \times 10^4 \qquad 4.21 \times 10^{-3}$$

If the decimal is moved to the left, the exponent is positive; if it is moved to the right, the exponent is negative. Writing numbers in exponential notation is convenient, since numerous zeros do not have to be written down and the exact number of significant digits can be expressed. Of course, an exponential notation number can be converted to the usual form by moving the decimal point the number of positions indicated in the power of 10:

$$6.00 \times 10^6 = 6,000,000$$
$$2.72 \times 10^{-4} = 0.000272$$

To convert a measurement in a prefixed unit to the base unit, we just multiply by the factor corresponding to the meaning of the prefix. The distance 15 km expressed in terms of meters is found by multiplying by the factor 10^3 m per kilometer:

$$15 \text{ km} \left(\frac{10^3 \text{ m}}{1 \text{ km}}\right) = 15,000 \text{ m or } 15 \times 10^3 \text{ m}$$

Note that the kilometer units cancel, leaving the meter unit. You might wonder how it is possible to cancel the units. Actually, units can be treated as algebraic terms and can be canceled and multiplied. In the above example, the kilometer units appear in the numerator and denominator and thus cancel. The numerical parts of the conversion factors are used to adjust the numerical part of the measurement to the proper magnitude associated with the desired units.

The distance 2.5 mm expressed in terms of meters is found by multiplying by the factor 1 m per 10^3 mm:

$$2.5 \text{ mm} \left(\frac{1 \text{ m}}{10^3 \text{ mm}}\right) = 0.0025 \text{ m or } 2.5 \times 10^{-3} \text{ m}$$

(See Appendix I for a discussion of calculations using powers of 10.)

To express 200,000 m in terms of kilometers, we note that for each kilometer there are 10^3 m. Thus, multiplying the distance in meters by this factor should convert the measurement to kilometers:

$$200,000 \text{ m} \left(\frac{1 \text{ km}}{10^3 \text{ m}}\right) = 200,000 \times 10^{-3} \text{ km or } 200 \text{ km}$$

Note that the decimal point is moved three places to the left.

To express the measurement 0.056 m in units of centimeters, we note that there are 10^2 cm in 1 m. Thus, multiplying the distance in meters by this factor will convert the measurement to centimeters:

$$0.056 \text{ m} \left(\frac{10^2 \text{ cm}}{1 \text{ m}}\right) = 5.6 \text{ cm}$$

Note that the decimal point is moved two places to the right.

In the above discussion conversion factors were used to convert a measurement from one set of units to another. In fact, we use many conversion factors in our daily lives. For instance, suppose we want to know how long it will take to make a car trip of 250 miles if we drive at an average speed of 50 miles per hour (50 mi/hr).* Using the speed as a conversion factor, we can determine the time required. The calculation would be

$$250 \text{ mi} \left(\frac{1 \text{ hr}}{50 \text{ mi}}\right) = 5.0 \text{ hr}$$

Of course, we do not usually calculate the answer to such a problem in this formal manner, but the point is that we often use conversion factors.

As another example, consider the fact that apples are normally priced

---

* A slash (/) should be read as "per". Thus (50 mi/hr) reads 50 miles per hour.

in terms of cost per pound. Such a price is really a conversion factor that allows us to determine the cost of a specific number of pounds of apples. Let us suppose apples cost 20 cents per pound (20¢/lb), and we want to purchase 2.6 lb. The cost would be

$$2.6 \text{ lb} \left(\frac{20\text{¢}}{1 \text{ lb}} \right) = 52\text{¢}$$

There are many other conversion factors that we use in our daily lives. Can you name a few?

Since the use of conversion factors is important in chemistry, let us consider the reasons why these factors can be used. (See Appendix II.) Conversion factors express a relation between two units. For example, we know that the relation between inches and feet is

$$1 \text{ ft} = 12 \text{ in.}$$

From this relation we can obtain a conversion factor to convert from inches to feet and a factor to convert from feet to inches. The first factor is obtained by dividing both sides of the relationship by 12 in. This gives

$$\left(\frac{1 \text{ ft}}{12 \text{ in.}} \right) = 1$$

The second factor can be obtained by dividing both sides by 1 ft. This gives

$$1 = \left(\frac{12 \text{ in.}}{1 \text{ ft}} \right)$$

Which factor we use in a conversion depends on whether we want to convert from feet to inches or vice versa. Notice that both factors are actually equivalent to 1 and, thus, when we multiply a quantity by such a factor we are essentially multiplying by 1. In this manner we do not fundamentally alter the quantity; we just convert it so that it is expressed in terms of another unit. Factors like those above, which are equivalent to 1, are called **unit factors.**

Example 1-1   Convert 8.00 ft to units of inches.
This can be done by multiplying the measurement by the unit factor relating inches to feet.

$$8.00 \text{ ft} \left(\frac{12 \text{ in.}}{1 \text{ ft}} \right) = 96.0 \text{ in.}^*$$

Notice that we choose the factor in a logical manner that allows the cancellation of the unwanted units and produces an answer in terms of the desired unit. Notice also that the numerical part of the factor is used in the calculation of the numerical portion of the answer.

Since conversion factors can be considered to be unit factors, you can see that it is possible to multiply a quantity by any number of factors so that a desired result is obtained.

Example 1-2   Convert 24 hr to units of seconds.
One approach to this problem is to convert the hours to minutes and then the minutes to

---

* For convenience and clarity the cancellation of units with slashes will not be shown throughout the text.

seconds. This will require the unit factor relating minutes to hours,

$$\left(\frac{60 \text{ min}}{1 \text{ hr}}\right)$$

and the factor relating seconds to minutes,

$$\left(\frac{60 \text{ sec}}{1 \text{ min}}\right)$$

The problem is solved using these factors by multiplying the 24 hr by the 60 min/hr factor and then multiplying this product by the 60 sec/min factor.

$$24 \text{ hr} \left(\frac{60 \text{ min}}{1 \text{ hr}}\right)\left(\frac{60 \text{ sec}}{1 \text{ min}}\right) = 8.6 \times 10^4 \text{ sec}$$

(See Appendix I for a discussion of calculations with exponential numbers.)

Notice that the desired units result when we algebraically cancel the hour and minute units. If we had used the wrong factors the units would not have worked out correctly.

The Number of Digits in a Calculated Result. The number of digits allowed in a result obtained when factors are multiplied and divided depends on the relative uncertainties of the factors. Actually, the uncertainty in each factor contributes to the uncertainty in the result. A general rule to determine the number of digits in a result obtained by multiplication and division is that the result can have no more significant digits than the factor with the least number of significant digits. Unfortunately, there are exceptions to this rule, but, for our purposes, it will usually indicate the proper number of digits needed in the answer.
Some examples of this rule are given below.

$$2.5 \left(\frac{27.2}{3.27}\right) = 21$$
The answer should have two significant digits, since the least accurate factor (2.5) has only two digits.

$$10.2 \left(\frac{52.7}{262}\right)\left(\frac{1.052}{99.8}\right) = 0.0216$$
The answer should have three digits, since the least accurate factors have three digits.

Note that the position of the decimal point does not determine the number of significant digits in a multiplication or division calculation. Moreover, most unit conversion factors can be expressed to more digits than the measurement data that is being converted. Consequently the number of digits in the measurement data normally gives the number of significant digits in the answer.

1-9   English to Metric Conversions

In the United States we often use the **English system** of units to make measurements. However, since the SI system is the commonly used technical system of units, it is necessary to convert measurements made

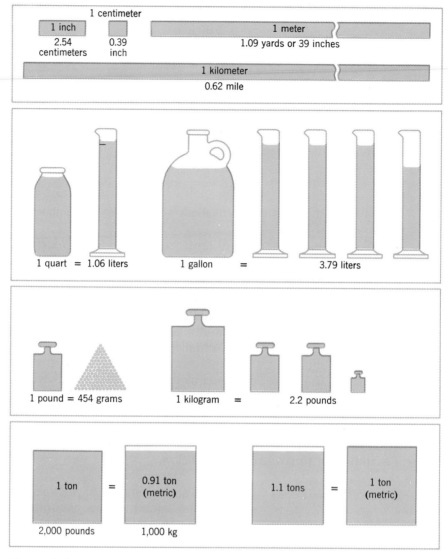

Figure 1-4   Comparison of SI(metric) and English units. From *Understanding Chemistry: From Atoms to Attitudes* by T. R. Dickson, Wiley, N.Y., 1974.

in English units to metric units, or vice versa. To accomplish such conversions it is necessary to use conversion factors that relate the units in each system. See Figure 1-4 for a comparison of English and metric units. Appendix V lists some of the most common conversion factors. To convert a measurement we just multiply by the proper factor. The factor that is used eliminates one unit and replaces it with a desired unit. To express the length 10.0 in. in terms of centimeters, we note that there are 2.54 cm/1 in. Thus, we multiply the 10.0 in. by this factor.

$$10.0 \text{ in.} \left(\frac{2.54 \text{ cm}}{1 \text{ in.}} \right) = 25.4 \text{ cm}$$

To express 80.0 kg in units of pounds, we multiply by the factor 2.2 lb/1 kg.

$$80.0 \text{ kg} \left(\frac{2.2 \text{ lb}}{1 \text{ kg}} \right) = 176 \text{ lb}$$

To express 50 yards in units of meters we multiply by the factor 0.914 m/yd.

$$50 \text{ yd} \left(\frac{0.914 \text{ m}}{1 \text{ yd}} \right) = 46 \text{ m}$$

To express the speed 100 km/hr in miles per hour, we multiply the speed by the fact that there are 0.62 miles/1 km.

$$\left(\frac{100 \text{ km}}{1 \text{ hr}} \right) \left(\frac{0.62 \text{ mi}}{1 \text{ km}} \right) = \left(\frac{62 \text{ mi}}{1 \text{ hr}} \right)$$

Example 1-3   Express the distance 10.0 ft in units of centimeters.
Since we are given a measurement in feet that we want to express in centimeters it is more convenient to convert the feet to inches and then to centimeters. That is, the conversion sequence can be

$$\text{ft} \rightarrow \text{in.} \rightarrow \text{cm}$$

First we multiply the 10.0 ft by the feet-to-inches factor.

$$10.0 \text{ ft} \left(\frac{12 \text{ in.}}{1 \text{ ft}} \right)$$

Next we multiply this product by the inch to centimeter factor given in Appendix V.

$$10.0 \text{ ft} \left(\frac{12 \text{ in.}}{1 \text{ ft}} \right) \left(\frac{2.54 \text{ cm}}{1 \text{ in.}} \right) = 305 \text{ cm}$$

Example 1-4   Express the distance 500 m in units of miles.
The conversion sequence can involve the conversion of the meters to kilometers and the kilometers to miles.

$$\text{m} \rightarrow \text{km} \rightarrow \text{mi}$$

First we multiply the 500 m by the meters-to-kilometers factor based on the meaning of the prefix.

$$500 \text{ m} \left(\frac{1 \text{ km}}{10^3 \text{ m}} \right)$$

Next we multiply this product by the kilometer-to-mile factor given in Appendix V.

$$500 \text{ m} \left(\frac{1 \text{ km}}{10^3 \text{ m}} \right) \left(\frac{0.621 \text{ mi}}{1 \text{ km}} \right) = 0.310 \text{ mi}$$

1-10   Volume Measurements

We often refer to measurements of volumes and areas in everyday life. For example, we refer to the engine displacement in an automobile in terms of cubic inches and carpeting in terms of square yards or square feet. Areas have dimensions of square length. Volumes represent amounts of three-dimensional space occupied and thus have dimensions of cubic length (i.e., cubic inches).

What units do we use when expressing volumes and areas? Since areas involve square length, it is best to use the unit of length raised to the power of 2 as the unit of area. Thus, we could express area in units of square feet, ft²; square inches, in.²; square centimeters, cm²; or square meters, m². Similarly, the unit of length raised to a power of 3 can be used as the unit of volume. For example, cubic inches, in.³; cubic feet, ft³; cubic centimeters, cm³; or cubic meters, m³. This is another example of how units can be treated algebraically. For example, if an object has a volume of 5.00 ft³ (read 5 cubic feet), the units correspond to the unit feet cubed (ft ft ft = ft³).

Squaring a number means multiplying the number by itself. The square is represented by a superscript 2 following the number: $3^2 = (3)(3) = 9$. Cubing a number means multiplying the number by itself three times. The cube is represented by a superscript 3 following the number: $3^3 = (3)(3)(3) = 27$. It is also possible to express the square or cube of a unit when measurements of the same units are multiplied. Thus, m² means square meter and represents (m)(m); cm³ means cubic centimeter and represents (cm)(cm)(cm).

When working with volumes of liquids we do not always use units of ft³ or in.³, but we use units of pints, quarts, or gallons. In other words, we use special units for volume.

In the metric system a special unit of volume has been defined for the measurement of volumes of substances. This unit is called the **liter, ℓ,** and is defined as a volume equal to 1000 cm³ (see Figure 1-5),

$$1 \ell = 10^3 \text{ cm}^3$$

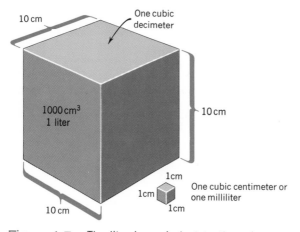

Figure 1-5   The liter is equivalent to the volume occupied by a cube measuring 10 cm (3.94 in.) on an edge. The liter is equivalent to 1.057 qt.

One liter is slightly larger than a quart; one liter equals 1.057 quarts. In the metric system one liter equals one cubic decimeter.

The metric prefixes can be used with the liter. If 1 ℓ of volume is divided into 1000 equal parts, each part is 1 milliliter (ml). There are 1000 ml in 1 ℓ. Note that since 1 ℓ is also equivalent to 1000 cubic centimeters, then the milliliter and cubic centimeter correspond to the same amount of volume:

$$1 \text{ ml} = 1 \text{ cm}^3$$

Since the milliliter and cubic centimeter have the same meaning, they are often used interchangeably. Notice that from the definition of the liter we are able to express the unit factors.

$$\left(\frac{1\ \ell}{10^3\ \text{cm}^3}\right), \left(\frac{10^3\ \text{cm}^3}{1\ \ell}\right), \left(\frac{1\ \ell}{10^3\ \text{ml}}\right) \quad \text{and} \quad \left(\frac{10^3\ \text{ml}}{1\ \ell}\right)$$

These factors can be used to convert volumes from cubic centimeters or milliliters to liters or vice versa.

Example 1-5   Express 0.250 ℓ in units of milliliters.
The volume can be multiplied by the factor 10^3 ml/ℓ to obtain the desired result.

$$0.250\ \ell \left(\frac{10^3\ \text{ml}}{1\ \ell}\right) = 250 \text{ ml}$$

Example 1-6   What is the area of a rectangular surface that has a base of 20.0 m and a height of 10.00 m?
The area is given by the product of the base and height:

$$(20.0 \text{ m})(10.00 \text{ m}) = 200 \text{ m}^2$$

Sometimes it is necessary to convert multiple units in one system to multiple units in another system. In such cases, the normal conversion factors must be raised to the appropriate power. For example, if we want to convert a measurement in square inches to units of square centimeters, we can use the square of the factor relating centimeters to inches:

$$\left(\frac{2.54 \text{ cm}}{1 \text{ in.}}\right)^2 = \left(\frac{2.54 \text{ cm}}{1 \text{ in.}}\right)\left(\frac{2.54 \text{ cm}}{1 \text{ in.}}\right) = \left(\frac{2.54^2 \text{ cm}^2}{1 \text{ in.}^2}\right) = \left(\frac{6.45 \text{ cm}^2}{1 \text{ in.}^2}\right)$$

Since this is a unit factor, when we square it the result is a unit factor and can be used as a factor for relating the square units.

Example 1-7   How many cm² are in a 200-in.² rectangular area?
This can be found by multiplying the 200 in.² by a factor to convert square inches to square centimeters. This factor is obtained by squaring the basic conversion factor as shown above.

$$200 \text{ in.}^2 \left(\frac{2.54 \text{ cm}}{1 \text{ in.}}\right)^2 = 200 \text{ in.}^2 \left(\frac{6.45 \text{ cm}^2}{1 \text{ in.}^2}\right) = 1.29 \times 10^3 \text{ in.}^3$$

Notice that when a conversion factor is raised to a power, all of the numbers and units in the factor are raised to that power. Furthermore, note once again how the undesired units cancel and the desired unit is obtained.

1-11   Measuring Instruments

The basic properties associated with matter are measured with devices that are carefully calibrated by a comparison with defined standards. Time measurements are usually carried out by use of mechanical or electrical clocks that are designed so that the movement of hands or digits corresponds to the passage of definite amounts of time. Length measurements are found by the use of straight pieces of metal or wood that have been calibrated by comparison to a standard length. For example, on a meter stick the distances between the markings are established by the standard meter reference unit. A meter stick is usually marked so that the distance between subsequent lines corresponds to 1/1000th of a meter. In other words, the distances between subsequent lines correspond to millimeter distances. Usually every tenth marking is numbered. Thus, the distance between numbered lines corresponds to 10 mm or 1 cm (1 cm = 10 mm). Lengths of objects can be measured by comparison to the standard meter stick.

The mass of an object is measured by comparing the mass of the object to standard reference masses. These are usually pieces of metal that have been calibrated by comparison to the standard unit of mass. Such masses are called weights. Mass comparisons, called weighing, are carried out by use of a **balance.** When we weigh an object on a balance we determine the mass of the object. Fundamentally, a balance consists of an upright with a sharp edge called a knife edge, (See Figure 1-6.) A metal beam is balanced on this knife edge, so that the mass of the beam on either side of the knife edge is the same. If an object of unknown mass is placed on one side of the beam, the beam will no longer be in a horizontal, balanced position. Now, if the correct number of calibrated masses or weights is placed on the other end of the beam, the beam will regain the horizontal, balanced position. Once this position has been regained, the mass of the object must equal the total mass of the calibrated masses used. Most balances function on a principal that is

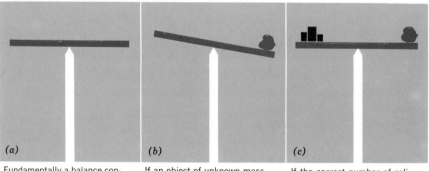

(a)   (b)   (c)

| (a) Fundamentally a balance consists of an upright with a sharp edge called a knife edge. A metal beam is balanced on this knife edge. The beam is balanced so that the mass of the beam on either side of the knife edge is the same. | (b) If an object of unknown mass is placed on one side of the beam, the beam will no longer be in an horizontal balanced position. The beam will be unbalanced. | (c) If the correct number of calibrated masses is placed on the other end of the beam, the beam will be restored to the horizontal balanced position. Once the balanced position has been restored, the mass of the object will equal the total mass of the calibrated masses used. |

Figure 1-6   The determination of the mass of an object.

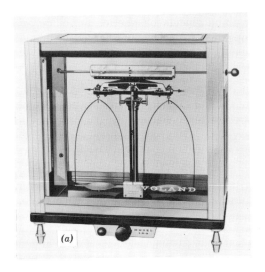

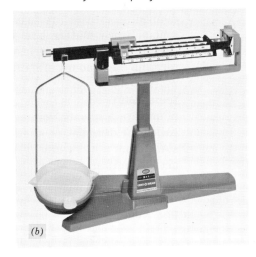

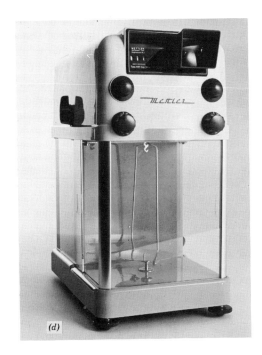

Figure 1-7   Some typical balances. (*a*) Two pan balance. (*b*) Triple beam balance. (*c*) Top loading balance. (*d*) Automatic single pan balance.

quite similar to this simplified view of weighing. However, balances vary quite a bit in design, as is shown in Figure 1-7.

1-12   Density and Property Conversion Factors

The conversion factors discussed above provide for the conversion of a measurement from one prefixed unit to another or from one system of units to another. Many physical relationships involve the definition of a quantity as a function of one or more fundamental properties. These defined quantities are used to describe objects or phenomena. For example, since matter has the properties of occupying space and possessing mass, a good way to distinguish between types of matter is to determine the amount of mass there is in a specific volume of each type of matter. As we shall see later, the volumes of objects can change with temperature so it is necessary to make the measurements of the amount of mass in a specific volume at a specific temperature. Such a property is called the **density,** D, of a substance and is defined as the mass of a substance in a unit volume of the substance at a specified temperature. The actual units used for mass and volume depend on the system of measurement used. Most often the units used are grams for mass and cubic centimeters or milliliters for volume. Density can be algebraically expressed as

$$D = \frac{m}{v} = \left(\frac{\text{number of grams}}{\text{volume}} \right)$$

Thus the density is the number of grams of a substance per cubic centimeter or milliliter of the substance. Usually when densities are stated, the temperature at which the density was determined is also given. The densities of a few common substances are given in Table 1-3. The density of a substance is determined by measuring the mass of a sample of the substance and volume occupied by the sample. Then, by dividing the mass by the volume, the density is expressed as the amount of mass per

Table 1-3   The Densities of Some Common Chemicals

Liquids	Density at 20°C
Water	1.00 g/ml
Ethyl alcohol	0.79 g/ml
Carbon tetrachloride	1.59 g/ml
Mercury	13.55 g/ml

Solids	Density at 20°C
Magnesium	1.74 g/ml
Aluminum	2.70 g/ml
Iron	7.87 g/ml
Gold	19.3 g/ml
Sulfur	2.07 g/ml
Salt	2.16 g/ml
Sugar	1.59 g/ml
Ice (0°C)	0.92 g/ml

Gases	Density at 0°C
Air (dry)	1.29 g/ℓ
Oxygen	1.42 g/ℓ
Nitrogen	1.25 g/ℓ
Carbon dioxide	1.96 g/ℓ
Chlorine	3.17 g/ℓ

unit of volume. The calculation of density in this manner is similar to many calculations we are used to. For example, when we want to determine how many miles per gallon of gasoline our car is using, we divide the total miles traveled by the number of gallons used, and we obtain an expression of the number of miles per unit volume of gas—the number of miles per gallon.

Example 1-8   What is the density of water at 20°C if a 20.0 g sample of the substance is found to have a volume of 20.0 cm³?
The mass can be divided by the volume to give the mass per unit volume—the density.

$$D = \left(\frac{20.0 \text{ g}}{20.0 \text{ cm}^3}\right) = \left(\frac{1.00 \text{ g}}{1 \text{ cm}^3}\right) \text{ or } 1.00 \text{ g/cm}^3$$

The density of a specific substance is a descriptive quantity that will always be the same, no matter how much of the substance you are dealing with. The density of a substance expresses the relationship between the mass and the volume of the substance. Consequently, the density can be used as the **property conversion factor** (a unit factor) to convert from the volume of a substance to the mass of that substance using the density as a factor or from mass to volume using as a factor the inverse or reciprocal of the density (the density "turned over"). Two quantities are called reciprocal or inverse of one another if they give a product of 1:

2 and 1/2 are the inverse of one another since $2 \times 1/2 = 1$; 3/4 and 4/3 are reciprocals since $3/4 \times 4/3 = 1$.

The reciprocal of a fraction or factor is found by switching the positions of the numerator and denominator or "turning the quantity over." Thus, the reciprocal of the density (5 g/1 cm³) is:

$$\left(\frac{1 \text{ cm}^3}{5 \text{ g}}\right)$$

To use a conversion factor like density you must decide the units of the quantity that you want to find and then use the factor or the inverse of the factor to convert the given data to the desired answer. That is, if you know the volume of a sample of a substance, the density can be used to convert to mass:

$$\text{volume} \left(\frac{\text{mass}}{\text{volume}}\right) = \text{mass}$$

On the other hand, if you know the mass of a sample of a substance, the inverse of the density can be used to convert to volume:

$$\text{mass} \left(\frac{\text{volume}}{\text{mass}}\right) = \text{volume}$$

Example 1-9   What is the mass of 20.0 cm³ of a substance that has a density of 2.50 g/cm³.

The volume can be multiplied by the density to convert the volume of the sample to the mass of the sample:

$$20.0 \text{ cm}^3 \left(\frac{2.50 \text{ g}}{1 \text{ cm}^3}\right) = 50.0 \text{ g}$$

Example 1-10   What is the volume of 10.0 g of a substance having a density of 0.500 g/cm³?

The mass can be multiplied by the inverse of the density to convert the mass of the sample to volume:

$$10.0 \text{ g} \left(\frac{1 \text{ cm}^3}{0.500 \text{ g}}\right) = 20.0 \text{ cm}^3$$

Sometimes a property conversion factor is used along with unit conversion factors for changing from one property in one set of units to another property expressed in different units.

Example 1-11   What is the volume in liters of 600 g of a substance having a density of 0.500 g/cm³?

The mass can be converted to volume in cm³ by use of the inverse of the density; then the volume can be converted to units of liters by use of the appropriate factor.

$$600 \text{ g} \left(\frac{1 \text{ cm}^3}{0.500 \text{ g}}\right) \left(\frac{1 \ell}{10^3 \text{ cm}^3}\right) = 1.20 \ \ell$$

The densities of common substances vary greatly. Some substances such as gases are of very low density, and some metals have very high densities. When two liquids that do not intermix are poured together, the liquid of lower density forms a layer on top of the liquid of higher density. For example, the less-dense salad oil floats on top of the more-dense vinegar in salad dressing. If a solid is more dense than a liquid, a chunk of it will sink in the liquid. On the other hand, if a solid is less dense than a liquid, it will float. For instance, a chunk of iron will sink in water, but a piece of low density wood will float.

1-13   Temperature and Heat

As noted earlier, the various forms of energy can be interconverted. One form of energy that we experience often in our environment is **heat.** We associate heat with hot bodies, and we shall find out later that the heat energy of a body is mainly the result of the motion of the very tiny constituent particles of the body. Heat is a convenient form of energy because it can be transferred from one body to another as a result of contact between the bodies. We say that when a body becomes hotter it gains heat energy. We say that a body is at a certain **temperature,** and when it gains heat energy it changes to a higher temperature.

In order to state the temperatures of bodies, it is necessary to establish a relative scale of comparison called a **temperature scale.** A temperature scale is based on the definition of certain states as corresponding to certain temperatures. Then the difference between one state and another can be measured with a thermometer, and the difference can be divided into a certain number of parts called degrees. For example, the **Celsius scale** (formerly called the centigrade scale) was originally defined by establishing the freezing point of water as zero degrees (0°) and the boiling point of water as 100 degrees (100°). The Celsius degree, °C, was defined as 1/100 of the temperature difference between the freezing point and boiling point of water.

There is an important difference between the concepts of heat and temperature. The amount of heat associated with a body depends on the amount of material involved while the temperature of the body may be the same for different amounts. To illustrate the difference, consider that much more heat is associated with a bonfire than a lighted match, while the temperatures may be the same.

When heat energy is absorbed by a body, the temperature of the body will usually increase. For instance, if we place a pan of water on the stove and heat it, the temperature increases. In fact, the more we heat the water the greater the increase in temperature. This, of course, is not true when the water is boiling, but we shall consider this later. The point here is that the amount of heat energy absorbed by the water is directly related to the amount of water and to the number of degrees increase in the temperature of the water. So, it would be possible to measure the amount of heat absorbed by measuring the increase in temperature of a given amount of water. In fact, this is one way a unit for heat can be established. The unit of heat commonly used is the **calorie** (**cal**) which can be defined as the amount of heat required to raise the temperature of 1 g (gram) of water from 14.5°C to 15.5°C. In this book, amounts of heat will be expressed in units of the calorie (cal). Do not confuse the calorie with the unit of energy used by dieticians. The dietetic Calorie is equal to a kilocalorie (kcal).

Three temperature scales are in common use today. The **Celsius scale** and the **Kelvin scale** are used in science, and the **Fahrenheit scale** is used as a nontechnical temperature scale. The Fahrenheit scale was established centuries ago using reference thermal states that were not reliable. Today the Fahrenheit scale is a secondary scale and is basically defined with reference to the scientific temperature scales. The symbol °F is used to represent a degree on the Fahrenheit scale. The Celsius scale mentioned above is a common scientific temperature scale. In the nineteenth century an English scientist, Lord Kelvin, established a more fundamental temperature scale that used the lowest possible temperature as a reference point for beginning of the scale. The lowest possible temperature was established as 0 K (zero Kelvin). This temperature in terms of the other temperature scales is −273.15°C or −459.7°F. The temperature scale established by Kelvin has become known as the **Kelvin** or **thermodynamic scale.** The size of the degree Kelvin, K, was chosen to be the same as a Celsius degree. This was

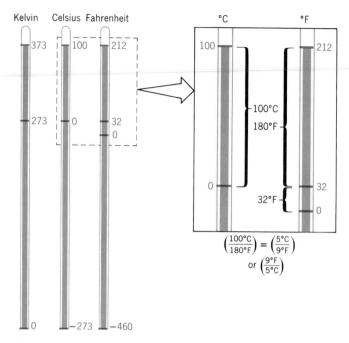

Figure 1-8   The relations between the Kelvin, Celsius, and Fahrenheit temperature scales.

done so that there would be a simple relation between the Kelvin and the Celsius scales. The three temperature scales are compared in Figure 1-8.

1-14   Temperature Conversion

The unit conversion factors discussed previously were used in calculations to directly convert from one unit to another. They can be used in this manner because the zero points for most units of a quality are the same. For example, zero length is the same whether measurements are made in meters or feet, and zero mass is the same for mass measured in grams or kilograms. In other words, most units have a natural zero point from which measurements can begin. However, the various temperature scales do not have the same zero points. (See Figure 1-8.) So, when a temperature measurement is converted from one scale to another, the differences in the zero points must be considered. The actual size of the Kelvin and Celsius degree are the same, consequently, temperatures expressed in these scales can be interconverted by adjusting the zero points. As can be seen in Figure 1-8, the zero points on the two scales differ by 273°. Thus, 273 must be subtracted from a Kelvin temperature to convert it to a Celsius temperature (°C = K − 273).

Example 1-12   Convert 373 K to the Celsius scale.

$$373 \text{ K} - 273 = 100 \text{°C}$$

To convert a Celsius temperature to the Kelvin scale, 273 must be added to the Celsius temperature (K = °C + 273).

Example 1-13   Convert − 100°C to the Kelvin scale.

$$-100°C + 273 = 173 \text{ K}$$

It is sometimes necessary to convert a temperature expressed in the Celsius scale to the Fahrenheit scale or vice versa. As can be seen in Figure 1-8, the Fahrenheit degree is smaller than the Celsius degree. Between the freezing point and the boiling point of water there are 100°C and 180°F. Thus, the relation of degrees in the two scales is

$$\left(\frac{100°C}{180°F}\right) = \left(\frac{5°C}{9°F}\right) \quad \text{or} \quad \left(\frac{180°F}{100°C}\right) = \left(\frac{9°F}{5°C}\right)$$

To convert a temperature in one scale to the other scale, the above ratios can be used as conversion factor. However, as is seen in Figure 1-8, the zero points on the two scales do not coincide and, in fact, differ by 32°F. Consequently, the zero points must be adjusted in the conversion process.

To convert a temperature measured in degrees Fahrenheit to degrees Celsius, we first subtract 32°F to adjust the zero points and then multiply by the factor 5°C/9°F.

Example 1-14   Express 68°F in the Celsius scale.
First subtract 32°F to adjust the zero point of the Celsius scale.

$$68°F - 32°F = 36°F$$

Then multiply by the factor 5°C/9°F.

$$\left(\frac{5°C}{9°F}\right) 36°F = 20°C$$

To convert a temperature measured in degrees Celsius to degrees Fahrenheit, we multiply by the factor 9°F/5°C and then add 32°F to adjust to the Fahrenheit zero point.

Example 1-15   Express 20°C in the Fahrenheit scale.
First multiply by the factor 9°F/5°C.

$$\left(\frac{9°F}{5°C}\right) 20°C = 36°F$$

Then add 32°F to adjust to the zero point of the Fahrenheit scale.

$$36°F + 32°F = 68°F$$

Questions and Problems

1. What is the difference between the terms mass and weight?
2. Reread section 1-4 and list the forms of energy mentioned.
3. What is a law? What is a theory or model?
4. List the four qualities of our environment for which units are defined in the SI system. Give the names of the basic SI units of these qualities.
5. One centimeter is how many milliliters?
6. One foot is how many centimeters?
7. A person has the following body measurements: 38 in., 22 in., 36 in. What would these measurements be in centimeters?
8. In the Olympics sprinters run the 100-m dash. How many yards is 100 m?
9. What is a liter?
10. How many cubic centimeters are in one cubic meter?
11. How many liters are in one cubic meter?
12. If a room measures 10 ft by 12 ft, how many square meters of carpet are needed to cover it?
13. If a person weighs 100 lb, what is their mass in kilograms?
14. One gallon is how many liters?
15. Give the definition of density.
16. Is it possible for two substances to have the same density? Why?
17. When carbon tetrachloride is poured into water it forms a layer that lies beneath the water. What is true concerning the density of carbon tetrachloride compared to the density of water?
18. A graduated cylinder has a mass of 52.77 g. When 8.80 ml of a liquid are placed in the cylinder, the mass of the liquid plus the cylinder is 74.93 g. Calculate the density of the liquid.
19. A copper penny has a mass of 3.17 g. If 10 pennies of this mass occupy a total volume of 3.54 ml, what is the density of copper?
20. A 1.600 kg, rectangular block of wood measures 25.0 cm by 20.0 cm by 3.00 cm. Will this wood float in water?
21. If the density of a person is 1.20 g/ml, what is their volume in liters if they weigh 100 lb?
22. A coin has a mass of 25.0 g and a volume of 1.35 ml. Is the coin made of pure gold? (The density of gold is 19.3 g/ml.)
23. The earth has a mass of 5.98×10^{24} kg and a radius of 6.37×10^{6} m. Assuming that the earth is a sphere calculate the density of the earth. The volume of a sphere is: $V = 4/3\pi r^{3}$ where π is 3.14 and r is the radius.
24. The density of air at 20°C is 1.29 g/ℓ. What is the mass of a 500 ml breath of air?
25. A water bed in the form of a rectangular solid measures 6 ft by 6 ft by 1 ft. Using a density of 1.00 g/ml for water determine how many grams of water the bed contains. How many pounds of water does the bed contain?

26. Normal body temperature is 98.6°F. What is normal body temperature in degrees Celsius? If a person has a temperature of 102°F, what is their temperature in degrees Celsius?

27. If the air temperature is 70°F, what is the temperature in degrees Celsius?

28. If you were told that the air temperature was 90°C, would you question such a temperature?

29. If you wanted to set your thermostat for 65°F, what should the Celsius setting be?

30. Convert the following temperatures:
    (a) 37°C to K
    (b) 273 K to °C
    (c) −53°C to °F
    (d) 25°C to °F
    (e) 373 K to °F

31. If the price of gasoline is 75¢/gal, what is the cost per liter?

32. In Europe a speed limit sign reads 80 km/hr. What is this speed limit in miles per hour?

33. Suppose the fuel use of a car is 20 mi/gal. Express this in terms of kilometers per liter.

34. A football field measures 100 yd long by 160 ft wide. What are these dimensions in meters?

35. A horse runs 6.0 furlongs in 4.0 min. What is the speed of the horse in miles per hour? (1 furlong = 220 yd, 1760 yd = 1 mi, 60 min = 1 hr.)

36. One light year is the distance that light will travel in one year. If the speed of light is 3.00×10^8 m/sec, how many miles are there per light year? (1 mi = 1.61 km, 1 min = 60 sec, 1 hr = 60 min, 1 day = 24 hr, 1 year = 365 days.)

37. The fuel use of a car is 20 mi/gal. Determine the fuel use in terms of furlongs per liquid ounce. (1 furlong = 220 yd, 1760 yd = 1 mi, 16 oz = 1 pt, 2 pt = 1 qt, 4 qt = 1 gal.)

38. Assuming that the volume of the ocean is about 1.4×10^9 km³, and the density of seawater is 1.02 g/cm³, calculate the approximate mass of the ocean in grams.

39. In the land of Who, the dimensions of objects are measured in units of bleks and substance in units of bleeps. They define Grossity as the number of bleeps per unit blek. If a sample of a Who object has a Grossity of 2.0 bleeps per blek, how many bleks are in 50 bleeps of the object?

2

Chapter 2: Chemical Elements and Compounds

Objectives

The student should be able to:

1 Give the names of the common elements given the symbols or the symbols given the names.

2 Describe the atomic theory of matter. (Dalton's theory)

3 Determine the molar mass or the number of grams per mole of an element using the periodic table or a list of atomic weights.

4 Use the molar mass or the number of grams per mole of an element to convert a given mass of the element to the number of moles of that element or to convert the number of moles to the mass.

5 Describe the meaning and give a molar interpretation of the formula of a compound.

6 Deduce the molar mass or the number of grams per mole of a compound given the formula.

7 Use the molar mass or the number of grams per mole of a compound to convert a given mass of the compound to the number of moles or to convert the number of moles to the mass.

8 Calculate the percentage by mass composition of a compound.

9 Determine the empirical formula of a compound.

10 List, with the aid of the periodic table, the elements that are gases, the elements that are liquids, and the elements that occur in the form of diatomic molecules.

11 Balance a simple chemical equation given the formulas of the reactants and products.

Terms to Know

Elements
Compounds
Law of conservation of matter
Law of constant composition
Atomic weight
Isotopes
Mole
Avogadro's Number

Chemical reaction
Catalyst
Lithosphere
Atmosphere
Hydrosphere
Biosphere
Ecosphere

2

Chemical Elements and Compounds

2-1 The States of Matter

As you look around at the water, air, and earth, you see that matter can take on three different physical forms. These forms are called physical states or phases. The three **states of matter** are the **solid** state, the **liquid** state, and the **gas** or vapor state. (See Figure 2-1). Matter in the solid state, such as a rock or a piece of metal, has a definite volume and a definite shape or firmness associated with it. A liquid, such as water or oil, also occupies a specific volume, but it requires a container. A given volume of a liquid will take the shape of the container that holds it. A gas has neither a definite shape nor volume, and, thus, it must be contained in a sealed vessel for storage. By considering an air-filled balloon, you can see that a gas occupies the entire volume into which it is placed.

2-2 The Nature of Matter

Chemists have always been interested in finding out what things are made of. To do this they observe the properties of the various forms of matter that is found in our environment. These characteristics or properties serve to distinguish one type of matter from another. Most matter that we see is made up of mixtures of materials. For example, as we look closely at soil we see the different parts. Concrete obviously has different parts. Food is made up of the varied forms of plant and animal cells. The grain in a piece of wood reveals that wood is a mixture. Such

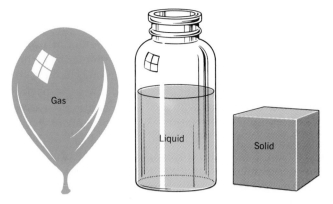

Figure 2-1   The three common states of matter.

mixtures are called **heterogeneous mixtures.** Heterogeneous means made up of different parts.

It is possible to separate mixtures into simpler parts. The components of mixtures are pure substances. (See Figure 2-2.) **Pure substances** are forms of matter that have the same properties throughout and have definite chemical compositions. Actually, there are two kinds of pure substances: elements and compounds. The chemical elements are the fundamental forms of matter and are components of all matter. **Elements** are pure substances that cannot be separated into simpler substances. Compounds have definite properties and composition, but, under certain circumstances, they can be separated into two or more chemical elements. In other words, **compounds** are pure substances

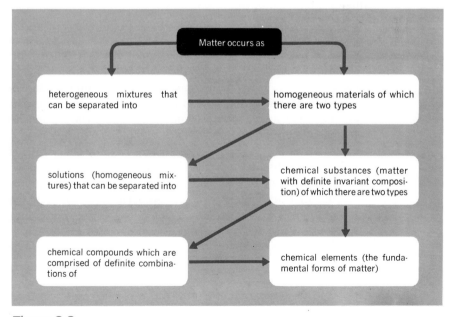

Figure 2-2

made up of combinations of chemical elements. For instance, the compound water is composed of the chemical elements hydrogen and oxygen. Chemistry is concerned with the study of the properties of the elements and with the ways in which the elements form compounds.

Chemical compounds are composed of elements, but a given compound will always contain the same elements in definite proportions by mass. For instance, the compound water always contains hydrogen and oxygen, and any sample will contain 8 g of oxygen to every 1 g of hydrogen. Compounds are characterized by this kind of definite composition. You might wonder if mixing sugar in water produces a compound since the resulting mixture appears to be pure and homogeneous. Mixtures of this kind do appear to be homogeneous but do not have definite composition. That is, sugar and water mixtures can contain variable amounts of sugar in water. Since such variable compositions occur, these mixtures are not compounds. Homogeneous mixtures of variable composition are called **solutions.** Solutions are mixtures of substances that dissolve in one another to become single-phase (i.e., solid, liquid, or gas) mixtures. Hold on, you may say, we can see the parts of a heterogeneous mixture, but how can a solution that looks pure be a mixture. First of all you have to realize that matter can be broken up into tiny pieces and mixed together. Some mixtures, like milk, appear to the eye to be pure substances but are actually heterogeneous mixtures. The difference between heterogeneous mixtures and solutions is that solutions involve very small particles of matter (far below even microscopic observation) that are intimately mixed, and heterogeneous mixtures involve larger aggregates of such particles. Solutions are quite common and include such liquids as wine and ocean water.

Solutions that are made by dissolving one substance in another are called binary solutions. In a binary solution, one of the substances is called a solvent, and the other is called a solute. For solutions involving the mixing of substances of two different phases, the **solvent** is considered to be the substance that is of the same phase as the resulting solution. For example, when solid sugar is dissolved in liquid water to form a liquid solution, the water is called the solvent, and the sugar is called the **solute.** If two substances are of the same phase, the solvent is usually considered to be the component present in the greater amount. Since solutions are a special type of mixture, the composition of solutions depends on the amount of each component that is used to make the solution. Often there are limits to the amount of solute that can be mixed with a solvent. The amount of solute that will dissolve in a given amount of solvent at a specific temperature is called the **solubility** of the solute in that solvent at that temperature. The solubilities of substances in liquid water are sometimes noted as distinguishing properties of the substance. Substances that do not dissolve in a solvent are said to be insoluble in that solvent. Solutions are discussed in greater detail in Chapter 10.

In the environment, a substance is in the solid, liquid, or gaseous state. Substances can be changed from one state to another. Such changes are called phase changes and can be made to occur by changing the environment of the substance. For example, by heating ice (solid water) it can be made to **melt,** change from the solid to the liquid state. By heating, liquid water can be **vaporized,** changed from the liquid state to the vapor or gaseous state. Under certain conditions some solids can be converted directly from the solid state to the gaseous state. You have probably observed this with a piece of dry ice (solid carbon dioxide). This phase change is called **sublimation.** Cooling or removal of heat can also cause certain phase changes to occur. A gas can be cooled and caused to change to the liquid state. This is what happens when a window becomes fogged with water droplets that are formed from the water vapor in the air. This phase change is called **condensation.** A liquid can be cooled and caused to change to the solid state. This is, of course, what happens when water freezes to form ice. Such a phase change is called **solidification, crystallization,** or **freezing.** The conditions under which certain phase changes occur can be used to characterize and identify substances. The temperature at which a solid melts is called the **melting point** of the solid. For instance, ice melts at about 0°C. Similarly, the temperature at which a liquid solidifies under specific conditions is called the **freezing point** of the liquid. For a pure substance the freezing and melting points are the same. The temperature at which a liquid under specific conditions boils is called the **boiling point** of the liquid. Liquid water boils at about 100°C. The observation of freezing, melting, and boiling points is often carried out in the laboratory for purposes of describing chemical substances.

2-3   The Elements

Since the elements are the fundamental building units of matter, it is important to understand their natures and differences. Much time and effort have been devoted to the isolation, purification, and description of the chemical elements. To date, 105 different elements have been identified. Most of these elements occur in nature as constituents of compounds or in the uncombined form. Some of the elements are not found in nature but have been synthesized by nuclear scientists. These elements are discussed in Chapter 15. Many of the elements that occur in compounds are separated from these compounds and used in numerous ways. For example, metals such as iron and aluminum are obtained (refined) from compounds contained in ores.

Each of the **elements** has a unique **name** and **symbol.** Since the elements have been isolated and identified over a period of centuries, these names and symbols have historical origin. An alphabetical list of the element names and corresponding symbols is given in Table 2-1. As you look over the list, note that most of the symbols are derived from the names of the elements. However, in some cases the symbols are derived from Latin names. For instance, iron has the symbol Fe, and mercury has the symbol Hg. Knowledge of the names of the elements and the symbols that represent them is fundamental to any discussion

Table 2-1   The Elements

Name	Symbol
Actinium (Gr. aktis, aktinos, beam or ray)	Ac
Aluminum (L. alumen, alum)	Al
Americium (the Americas)	Am
Antimony (L. antimonium, stibium, mark)	Sb
Argon (Gr. argon, inactive)	Ar
Arsenic (L. arsenicum, Gr. arsenikon, yellow orpiment — identified with arsenikos, male, from the belief that metals were different sexes — Arab. az-zernikh, the orpiment from Persian zerni-zar, gold)	As
Astatine (Gr. astatos, unstable)	At
Barium (Gr. barys, heavy)	Ba
Berkelium (Berkeley, home of Univ. of Calif.)	Bk
Beryllium (Gr. berryllos, beryl; also called Glucinium or Glucinum, Gr. glykys, sweet)	Be
Bismuth (Ger. Weisse Masse, white mass; later Wismuth and Bisemutum)	Bi
Boron (Ar. Buraq, Pers. Burah)	B
Bromine (Gr. bromos, stench)	Br
Cadmium (L. cadmia; Gr. kadmeia — ancient name for calamine, zinc carbonate)	Cd
Carbon (L. carbon, charcoal)	C
Cerium (named for the asteroid Ceres)	Ce
Cesium (L. caesius, sky blue)	Cs
Chlorine (Gr. chloros, greenish-yellow)	Cl
Chromium (Gr. chroma, color)	Cr
Cobalt (Kobold, from the German, goblin or evil spirit)	Co
Copper (L. cuprum, from the island of Cyprus)	Cu
Curium (Pierre and Marie Curie)	Cm
Dysprosium (Gr. dysprositos, hard to get at)	Dy
Einsteinium (Albert Einstein)	Es
Erbium (Ytterby, a town in Sweden)	Er
Europium (Europe)	Eu
Fermium (Enrico Fermi)	Fm
Fluorine (L. and F. fluere, flow or flux)	F
Francium (France)	Fr
Gadolinium (gadolinite — a mineral named for Gadolin, a Finnish chemist)	Gd
Gallium (L. Gallia, France)	Ga
Germanium (L. Germania, Germany)	Ge
Gold (Sanskrit Jval; anglo-Saxon gold; L. aurum, shining dawn)	Au
Hafnium (Hafnia, Latin name for Copenhagen)	Hf
Helium (Gr. helios, the sun)	He
Holmium (L. Holmia, for Stockholm)	Ho
Hydrogen (G. hydro, water, and genes, forming)	H
Indium (from the brilliant indigo line in its spectrum)	In
Iodine (Gr. iodes, violet)	I
Iridium (L. iris, rainbow)	Ir
Iron (Anglo-Saxon, iron; L. ferrum)	Fe
Krypton (Gr. kryptos, hidden)	Kr
Lanthanum (Gr. lanthanein, to lie hidden)	La

Table 2-1   (Continued).

Name	Symbol
Lawrencium (Ernest O. Lawrence, inventor of the Cyclotron)	Lr
Lead (Anglo-Saxon lead; L. plumbum)	Pb
Lithium (Gr. lithos, stone)	Li
Lutetium (Lutetia, ancient name for Paris)	Lu
Magnesium (Magnesia, district in Thessaly)	Mg
Manganese (L. magnes, magnet)	Mn
Mendelevium (Dmitri Mendeleev)	Md
Mercury (Planet Mercury; hydrargyrum, liquid silver)	Hg
Molybdenum (Gr. molybdos, lead)	Mo
Neodymium (Gr. neos, new and didymos, twin)	Nd
Neon (Gr. neos, new)	Ne
Neptunium (planet Neptune)	Np
Nickel (Ger. Nickel, Satan or "Old Nick" and from kupfernickel, Old Nick's copper)	Ni
Niobium (Niobe, daughter of Tantalus)	Nb
Nitrogen (L. nitrum, Gr. nitron, native soda; genes, forming)	N
Nobelium (Alfred Nobel)	No
Osmium (Gr. osme, a smell)	Os
Oxygen (Gr. oxys, sharp, acid and genes, forming; acid former)	O
Palladium (named after the asteroid Pallas; Gr. Pallas, goddess of wisdom)	Pd
Phosphorus (Gr. phosphoros, light-bearing; ancient name for the planet Venus when appearing before sunrise)	P
Platinum (Sp. platina, silver)	Pt
Plutonium (Planet Pluto)	Pu
Polonium (Poland, native country of Mme. Curie)	Po
Potassium (English, potash—pot ashes; L. kalium; Arab. quali, alkali)	K
Praseodymium (Gr. prasios, green, and didymos, twin)	Pr
Promethium (Prometheus, who, according to mythology, stole fire from heaven	Pm
Protactinium (Gr. protos, first)	Pa
Radium (L. radius, ray)	Ra
Radon (from radium)	Rn
Rhenium (L. Rhenus, Rhine)	Re
Rhodium (Gr. rhodon, rose)	Rh
Rubidium (L. rubidius, deepest red)	Rb
Ruthenium (L. Ruthenia, Russia)	Ru
Samarium (Samarskite, a mineral)	Sm
Scandium (L. Scandia, Scandinavia)	Sc
Selenium (Gr. Selene, moon)	Se
Silicon (L. silex, silicis, flint)	Si
Silver (Anglo-Saxon; Seolfor, siolfur; L. argentum)	Ag
Sodium (English, soda; Medieval Latin, sodanum, headache remedy; L. natrium)	Na
Strontium (Strontian, town in Scotland)	Sr
Sulfur (Sanskrit, sulvere; L. sulphurium)	S
Tantalum (Gr. Tantalos, mythological character—father of Niobe)	Ta
Technetium (Gr. technetos, artificial)	Tc
Tellurium (L. tellus, earth)	Te

Table 2-1   (Continued).

Name	Symbol
Terbium (Ytterby, village in Sweden)	Tb
Thallium (Gr. thallos, a green shoot or twig)	Tl
Thorium (Thor, Scandinavian god of war)	Th
Thulium (Thule, the earliest name for Scandinavia)	Tm
Tin (Anglo-Saxon, tin; L. stannum)	Sn
Titanium (L. Titans, the first sons of the Earth, mythology)	Ti
Tungsten (Swedish, tung sten, heavy stone)	W
Uranium (Planet Uranus)	U
Vandium (Scandinavian goddess, Vanadis)	V
Xenon (Gr. xenon, stranger)	Xe
Ytterbium (Ytterby, village in Sweden)	Yb
Yttrium (Ytterby, village in Sweden)	Y
Zinc (Ger. Zink, or obscure origin)	Zn
Zirconium (Arabic zargum, gold color)	Zr

Adapted from *Handbook of Chemistry and Physics*. Chemical Rubber Publishing Co.

of chemistry. It is necessary to memorize some of the element names and symbols. A list of the most important ones is given in Table 2-2.

Many elements are found on earth in very small amounts while some occur in much greater amounts. The distribution of the elements on earth is discussed in Section 2-14.

Table 2-2   Some Important Elements

Element	Symbol	Element	Symbol
Aluminum	Al	Iron	Fe
Antimony	Sb	Lead	Pb
Arsenic	As	Lithium	Li
Barium	Ba	Magnesium	Mg
Beryllium	Be	Manganese	Mn
Bismuth	Bi	Mercury	Hg
Boron	B	Nickel	Ni
Bromine	Br	Nitrogen	N
Cadmium	Cd	Oxygen	O
Calcium	Ca	Phosphorus	P
Carbon	C	Platinum	Pt
Cesium	Cs	Potassium	K
Chlorine	Cl	Silicon	Si
Chromium	Cr	Silver	Ag
Cobalt	Co	Sodium	Na
Copper	Cu	Strontium	Sr
Fluorine	F	Sulfur	S
Gold	Au	Tin	Sn
Helium	He	Tungsten	W
Hydrogen	H	Uranium	U
Iodine	I	Zinc	Zn

2-4   The Nature of Elements

Since the beginnings of reflective thought humans have wondered about the nature of matter at a submicroscopic, unseeable level. If you imagine dividing a piece of pure substance into smaller and smaller bits, will you soon reach a characteristic particle that cannot be further broken down, or could this subdivision continue indefinitely? The Greek philosophers pondered this question. Some philosophers, realizing that they could feel the wind blow and that bricks in the street seemed to wear away in unseeable pieces, considered matter to be made up of tiny units or particles. The Greek philosopher Democritus (about 400 B.C.) proposed that matter was composed of tiny, indestructible particles that he called **atoms,** from the Greek word for indestructible. However, the dominant philosophers such as Aristotle and Plato rejected this idea and considered matter to be continuous and composed of only one substance. The Aristotelian view prevailed for centuries. In the Middle Ages, the alchemists investigated and experimented with matter with the intention of trying to convert other elements into gold. This feat was never accomplished, but many compounds and elements were discovered in the process that later contributed to the understanding of matter.

By the 1600s alchemy was well on the decline, and the pioneers of modern chemistry, the first chemists, began to develop the science of chemistry. These pioneers included such men as Robert Boyle (1627–1691), Joseph Priestly (1733–1804), Antoine Lavoisier (1743–1794), and John Dalton (1766–1844). In the next few hundred years significant development of chemistry occurred. Chemists actually experimented with substances and were able to identify many elements and compounds. They were able to decompose compounds into the constituent elements and to form compounds from the elements. Furthermore, with the development of weighing methods, they were able to weigh samples of elements and compounds. This allowed the precise observation of the relative masses of elements combined in compounds and provided evidence for the development of a theory concerning the submicroscopic nature of matter.

Let us consider some of this evidence. It was realized that when elements were combined to form compounds or compounds were decomposed to form elements, no mass was lost or gained. For example, when 56 g of iron were compounded with 32 g of sulfur, the resulting compound had a mass of 88 g. This realization indicated that in a chemical process mass is conserved, and this was stated as the **law of conservation of matter:**

Matter is not created or destroyed in a chemical process.

This idea was surprising to many since they thought that, for example, when a candle burned, its matter was lost. It can be shown that when a candle is burned in a closed container of air the mass of the candle and air in the container is not changed. Actually, the products of the burning candle are given off as gases.

Another observation that developed was the fact that a specific compound always contained the same elements. This idea was mentioned previously, but let us consider it in more detail. The common compound salt or sodium chloride will be used as an example. When salt was produced in the laboratory or obtained from a natural source, it was always found to be a compound of the elements sodium and chlorine. Furthermore, pure compounds were always found to have the same amounts of each element. For instance, 100 g of salt would always contain 39.4 g of sodium and 60.6 g of chlorine (39.4 percent sodium and 60.6 percent chlorine). The observation became known as the **law of constant composition:**

A compound always contains the same elements, and they are present in the definite proportions by mass.

An additional observation was that a given element tended to form specific kinds of compounds with other elements. Sometimes, two or more elements formed similar compounds with another element. Sodium and potassium were found to form similar compounds with chlorine. Also, on occasion, it was found that two elements could form more than one kind of compound. Hydrogen and oxygen commonly form the compound water but can also form the compound hydrogen peroxide. Water contains 88.8 percent oxygen and 11.2 percent hydrogen by mass, and hydrogen peroxide contains 5.9 percent hydrogen and 94.1 percent oxygen.

All these observations indicated that elements must consist of some kind of building units that could join to form compounds.

2-5   The Atomic Theory

In 1803, John Dalton, drawing from the work of many early scientists, proposed a theory or model of the submicroscopic nature of matter. This is the **atomic theory,** which can be stated as follows (see Figure 2-3):

1. Elements are composed of tiny, fundamental particles of matter called atoms. (Dalton used the term atom as recognition of the brilliant suggestion of Democritus 2000 years before.)
2. Atoms of the same element are the same but differ from atoms of other elements.
3. Atoms enter into combinations with other atoms to form compounds.

The atomic theory was and is very important since it provides a mental picture of matter. An atom can be pictured as a tiny spherical particle. An **atom** is the smallest representative particle of an element. Each kind of element has a unique kind of atom, which has mass. Atoms are indeed very tiny particles. A penny contains about

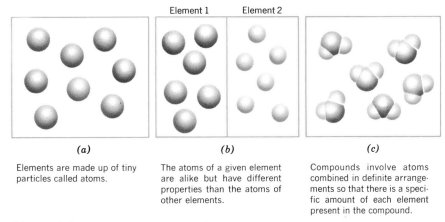

Element 1   Element 2

| (a) | (b) | (c) |

Elements are made up of tiny particles called atoms.

The atoms of a given element are alike but have different properties than the atoms of other elements.

Compounds involve atoms combined in definite arrangements so that there is a specific amount of each element present in the compound.

Figure 2-3   Illustration of the atomic theory.

30,000,000,000,000,000,000,000 (3×10^{22}) copper atoms. The masses of the atoms of the various elements differ from one another, but atoms have masses in the range of 10^{-24}–10^{-23} g. As you look at a sample of a compound, such as a salt, you can imagine that it consists of numerous combined atoms of sodium and chlorine. Even as you view a rock or a flower, you can imagine that it consists of a complex combination of atoms. The natures of such atomic combinations are discussed in Chapter 4.

Let us consider how the atomic theory explains some of the observations mentioned previously. The law of constant composition becomes clear when it is realized that compounds involve definite combinations of atoms. That is, a given compound will always include the same atoms in the same amounts. (See Figure 2-4.) For instance, water always contains two combined atoms of hydrogen for every combined atom of oxygen. Since the atoms of elements differ from one another, they form different kinds of compounds with other elements. In some cases it is possible to have more than one combination of the atoms of two elements. (See Figure 2-4.)

The atomic theory served as the foundation for the development of chemistry and is still, at this time, one of the most fundamental models in chemistry. Once the idea of the atom and the idea that elements were unique had become established, chemists searched nature for the various elements. Over the years, numerous elements were discovered, and now 105 elements are known.

2-6   Atomic Weights

The atom can be considered to be the smallest unit or particle of an element. Compounds involve these atoms from different elements combined according to some definite pattern. The atomic theory allows us to view the elements as consisting of these characteristic particles that can combine with one another to form compounds.

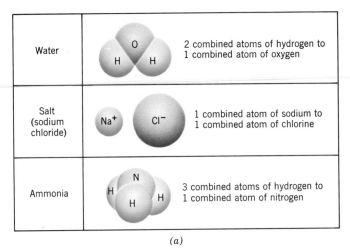

(a)

A compound always contains the same elements in the same relative amounts.

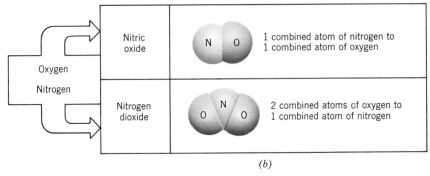

(b)

Two elements may form more than one kind of compound.

Figure 2-4   Compounds viewed using the atomic theory. From *Understanding Chemistry: From Atoms to Attitudes* by T. R. Dickson, Wiley, N.Y., 1974.

When an attempt is made to isolate an individual atom of an element, it is found that such particles are so small that it is not possible to observe them by any available method. Consequently, it is not possible to make the same types of measurements on atoms that we are able to make on larger samples of substances that are aggregates of atoms. Samples of substances with which we normally deal consist of a fantastic number of atoms. Atoms are extremely small particles. How, then, is it possible to describe atoms so that we can distinguish between atoms of different elements? Fortunately there are methods by which the relative masses of the atoms of elements can be determined.

When the masses of the atoms of a given element are compared, it is found that all of its atoms do not have the same mass. In fact, it is found that definite percentages of the atoms of a given element have specific masses. For example, 75.5 percent of the atoms of naturally occurring chlorine have one mass, while 24.5 percent of the atoms have another mass. Atoms of the same element that have different masses are called **isotopes** of that element. Of the elements that exist in nature, 22 have no naturally occurring isotopes and, thus, exist in the form of one type of atom. The other elements have two or more isotopes. For instance, hydrogen has 2 natural isotopes, oxygen has 3, carbon has 2 and tin has 10. (See Figure 2-5.) Even though an element has isotopes, all of the atoms of that element behave as atoms of that element. Extensive studies of naturally occurring elements have shown that the percentages of the isotopes for a given element are generally constant. No matter where we obtain a sample of that element on earth, it will generally contain the same percentage of each isotope. The percentage distributions of isotopes of natural elements are used in the expression of the relative masses of the atoms of elements.

Percentage. Percentage is a common way of expressing what fraction one quantity is of another. Percentage means parts per hundred and is represented by %. Thus, 20% means 20 parts per 100 parts or 20/100. A percentage can be expressed as a decimal fraction by dividing by 100. So, 20% as a decimal fraction is 0.20. To calculate what percentage one quantity, a, is of another quantity, b, a is divided by b and multiplied by 100:

$$\text{percentage} = \left(\frac{a}{b}\right) 100$$

Percentages can use various bases of comparisons. For example, if it is found that, in a large group of people, 3 out of every 100 are left handed, we say that 3% are left handed. The basis of comparison is numbers of people. Very often percentages are expressed on a basis of mass or weight. A deposit of iron ore may contain 20 g of iron in each 100 g of ore. Thus, we say that the ore contains 20% iron by mass. It is essential to state the basis of comparison when giving a percentage. When this is not done, the context in which the percentage is used must indicate the basis of comparison.

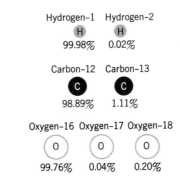

Hydrogen–1   Hydrogen–2
99.98%   0.02%

Carbon–12   Carbon–13
98.89%   1.11%

Oxygen–16 Oxygen–17 Oxygen–18
99.76%   0.04%   0.20%

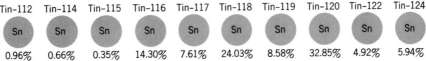

Tin–112	Tin–114	Tin–115	Tin–116	Tin–117	Tin–118	Tin–119	Tin–120	Tin–122	Tin–124
0.96%	0.66%	0.35%	14.30%	7.61%	24.03%	8.58%	32.85%	4.92%	5.94%

Figure 2-5   Some typical isotopes. The isotopes are given numbers that indicate the approximate relative masses. The percentages indicate how much of the naturally occurring element is composed of each isotope.

Let us now consider how the masses of atoms of elements are expressed. Atoms are such tiny particles that it is not possible to weigh individual atoms. Nevertheless, it is possible to determine the relative masses of elements by special chemical and instrumental methods. The relative masses of the atoms of all naturally occurring elements have been determined through many years of experimental work by chemists. A relative mass expresses how the mass of an atom of one element compares to the mass of an atom of another element. For example, a typical sulfur atom has a mass that is twice the mass of a typical oxygen atom.

To illustrate the significance of relative masses, consider two persons with relative masses of 1.2 to 1. This means the first person has a mass that is 1.2 times the second. Relative masses allow for the use of any units of mass. For instance if the second person has a mass of 50 kg, the first must have a mass of 60 kg (1.2×50 kg $= 60$ kg). Furthermore, if we want to work in pounds the lighter person is 110 lb, and the other must weigh 132 lb (1.2×110 lb $= 132$ lb). Once we know the relative mass we can express the masses in any unit if we know the mass of one of the individuals. In fact we can devise a new unit if desired. For instance, suppose we arbitrarily assigned the lighter person a mass of 10 people weight units or pwu. The other person's mass would be 12 pwu (1.2×10 pwu $= 12$ pwu). In fact anyone could express their mass in pwu by finding their mass relative to the standard person (110 lb) and multiply by 10 pwu. What is your mass in pwu?

Since atomic masses are very small when expressed in units of grams, a unit of mass that is more convenient for expressing masses on an atomic level has been devised. This unit of mass used for atomic masses is arbitrary and is established only for convenience. The masses of normal objects can be expressed in grams, but to express the masses of atoms, an **atomic mass unit** (**amu**) has been established. This unit is based on the following definition.

An atom of the most common isotope of carbon (referred to as carbon-12) is considered to have a mass of exactly 12 atomic mass units (amu). Therefore, one amu is 1/12 the mass of an atom of carbon-12.

The figure 12 amu was chosen so that no isotopes of an element would have a relative mass of less than 1 amu. The masses of the isotopes of the elements can be expressed in terms of atomic mass units. For example, if an isotope has a mass that is 2.56 times as great as the mass of a carbon-12 atom, then its mass is 30.72 amu (12 amu $\times$ 2.56).

When comparing the masses of elements, it is not convenient to compare the masses of all of the isotopes of each element. It is more convenient to refer to the masses of the hypothetical average atoms of the elements. The mass of an average atom of an element is called the atomic weight of the element. The **atomic weight** can be defined as follows.

The atomic weight of an element is the weighted average mass of the isotopes of the naturally occurring element expressed in atomic mass units (amu).

A weighted average is just the average calculated by considering the relative amounts. For example, the average of a set of numbers can be found by adding the numbers and dividing by the number of values.

$$\begin{array}{c} 2 \\ 2 \\ 3 \\ \underline{5} \\ 12 \end{array} \qquad \frac{12}{4} = 3 \text{ (average value)}$$

This list of numbers could be expressed as

$$\begin{array}{cc} 50\% & 2 \\ 25\% & 3 \\ 25\% & 5 \end{array}$$

To find the average in this case we multiply each number by its fractional abundance (%/100) and then add these products.

$$\begin{array}{l} 2 \ (.50) = 1.00 \\ 3 \ (.25) = 0.75 \\ 5 \ (.25) = \underline{1.25} \\ \qquad\qquad 3.00 \text{ (average value)} \end{array}$$

For example, the atomic weight of silicon could be calculated using the following data concerning the naturally occurring isotopes of silicon.

Isotope	Mass Relative to Carbon-12	Percentage Natural Abundance
Silicon-28	27.98 amu	92.21
Silicon-29	28.98 amu	4.70
Silicon-30	29.97 amu	3.09

The weighted average mass is calculated by multiplying each isotopic mass by the corresponding fractional abundance (percentage abundance/100) and then adding the products.

$$\begin{array}{l} (.9221)(27.98 \text{ amu}) = 25.80 \text{ amu} \\ (.0470)(28.98 \text{ amu}) = \ 1.36 \text{ amu} \\ (.0309)(29.97 \text{ amu}) = \ \underline{\ .93 \text{ amu}} \\ \qquad\qquad\qquad\qquad\qquad 28.09 \text{ amu} \end{array}$$

For an element with no isotopes, the atomic weight is the mass of an atom of the element. For an element with isotopes the atomic weight expresses the average mass of an atom of an element considering the contribution of each isotope. Atomic weights are very useful in chemistry and will be frequently referred to and used in calculations. A list of the atomic weights of the elements is given inside the back cover of the book.

2-7   The Mole

When working with elements, we never deal with individual atoms, and, in the laboratory, we normally work with amounts of elements and compounds that are measured in grams. Thus, the amu is not a very convenient unit for normal work. That is, it is more reasonable to work with grams of elements rather than atomic mass units. It is necessary, then, to be able to deal with the relative masses of elements in units of grams. However, several grams of an element would contain around one septillion (10^{24}) atoms. To solve this problem chemists have devised a way in which the mass of a sample of an element can be related to the number of atoms in the sample. To relate the relative masses of elements in units of grams, the following definition is established.

A **mole** of an element is the amount in grams that contains the same number of atoms that is contained in exactly 12 g of carbon-12.

As an explanation of this definition consider that the atomic weights express the relative masses of the atoms of elements in terms of atomic mass units. The mole allows for the use of relative masses on a gram basis. This can be illustrated as follows:

$$\left(\frac{12 \text{ amu}}{1 \text{ atom carbon-12}}\right) \qquad \left(\frac{16 \text{ amu}}{1 \text{ atom O}}\right)$$

$$\left(\frac{12 \text{ g}}{\text{certain number of carbon-12 atoms}}\right) \qquad \left(\frac{16 \text{ g}}{\text{certain number of O-atoms}}\right)$$

The number of carbon atoms is the same as the number of oxygen atoms.

Twelve grams of carbon-12 will contain a specific number of carbon-12 atoms. Since the relative mass of carbon-12 to oxygen is 12 to 16, then 16 g of oxygen will contain the same number of atoms as 12 g of carbon-12. This number of atoms is a mole of atoms. The amount of any element corresponding to a mole will be the number of grams numerically the same as the atomic weight. This is the basic reason for the definition of the mole. That is, an amount of an element in grams numerically the same as the atomic weight will contain a mole of atoms of that element (e.g., 1 mole sodium = 23.0 g, 1 mole oxygen = 16.0 g)

A mole of an element consists of a definite number of atoms. This number has been determined by experimental methods and is called Avogadro's number. **Avogadro's number** is expressed as

$$\left(\frac{6.02 \times 10^{23} \text{ atoms}}{1 \text{ mole}}\right)$$

Avogadro's number provides an alternate view of the mole. A mole of an element is considered to be the mass of Avogadro's number of atoms of that element. Avogadro's number is extremely large. Consider that if there are about 3 billion (3×10^9) people on earth, there are 200 trillion (2×10^{14}) times as many atoms in a mole. One mole of an element contains Avogadro's number of atoms, which is 6.02×10^{23} atoms.

The **number of grams per mole** or **molar mass** of each element can be determined from the atomic weights. The molar mass of an element can be used to determine the number of atoms in any sample of the element if the mass is known. As an analogy that illustrates this idea, consider a situation in which we want to count a large number of bolts of a given size. Rather than actually count the bolts, we can devise a way to determine the number of bolts in a known mass of bolts. This can be done by determining the number of grams of bolts per dozen bolts. For instance, suppose we find that there are 40 grams per dozen bolts. This can be expressed in a factor form as:

$$\left(\frac{40 \text{ g}}{1 \text{ doz bolts}}\right) \quad \text{or} \quad \left(\frac{1 \text{ doz bolts}}{40 \text{ g}}\right)$$

Now suppose we had an 800-g sample of bolts. The number of dozen bolts in the sample can be deduced by multiplying the 800 g by the factor 1 doz bolts/40 g.

$$800 \text{ g} \left(\frac{1 \text{ doz bolts}}{40 \text{ g}}\right) = 20 \text{ doz bolts}$$

The point is that the factor can be used to determine the number of dozen bolts in any sample of bolts of known mass. Furthermore, the factor can be used to determine the number of grams needed to contain a specific number of dozen. For instance, suppose the number of grams per dozen of nuts to match the bolts is

$$\left(\frac{20 \text{ g}}{1 \text{ doz nuts}}\right)$$

If we want to determine the number of grams of nuts needed to supply enough nuts to match the 20 dozen bolts we can multiply 20-dozen nuts by the above factor.

$$20 \text{ doz nuts} \left(\frac{20 \text{ g}}{1 \text{ doz nuts}}\right) = 400 \text{ g}$$

Now let us consider how the molar mass of an element can be used in a similar fashion as the examples given above. The molar mass of an element is obtained from a list of atomic weights. (See the list inside the cover of the book or the periodic table.) The molar mass can be expressed as a factor giving the number of grams per mole of the element. For example, the molar mass of sodium is

$$\left(\frac{23.0 \text{ g}}{1 \text{ mole Na}}\right)$$

and the molar mass of sulfur is

$$\left(\frac{32.1 \text{ g}}{1 \text{ mole S}}\right)$$

These factors can be used to convert the mass of a sample of the element to the number of moles or vice versa. The number of atoms of an element is often expressed in terms of moles rather than the absolute number of atoms just as we often use the term dozen when referring to donuts or eggs rather than stating the actual number. In a sense the mole is the chemist's dozen.

Example 2-1   How many moles of sodium are in a 46.0 g-sample of sodium?
The sample mass is multiplied by the inverse of the molar mass to convert grams to moles.

$$46.0 \text{ g} \left(\frac{1 \text{ mole Na}}{23.0 \text{ g}} \right) = 2.00 \text{ moles Na}$$

Example 2-2   What is the mass of a sample of sulfur that contains 0.500 moles of sulfur?
The mass is found by multiplying the number of moles by the molar mass.

$$0.500 \text{ moles S} \left(\frac{32.1 \text{ g}}{1 \text{ mole S}} \right) = 16.0 \text{ g}$$

Using the mole concept, we can determine the number of moles of an element contained in a certain number of grams of that element. Once the number of moles has been found, the actual number of atoms can be determined using Avogadro's number. Thus, the mole concept provides for the determination of the number of atoms in a given mass of an element. Furthermore, if we have a certain number of moles of an element, we can easily determine the number of grams that we have since we know the number of grams per 1 mole of the element. The number of grams per mole of an element can be used as a property conversion factor to convert from grams of an element to the number of moles or from the number of moles to grams.

Example 2-3   How many iron atoms are in a 112-g sample of iron?
First we determine the number of moles of iron by multiplying the sample mass by the factor 1 mole Fe/55.8 g.

$$112 \text{ g} \left(\frac{1 \text{ mole Fe}}{55.8 \text{ g}} \right)$$

Next this product is multiplied by Avogadro's number to give the number of iron atoms.

$$112 \text{ g} \left(\frac{1 \text{ mole Fe}}{55.8 \text{ g}} \right) \left(\frac{6.02 \times 10^{23} \text{ atoms Fe}}{1 \text{ mole Fe}} \right) = 1.21 \times 10^{24} \text{ atoms Fe}$$

2-8   Formulas and the Molar Mass of Compounds

Atoms are basic chemical species. A species can be considered to be an identifiable chemical particle. Atoms are the building blocks of substances. Other, more complex species can be formed from atoms. Two or more atoms may bind together to form a chemical species called a molecule. For example, two atoms of hydrogen and one atom of oxygen are bound together in the form of a molecule of water. Another impor-

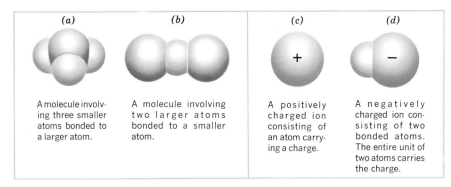

(a)	(b)	(c)	(d)
A molecule involving three smaller atoms bonded to a larger atom.	A molecule involving two larger atoms bonded to a smaller atom.	A positively charged ion consisting of an atom carrying a charge.	A negatively charged ion consisting of two bonded atoms. The entire unit of two atoms carries the charge.

Figure 2-6   Idealized representations of molecules and ions.

tant chemical species is the ion. An ion is an atom or a bonded group of atoms that carries an electrical charge. (See Figure 2-6.) The natures of these chemical species are discussed in Chapter 4. They are mentioned here to emphasize the fact that chemical compounds consist of aggregates of atoms that exist as stable units. That is, compounds are composed of chemically combined atoms in the form of molecules or ions. Compounds contain atoms combined in definite proportions. For example, water contains two combined atoms of hydrogen for every one combined atom of oxygen. Each pure compound contains a definite proportion of combined atoms of each element present in the compound. Since this is true, compounds can be represented by a **formula** indicating the kinds of elements present and the relative number of combined atoms of each element. Elements in the combined state are represented by their usual symbol. Thus, formulas of compounds include the symbol of each element present, and each symbol is followed by a subscript indicating the relative number of atoms. (See Figure 2-7.) For example, the formula for water is H_2O, which indicates that the compound water is made up of hydrogen and oxygen, and that the compound contains twice as many combined atoms of hydrogen as there are combined atoms of oxygen. Note that when the subscript is one it is not written as the number one but is omitted. Formulas are very useful. The formula of a compound tells us which elements make up the compound and the relative number of combined atoms of each element. At this point, we are only concerned with the fact that a compound can be represented by a formula. However, as we shall see, the real significance of the formula depends on the nature of the chemical species that makes up the compound. Incidentally a formula represents the chemical nature of a compound. Each compound has a unique chemical name. Nomenclature is discussed in Chapter 6. When you encounter a formula it can be read by just pronouncing the letters and numbers. A few examples are

H_2O read H-two-O
NaCl read N-A-C-L
K_2SO_4 read K-two-S-O-four
NH_3 read N-H-three

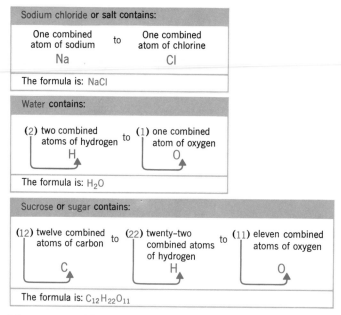

Figure 2-7   The formulas of some compounds.

The formula of a compound can be interpreted as indicating the relative number of moles of each element in the compound. For instance, the formula of water, H_2O, indicates that in the compound water there are twice as many moles of combined hydrogen as there are moles of combined oxygen. This just means that if we had a sample of water that contained Avogadro's number (6.02×10^{23}) of combined oxygen atoms, it would also contain two times Avogadro's number of combined hydrogen atoms. This is illustrated in Figure 2-8. Since a formula can be interpreted in this manner, it is possible to express the number of grams associated with a mole of a compound. A **mole of a compound** is the amount of the compound (in grams) that contains the number of moles of each element given by the subscripts in the formula. A mole of water would be the amount that contains 1 mole of combined oxygen atoms and 2 moles of combined hydrogen atoms (See Figure 2-8.) Why would we want to deal with a mole of a compound? Well, as was the case with elements, we normally work with amounts of compounds measured in grams. By the use of the molar interpretation of a formula, it is possible to refer to the number of grams per mole or molar mass of a compound.

Before illustrating the determination of the molar mass of a compound, consider the following analogy. Suppose we were working with a nut-bolt combination ("compound") consisting of one bolt and two nuts (BN_2). Furthermore, we will assume the same nuts and bolts discussed in Section 2-7. To determine the mass per dozen of the BN_2 combination we must consider how each component of the combination contributes to the mass. One dozen combinations contain 1 dozen bolts that make up part of the mass. This contribution is

$$1 \text{ doz B} \left(\frac{40 \text{ g}}{1 \text{ doz B}} \right) = 40 \text{ g}$$

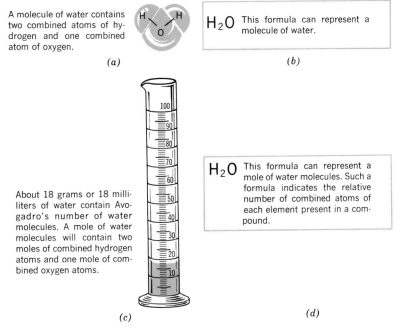

A molecule of water contains two combined atoms of hydrogen and one combined atom of oxygen.

(a)

H_2O This formula can represent a molecule of water.

(b)

About 18 grams or 18 milliliters of water contain Avogadro's number of water molecules. A mole of water molecules will contain two moles of combined hydrogen atoms and one mole of combined oxygen atoms.

H_2O This formula can represent a mole of water molecules. Such a formula indicates the relative number of combined atoms of each element present in a compound.

(c)

(d)

Figure 2-8 The relationship between a mole of a compound and the number of moles of the elements making up the compound.

One dozen combinations also contain 2 dozen nuts that contribute the following mass

$$2 \text{ doz N} \left(\frac{20 \text{ g}}{1 \text{ doz N}} \right) = 40 \text{ g}$$

The mass per dozen combinations is the sum of these two masses or 80 grams. This can be expressed as

$$\left(\frac{80 \text{ g}}{1 \text{ doz BN}_2} \right)$$

The **molar mass of a compound** is determined by multiplying the number of grams per mole of each element by the corresponding subscript in the formula and then adding up these products. The molar mass for each element can be obtained from an atomic weight table. For example, the number of grams per mole of water, H_2O, can be determined as follows:

$$H \quad 2 \left(\frac{1.008 \text{ g}}{1 \text{ mole H}} \right) = 2.016 \text{ g}$$

$$O \quad 1 \left(\frac{16.00 \text{ g}}{1 \text{ mole O}} \right) = \frac{16.00 \text{ g}}{18.016 \text{ g}} \quad \text{or} \quad \left(\frac{18.02 \text{ g}}{1 \text{ mole H}_2O} \right)$$

For simplicity, the calculations can be

$$
\begin{array}{ll}
H & 2(1.008 \text{ g}) = \quad 2.016 \text{ g} \\
O & \quad 16.00 \text{ g} = \underline{16.00 \text{ g}} \\
& \quad\quad\quad\quad\quad 18.016 \text{ g} \quad \text{or} \quad \left(\frac{18.02 \text{ g}}{1 \text{ mole H}_2O} \right)
\end{array}
$$

Normally we would carry an extra digit in the calculations and round off as shown. See Appendix I for round off rules.

Significant Digits in Addition-Subtraction Calculations. When a calculation is carried out, a question arises concerning how many significant digits there should be in the answer. Since the factors involved have specific numbers of significant digits, the calculated result obtained from the factors must have a specific number of significant digits. The number of digits in a result calculated by addition or subtraction depends on the position of the decimal point. The result can have no more significant digits to the right of the decimal than the factor with the least number of digits to the right of the decimal. Often, an extra digit is carried, and the result is rounded off to the correct number of digits.

$$
\begin{array}{ll}
16.00 & \text{(two digits past decimal)} \\
\underline{2.016} & \text{(three digits past decimal)} \\
18.016 & \rightarrow \text{answer is } 18.02 \text{, since the result may} \\
& \text{have only two digits past decimal}
\end{array}
$$

$$
\begin{array}{ll}
12.02 & \text{(two digits past decimal)} \\
6.048 & \text{(three digits past decimal)} \\
\underline{16.00} & \text{(two digits past decimal)} \\
34.048 & \rightarrow \text{answer is } 34.05 \text{, since the result may} \\
& \text{have two digits past decimal}
\end{array}
$$

Example 2-4   Determine the molar mass of ammonia, NH_3. The molar mass is found by multiplying the molar mass of each element by the subscript and adding up the products.

$$
\begin{array}{lll}
N & 14.01 \text{ g} = 14.01 \text{ g} \\
H & 3(1.008 \text{ g}) = \underline{3.024 \text{ g}} \\
& \phantom{3(1.008 \text{ g}) =} 17.034 \text{ g} & \text{or} \quad \left(\dfrac{17.03 \text{ g}}{1 \text{ mole } NH_3} \right)
\end{array}
$$

Example 2-5   Determine the molar mass of sodium sulfate, Na_2SO_4. The molar mass is found by multiplying the molar mass of each element by the subscript and adding up the products.

$$
\begin{array}{lll}
Na & 2(23.0 \text{ g}) = 46.0 \text{ g} \\
S & 32.0 \text{ g} = 32.0 \text{ g} \\
O & 4(16.0 \text{ g}) = \underline{64.0 \text{ g}} \\
& \phantom{4(16.0 \text{ g}) =} 142.0 \text{ g} & \text{or} \quad \left(\dfrac{142 \text{ g}}{1 \text{ mole } Na_2SO_4} \right)
\end{array}
$$

As was true with the elements, the number of grams per mole for a compound can be used as a conversion factor to convert from grams of the compound to number of moles and vice versa.

Example 2-6   How many moles of ammonia, NH_3, are contained in a 34.1 g sample of ammonia?

The number of moles of ammonia can be found by multiplying the sample mass by the inverse of the molar mass of ammonia.

$$34.1 \text{ g} \left(\frac{1 \text{ mole NH}_3}{17.03 \text{ g}}\right) = 2.00 \text{ moles NH}_3$$

Example 2-7   What is the mass of 3.00 moles of water?
The mass is found by multiplying the 3.00 moles H_2O by the molar mass of water.

$$3.00 \text{ moles H}_2\text{O} \left(\frac{18.02 \text{ g}}{1 \text{ mole H}_2\text{O}}\right) = 54.1 \text{ g}$$

2-9   Percentage-by-Mass Composition

Sometimes we find it convenient to express the makeup of a compound in terms of the percentage of each element present in the compound. Since the mass and number of moles are related, the formula of the compound, which gives the number of moles of each constituent element, can be used to determine the percentage by mass of each element present. This is referred to as the **percentage-by-mass composition** of the compound.

Since we can deduce the mass of a mole of a compound, and we know the mass per mole of each element in the compound, the percentage by mass of any element can be found. This is done by multiplying the number of grams per mole of the element by the subscript in the formula and dividing this by the number of grams per mole of the compound. This result should be multiplied by 100 to express a percentage.

Example 2-8   What is the percentage-by-mass composition of water, H_2O?

$$\text{percent H} = \left(\frac{2 \text{ moles H}}{1 \text{ mole H}_2\text{O}}\right)\left(\frac{1.008 \text{ g}}{1 \text{ mole H}}\right)\left(\frac{1 \text{ mole H}_2\text{O}}{18.02 \text{ g}}\right) 10^2 = 11.2\%$$

$$\text{percent O} = \left(\frac{1 \text{ mole O}}{1 \text{ mole H}_2\text{O}}\right)\left(\frac{16.00 \text{ g}}{1 \text{ mole O}}\right)\left(\frac{1 \text{ mole H}_2\text{O}}{18.02 \text{ g}}\right) 10^2 = 88.8\%$$

or after canceling the units the calculations are

$$\text{percent H} = 2 \left(\frac{1.008}{18.02}\right) 10^2 = 11.2\%$$

$$\text{percent O} = \left(\frac{16.00}{18.02}\right) 10^2 = 88.8\%$$

Of course, the sum of the percentages of the elements in a compound should equal 100%.

Example 2-9   What is the percentage-by-mass composition of sodium sulfate, Na_2SO_4?

$$\text{percent Na} = \left(\frac{2 \text{ moles Na}}{1 \text{ mole Na}_2\text{SO}_4}\right)\left(\frac{23.0 \text{ g Na}}{1 \text{ mole Na}}\right)\left(\frac{1 \text{ mole Na}_2\text{SO}_4}{142.0 \text{ g}}\right) 10^2 = 32.4\% \text{ Na}$$

$$\text{percent S} = \left(\frac{1 \text{ mole S}}{1 \text{ mole Na}_2\text{SO}_4}\right)\left(\frac{32.1 \text{ g}}{1 \text{ mole S}}\right)\left(\frac{1 \text{ mole Na}_2\text{SO}_4}{142.0 \text{ g}}\right) 10^2 = 22.6\% \text{ S}$$

$$\text{percent O} = \left(\frac{4 \text{ moles O}}{1 \text{ mole Na}_2\text{SO}_4}\right)\left(\frac{16.00 \text{ g}}{1 \text{ mole O}}\right)\left(\frac{1 \text{ mole Na}_2\text{SO}_4}{142.0 \text{ g}}\right) 10^2 = 45.0\% \text{ O}$$

or, after canceling the units, the calculations are

$$\text{percent Na} = 2\left(\frac{23.0}{142.0}\right)10^2 = 32.4\%$$

$$\text{percent S} = \left(\frac{32.1}{142.0}\right)10^2 = 22.6\%$$

$$\text{percent O} = 4\left(\frac{16.00}{142.0}\right)10^2 = 45.0\%$$

2-10 Empirical Formulas

A good question at this point is how the formulas of compounds can be determined. Often it is possible to experimentally determine the percentage-by-mass composition of a pure compound. This is done by decomposing a compound to find the amounts of the component elements or by synthesizing the compound and measuring the amounts of the elements needed. A formula is considered to be a symbolic expression of the molar relationship between the constituent elements of the compound. For example, the formula of water, H_2O, indicates that there are 2 moles of combined hydrogen for every 1 mole of combined oxygen in the compound. In other words, we know from the formula that the molar ratio is

$$\left(\frac{2 \text{ moles H}}{1 \text{ mole O}}\right)$$

If the percentage-by-mass composition of a compound is known, it is possible to determine the number of moles of each element present in a given mass of the compound. The molar ratios of the elements can be found from the number of moles of each element. The molar ratios obtained in this manner indicate the subscripts that are to be used in the formula of the compound. Such a formula is called an **empirical formula.** The term empirical means derived from experimental data.

Before giving an example of the determination of the empirical formula of a compound, let us consider an analogy concerning the nuts and bolts referred to in Sections 2-7 and 2-8. Suppose we had a sample of nut-bolt combinations ("compound") that contained 50%-by-mass nuts and 50%-by-mass bolts, and we want to determine the number of nuts per bolt ("formula") in the combination. This can be done by supposing we had 100 g of the combination that would contain 50 g of nuts and 50 g of bolts. Using the grams-per-dozen factors, we can find the number of dozen bolts and nuts and the ratio of these will give the number of nuts per bolt. The calculations that accomplish this are:

$$50 \text{ g}\left(\frac{1 \text{ doz N}}{20 \text{ g}}\right) = 2.5 \text{ doz N}$$

$$50 \text{ g}\left(\frac{1 \text{ doz B}}{40 \text{ g}}\right) = 1.25 \text{ doz B}$$

The ratio is

$$\left(\frac{2.5 \text{ doz N}}{1.25 \text{ doz B}}\right) = \left(\frac{2 \text{ doz N}}{1 \text{ doz B}}\right)$$

Thus, there are two nuts for every bolt (The "formula" is BN_2).

Now consider the determination of the empirical formula of a compound. What is the empirical formula of a compound that is found to contain 11.18%-by-mass combined hydrogen and 88.8%-by-mass combined oxygen? Since we know the percentage composition, and we want to deal with masses, we can say that if we had 100.0 g of the compound, it would contain 11.18 g of combined hydrogen and 88.8 g of combined oxygen. Using these masses, we find the number of moles of each element and the molar ratios.

$$11.18 \text{ g} \left(\frac{1 \text{ mole H}}{1.008 \text{ g}}\right) = 11.10 \text{ moles H}$$

$$88.8 \text{ g} \left(\frac{1 \text{ mole O}}{16.00 \text{ g}}\right) = 5.55 \text{ moles O}$$

Using the number of moles of each element, we express the molar ratios as

$$\left(\frac{11.10 \text{ moles H}}{5.55 \text{ moles O}}\right) = \left(\frac{2.00 \text{ moles H}}{1 \text{ mole O}}\right)$$

or

$$\left(\frac{5.55 \text{ moles O}}{11.10 \text{ moles H}}\right) = \left(\frac{0.500 \text{ moles O}}{1 \text{ mole H}}\right)$$

The first ratio is easier to interpret when deducing the formula from the molar ratios. From this molar ratio, it is seen that the formula for the compound is H_2O. That is, there are 2 moles of combined hydrogen per 1 mole of combined oxygen. It is always a good idea to divide by the smallest number of moles when finding the molar ratios so that the molar ratios are not less than 1. Thus, the first molar ratio expressed above would be the one conventionally used to deduce the formula.

Example 2-10 What is the formula for a compound that is found to have the following percentage-by-mass composition: 26.5% combined potassium, 35.4% combined chromium, and 38.1% combined oxygen?

Again, to express the amounts of the elements in terms of mass we can express the number of grams of each element that would be present in 100 g of the compound. Then the number of moles of each compound can be determined.

$$26.5 \text{ g} \left(\frac{1 \text{ mole K}}{39.1 \text{ g}}\right) = 0.678 \text{ moles K}$$

$$35.4 \text{ g} \left(\frac{1 \text{ mole Cr}}{52.0 \text{ g}}\right) = 0.681 \text{ moles Cr}$$

$$38.1 \text{ g} \left(\frac{1 \text{ mole O}}{16.00 \text{ g}}\right) = 2.38 \text{ moles O}$$

Using the smallest number of moles, which is the number of moles of potassium, in the denominator the molar ratios are

$$\left(\frac{0.681 \text{ moles Cr}}{0.678 \text{ moles K}}\right) = \left(\frac{1.00 \text{ moles Cr}}{1 \text{ mole K}}\right)$$

$$\left(\frac{2.38 \text{ moles O}}{0.678 \text{ moles K}}\right) = \left(\frac{3.51 \text{ moles O}}{1 \text{ mole K}}\right)$$

From the molar ratios, the formula is seen to be $KCrO_{3.5}$. We normally avoid writing fractional subscripts in formulas so we multiply each of the subscripts by two in order to obtain whole number subscripts. Thus, the best empirical formula is $K_2Cr_2O_7$. Whenever you are determining the formula of a compound from percentage composition and fractional subscripts are obtained from the molar ratios, the subscripts should be multiplied by an appropriate number that will give whole number subscripts.

Notice that the molar ratios calculated above had a few extra digits that we neglected. These extra digits probably arise from round-off errors. Normally, the subscripts in formulas are small whole numbers so we neglect these extra digits when deducing the formula from the molar ratios.

2-11   The Physical and Chemical Forms of the Elements

Since we are concerned with studying the properties and behavior of elements, it is important to know the nature of the naturally occurring elements and the nature of pure elements that can be separated from the complex compounds in which they occur. We can describe the nature of such elements by stating some of their physical properties under normal earth conditions. Under normal conditions, some of the elements are in the gaseous state. Table 2-3 indicates these elements. Two elements, mercury and bromine, occur as liquids under normal conditions, and gallium and cesium become liquids at slightly above-normal temperature conditions. All of the other elements occur as solids under normal conditions. The elements can be classified according to their

Table 2-3   Gaseous Elements

Element	Element Symbol	Formula of Gaseous Element
Hydrogen	H	H_2
Nitrogen	N	N_2
Oxygen	O	O_2
Fluorine	F	F_2
Chlorine	Cl	Cl_2
Helium	He	He
Neon	Ne	Ne
Argon	Ar	Ar
Krypton	Kr	Kr
Xenon	Xe	Xe
Radon	Rn	Rn

properties into three groups. A large portion of the elements have metallic properties—good electrical conductors, flexible enough to be shaped into various forms, possess metallic luster—and are called **metals.** Some elements do not possess these metallic properties. They are nonconductors and gases or brittle nonlusterous solids. These elements are called **nonmetals.** A few elements display properties of both metals and nonmetals and are called the **metalloids.** Note the positions of the metals, nonmetals, and metalloids in Figure 2-9. The metals occur as solids (except mercury) and are in a form in which the atoms are distributed in a definite geometrical pattern called a crystal. The atom is the representative particle in the crystal, so the metals are represented by the atomic symbols. Certain nonmetals occur in the form of monoatomic gases. That is, they occur in the gaseous state in which the particles comprising the gas are uncombined atoms. These elements called the noble gases are helium, neon, argon, krypton, xenon, and radon and are represented by the atomic symbols.

The nonmetals hydrogen, nitrogen, oxygen, fluorine, chlorine, bromine, and iodine occur in the form of **diatomic molecules**—molecules made up of two atoms of the element. These elements are represented in this state by the formulas H_2, N_2, O_2, F_2, Cl_2, Br_2, and I_2. Thus, when we want to refer to oxygen, for example, in its natural state, we can use the formula O_2, since oxygen occurs as a diatomic molecule in the natural state. The nonmetal sulfur occurs in solid form under normal conditions. This solid consists of molecules that are made up of eight atoms of sulfur. Sulfur in this state is represented as S_8. However, sometimes the molecular nature of sulfur in the natural state is not taken into consideration, and sulfur in this state is represented by the atomic symbol S. The other nonmetals and the metalloids generally occur under normal conditions in the form of complicated molecules involving many atoms. These elements are usually represented by the atomic symbols. The states and forms of the elements are summarized in Figure 2-10. When dealing with samples of pure elements at normal conditions, we normally use the symbols shown in the figure. For example, if we have a sample of iron, the representative symbol is Fe, and if we want to refer to a sample of iodine, we use the symbol I_2.

2-12   Chemical Reactions and Equations

The elements can combine to form compounds. Similarly, many compounds can be decomposed to form the constituent elements. These changes occur as chemical processes. Any chemical process in which certain substances are converted into other substances is called a **chemical reaction.** In a chemical reaction the initial substances, called **reactants,** are said to react chemically to form new substances called **products.** In a chemical reaction the chemical particles (atoms, ions, or molecules) comprising the reactants are transformed into new chemical compounds that are the products. Chemical reactions are occurring in and around us continuously. The digestion and metabolism of food involves chemical reactions. The rusting of metal and the burning of fuel

Nonmetals | Metalloids | Metals

IA	IIA	IIIB	IVB	VB	VIB	VIIB	VIII	VIII	VIII	IB	IIB	IIIA	IVA	VA	VIA	VIIA	NOBLE GASES
H H_2																H H_2	He
Li metal	Be metal											B	C	N N_2	O O_2	F F_2	Ne
Na metal	Mg metal											Al metal	Si	P	S S_8	Cl Cl_2	Ar
K metal	Ca metal	Sc metal	Ti metal	V metal	Cr metal	Mn metal	Fe metal	Co metal	Ni metal	Cu metal	Zn metal	Ga metal	Ge	As	Se	Br Br_2	Kr
Rb metal	Sr metal	Y metal	Zr metal	Nb metal	Mo metal	Tc metal	Ru metal	Rh metal	Pd metal	Ag metal	Cd metal	In metal	Sn metal	Sb	Te	I I_2	Xe
Cs metal	Ba metal	La metal	Hf metal	Ta metal	W metal	Re metal	Os metal	Ir metal	Pt metal	Au metal	Hg metal	Tl metal	Pb metal	Bi metal	Po	At	Rn
Fr metal	Ra metal	Ac metal															

Ce metal	Pr metal	Nd metal	Pm metal	Sm metal	Eu metal	Gd metal	Tb metal	Dy metal	Ho metal	Er metal	Tm metal	Yb metal	Lu metal
Th metal	Pa metal	U metal	Np metal	Pu metal	Am metal	Cm metal	Bk metal	Cf metal	Es metal	Fm metal	Md metal	No metal	Lr metal

Figure 2-9 A periodic table showing the metals, nonmetals, and metalloids.

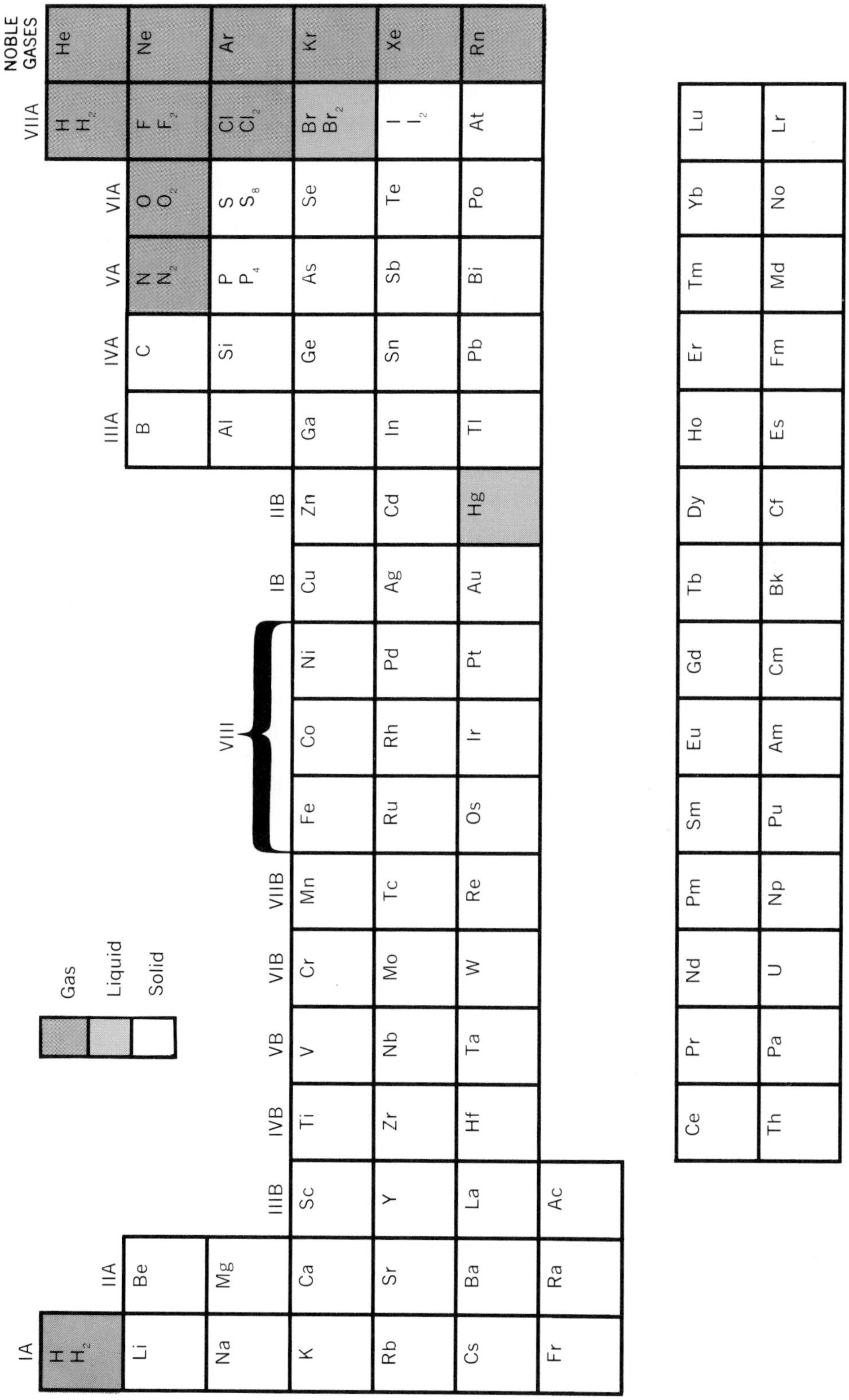

Figure 2-10 The states and forms of the elements.

are chemical reactions. Some reactions occur quickly and some quite slowly. The explosion of dynamite is a fast chemical reaction while the formation of a fossil involves extremely slow chemical processes. Chemical reactions account for many of the changes we see around us. A flame is the indication of a relatively simple chemical reaction, whereas the growth of a plant involves a complex series of chemical reactions.

Chemical reactions can be described by indicating the reactants and products. For instance, when we burn charcoal (e.g., charcoal briquettes in the barbeque), the main chemical reactants are carbon (C, the main component of charcoal) and oxygen (O_2, remember oxygen occurs as diatomic molecules) gas in the air. They react to form carbon dioxide as a product. Thus, we say that carbon reacts with oxygen to form carbon dioxide. Such a chemical reaction can be represented symbolically by a **chemical equation.** A chemical equation must include the proper formulas for the reactants and the products. Furthermore, the reactant formulas are separated by plus signs that are not intended to mean a mathematical plus but instead should be interpreted as meaning "and." The product formulas are also separated by plus signs. The reactants and products are separated by an arrow ($\rightarrow$), which is a symbol that means "react to produce, yield, or give." For the reaction between carbon and oxygen mentioned above, the chemical equation is written as follows:

$$C + O_2 \rightarrow CO_2$$

This equation says that carbon and oxygen react to produce carbon dioxide, or carbon plus oxygen react to yield carbon dioxide. Notice that the symbol used for solid carbon is the atomic symbol C, and the symbol used for gaseous oxygen is the formula for the diatomic molecule O_2.

The combined atoms comprising the substances are not created or destroyed in a chemical reaction, but chemical bonds are broken and new combinations result. Consequently, the total number of combined atoms of each element involved remains constant. Since atoms are not destroyed or created in a chemical reaction, the total mass of the reactants must equal the total mass of the products. Let us consider what happens when we use a flashbulb in a camera. The bulb contains some magnesium metal and some oxygen gas. A small electrical current passing through the magnesium initiates a violent chemical reaction between the magnesium and the oxygen. Even though this reaction is accompanied by the release of energy in the form of heat and light, no detectable mass is lost or gained in the reaction. That is, the mass of the flashbulb is the same before and after the reaction. (See Figure 2-11.) This observation is generally true for all chemical reactions. The fact that mass is conserved in a chemical reaction is called the law of conservation of matter. Recall that this idea was mentioned in Chapter 1.

2-13   Balancing Chemical Equations

The law of conservation of matter affects the way in which an equation is written. The number of atoms of each element in a reaction must be the same on either side of the equation. Consider the reaction that

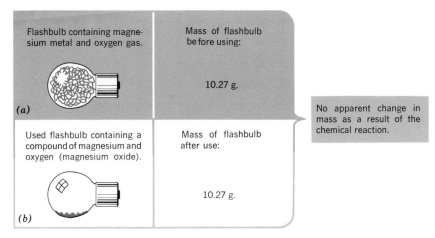

Figure 2-11

occurs when methane, CH_4 (the main component of natural gas), reacts with oxygen gas (i.e., burns in the air) to produce carbon dioxide, CO_2, and water, H_2O. The equation is written showing the formulas of the reactants and products separated by an arrow:

$$CH_4 + O_2 \rightarrow CO_2 + H_2O$$

However, we must be sure that the number of atoms of each element is the same on both sides of the equation since atoms are not created or destroyed in a reaction. We cannot change the formulas of the compounds in order to make the number of atoms of each element the same on each side of the arrow since this would not be representative of the proper compounds. We can, however, change the number of each species that appears in the equation. We do this by placing a numerical coefficient in front of the formula of any species to refer to a specific number of such species. As illustrated in Figure 2-12, to have four hydrogen atoms on each side of the above equation, two water molecules must be produced; and to have four oxygen atoms on each side, two oxygen molecules must react. This is denoted by placing the necessary number, called a **coefficient,** in front of the formula of the appropriate species. Thus, the correctly written equation is

$$CH_4 + 2O_2 \rightarrow CO_2 + 2H_2O$$

This equation can be read in four ways.

1. C-H-four plus two-O-two react to give C-O-two plus two-H-two-O.
2. Methane reacts with oxygen to produce carbon dioxide and water.
3. One molecule of methane and two molecules of oxygen react to produce one molecule of carbon dioxide and two molecules of water.
4. One mole of methane and 2 moles of oxygen react to produce 1 mole of carbon dioxide and 2 moles of water.

Unless a specific interpretation of an equation is desired the first method given above is sufficient for the reading of an equation.

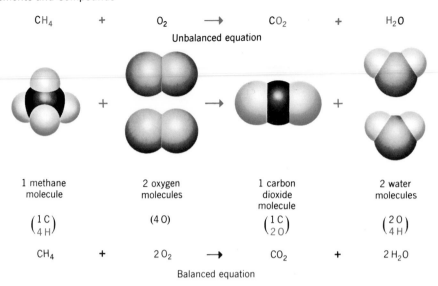

$$CH_4 \qquad + \qquad O_2 \qquad \longrightarrow \qquad CO_2 \qquad + \qquad H_2O$$

Unbalanced equation

1 methane molecule	2 oxygen molecules	1 carbon dioxide molecule	2 water molecules
$\begin{pmatrix} 1\,C \\ 4\,H \end{pmatrix}$	$(4\,O)$	$\begin{pmatrix} 1\,C \\ 2\,O \end{pmatrix}$	$\begin{pmatrix} 2\,O \\ 4\,H \end{pmatrix}$

$$CH_4 \qquad + \qquad 2\,O_2 \qquad \longrightarrow \qquad CO_2 \qquad + \qquad 2\,H_2O$$

Balanced equation

Figure 2-12   Balancing of the equation for the reaction between methane and oxygen to give carbon dioxide and water. From *Understanding Chemistry: From Atoms to Attitudes* by T. R. Dickson, Wiley, N.Y., 1974.

In any equation representing a chemical reaction, the coefficients must be adjusted to satisfy the requirement that the number of atoms of each element must be the same in the reactants and the products. An equation in which the number of atoms of each element is the same on both sides of the arrow is called a **balanced equation.** The process of adjusting the coefficients is called balancing the equation. Often such balancing is carried out by trial and error, that is, by trying various coefficients.

Chemical equations are used in chemistry to show a chemical change in a shorthand fashion. All chemical reactions could be stated in words, but the equation provides for the precise description of a chemical process in convenient symbolism. You will see many other chemical equations in this book.

Sometimes special conditions are needed to have a chemical reaction occur. Often these conditions are indicated above or below the arrow. For example, nitrogen, N_2, and hydrogen, H_2, react to form ammonia, NH_3, under high pressure. The equation

$$N_2 + 3H_2 \xrightarrow{\text{pressure}} 2NH_3$$

indicates that pressure is needed in order for the reaction to proceed. Some chemical reactions will not occur readily unless a specific substance is present in the reaction environment. Such a substance, the presence of which is necessary for the reaction to occur but is not permanently altered chemically in the reaction, is called a **catalyst.** The nitrogen and hydrogen reaction mentioned above only occurs readily in the presence of a catalyst such as iron, Fe. To denote this, we include iron above or below the arrow:

$$N_2 + 3H_2 \xrightarrow[\text{Fe}]{\text{pressure}} 2NH_3$$

This indicates that nitrogen and hydrogen under pressure and in the presence of the catalyst iron will react to form ammonia. Whenever needed, additional information concerning a chemical reaction is usually written above or below the arrow.

2-14   Chemical Elements in the Ecosphere

There are four distinct regions of our earth environment. The rocky and mountainous crust of the earth is called the **lithosphere** (Greek: litho = stone). The envelope of gases surrounding the earth is the **atmosphere** (Greek: atmos = vapor). The **hydrosphere** (Greek: hydro = water) includes the vast amount of water in lakes, rivers, oceans, and underground deposits, the water contained in the ice and snow of the earth, and the water that makes up the clouds and moisture in the atmosphere. Within the atmosphere and hydrosphere and upon the lithosphere dwell the various plants and animals that constitute the **biosphere** (Greek: bio = life). Since these four realms are interrelated and form our normal environment, they are referred to as a whole as the **ecosphere** (Greek: oikos = house). The components of the lithosphere, hydrosphere, and atmosphere are intermingled to a certain extent. The atmosphere contains water vapor and dust particles. The hydrosphere contains dissolved gases and particles of the lithosphere. Waters of the hydrosphere cover and permeate portions of the lithosphere. This harmonious interfacing of the three portions of the ecosphere provides for the existence and maintenance of the biosphere. Certain activities of humans are affecting the very composition and the natural balances of the subspheres.

The four portions of our environment contain elements in various states of combinations. Table 2-4 gives the percentages by mass of the

Table 2-4   The Percentages by Mass of the Elements in the Ecosphere

Element	Percentage	Element	Percentage
Oxygen	49.20	Chlorine	0.19
Silicon	25.67	Phosphorus	0.11
Aluminum	7.50	Manganese	0.09
Iron	4.71	Carbon	0.09
Calcium	3.39	Sulfur	0.06
Sodium	2.63	Barium	0.04
Potassium	2.40	Fluorine	0.03
Magnesium	1.93	Nitrogen	0.03
Hydrogen	0.87		
Titanium	0.58	Other elements	0.47

From *Understanding Chemistry: From Atoms to Attitudes,* by T. R. Dickson, Wiley, N.Y., 1974.

18 elements that make up about 99.5% of the lithosphere, atmosphere, hydrosphere, and biosphere. The lithosphere, that is made up mainly of rock and some soil, contains combined silicon, oxygen, aluminum, and many other metals. Most of the metals are combined in chemical compounds, and a few, such as gold, are found only occasionally as the free uncombined metal. Metals are distributed within the lithosphere in very small percentages. Sometimes, as a result of geological processes, certain chemical substances are localized in the lithosphere. These accumulations of substances are called **mineral deposits.** When such mineral deposits contain elements that humans need and mining is feasible, the mineral is called an **ore.** It is estimated that usable mineral deposits make up far less than 1% of the crust of the earth. Let us consider an example of an ore. Aluminum content in most rocks is about 8%. However, deposits of the mineral bauxite contain around 25% aluminum. Thus, such bauxite deposits are aluminum ore from which aluminum metal can be extracted by chemical processes. The minerals that are used the most by industrial societies are iron ore, copper ore, aluminum ore, and minerals that find use as fertilizers (phosphorous-, nitrogen-, and potassium-containing minerals).

The hydrosphere consists mainly of hydrogen and oxygen combined as water. However, many elements in various dissolved forms are found in seawater. These elements are present in small percentages. If we evaporated the water from 1 ℓ of typical seawater, about 35 g of chemical substances would remain. Much salt or sodium chloride, $NaCl$, is obtained from seawater. Significant amounts of magnesium and bromine are also extracted from seawater by chemical processes.

The atmosphere is the layer of air surrounding the earth. The atmosphere is most dense near the surface of the earth and decreases in density as it extends into space. The portion of the atmosphere extending out about 12 km is called the troposphere. This is where cloud formation and movement of air (wind) occurs. The portion of the atmosphere extending out from 12 km to about 50 km is called the stratosphere. The stratosphere is free of clouds and winds. The portion of the atmosphere above about 50 km is called the ionosphere.

Air is a homogeneous mixture of gases. The major components of dry air, in percent by volume, are

nitrogen	N_2	78.1%	neon	Ne	1.8×10^{-3}%
oxygen	O_2	21.0%	helium	He	5.3×10^{-4}%
argon	Ar	0.93%	krypton	Kr	1×10^{-4}%
carbon dioxide	CO_2	3×10^{-2}%			

The atmosphere also contains variable amounts of water vapor and dust. The amount of water vapor varies from small amounts in desert regions to large amounts in tropical regions. The water vapor condenses to form clouds, rain, and snow.

The biosphere includes all plant and animal life. It centers near the interface between the lithosphere and hydrosphere, but in the extremes extends to thousands of feet into the atmosphere (birds) and thousands of feet into the ocean depths (fishes). The major elements that structure living matter are carbon, oxygen, and hydrogen. These elements are combined in the chemical substances of life so that the average per-centage-by-mass composition of the biosphere is 52% oxygen, 39% carbon, and 6.7% hydrogen. In addition, there are over 10 elements that are found in the biosphere in small amounts. The most important of these elements are nitrogen, sulfur, and phosphorus.

Questions and Problems

1. Which element would you like to be?

2. Choose an element from a list, look it up in a chemistry reference book (see "The Elements" in the *Handbook of Chemistry and Physics,* Chemical Rubber Co.), and describe the nature, source, and uses of the element and when and by whom it was discovered. If any other elements interest you, look them up also.

3. Look up John Dalton's atomic theory in an encyclopedia or a history of science book. Describe the theory as Dalton developed it and describe the historical background of the theory.

4. Give a statement of the atomic theory.

5. Why is the atomic theory so fundamental and necessary to chemistry?

6. Define the following terms:
   (a) element
   (b) compound
   (c) isotopes
   (d) atomic weight
   (e) mole
   (f) chemical reaction

7. Give the symbol for each of the following elements:
   (a) antimony   (j) magnesium
   (b) barium   (k) nitrogen
   (c) bromine   (l) oxygen
   (d) calcium   (m) potassium
   (e) chromium   (n) silicon
   (f) copper   (o) sodium
   (g) gold   (p) sulfur
   (h) hydrogen   (q) tungsten
   (i) iron   (r) zinc

8. Give the name for each of the following elements for which the symbol is given:
   (a) Al   (e) B
   (b) As   (f) Cd
   (c) Be   (g) C
   (d) Bi   (h) Cs

(i) Cl

(j) Co

(k) F

(l) He

(m) I

(n) Pb

(o) Hg

(p) P

(q) Ag

(r) Sn

(s) U

(t) Fe

9. Calculate the number of moles of atoms of the indicated element contained in the following masses:

(a) 17.3 g C

(b) 47.9 g Ba

(c) 94.0 g Pt

(d) 8.81 g S

(e) 0.142 g Fe

(f) 59.6 g Cl

(g) 33.4 g P

(h) 432 g Ag

(i) 751 kg B

(j) 0.271 mg Al

10. Calculate the mass in grams corresponding to each of the following:

(a) 0.576 moles O

(b) 97.3 moles H

(c) 14.7 moles Fe

(d) 4.27 moles As

(e) 23.2 moles N

(f) 493 moles Cl

(g) 39.1 moles C

(h) 3.14 moles S

(i) 2.71 mmoles K (millimoles)

11. How many atoms of the indicated element are contained in the following masses?

(a) 52.6 g Fe

(b) 157 g S

(c) 2.13 mg K

(d) 6.43 g He

(e) 8.61×10^4 g Hg

(f) 1.32×10^{-12} g Au

12. Determine the number of grams per mole for each of the following compounds:

(a) sulfuric acid, H_2SO_4

(b) sodium hydroxide, NaOH

(c) lead carbonate, $PbCO_3$

(d) calcium iodide, CaI_2

(e) potassium dichromate, $K_2Cr_2O_7$

(f) sulfur, S_8

(g) magnesium chloride, $MgCl_2$

(h) germanium difluoride, GeF_2

(i) bromine, Br_2

(j) toluene, C_7H_8

(k) sucrose, $C_{12}H_{22}O_{11}$

(l) lysergic acid, $C_{16}H_{16}N_2O_2$

13. Calculate the number of moles of the indicated compound contained in the following (see problem 12):

(a) 42.3 g H_2SO_4

(b) 0.892 g NaOH

(c) 514 g $PbCO_3$

(d) 27.2 g CaI_2

(e) 5.02×10^3 g $K_2Cr_2O_7$

(f) 872 megagrams S_8

(g) 4.32×10^{-7} g $MgCl_2$

(h) 17.9 g GeF_2

(i) 746 g Br_2

(j) 1.92 g C_7H_8

(k) 12.4 g $C_{12}H_{22}O_{11}$

(l) 0.407 mg $C_{16}H_{16}N_2O_2$

14. Calculate the mass in grams for each of the following:

(a) 2.37 moles O_2

(b) 0.963 moles H_2

(c) 27.5 moles NH_3

(d) 10.72 moles NaCl

(e) 5.26 moles Cl_2

(f) 2.50 moles $KMnO_4$, potassium permanganate

(g) 82.4 moles CH_4, methane

(h) 2.47×10^{-3} moles H_2O

15. How many moles of combined hydrogen atoms are contained in 102 g of water?

16. How many moles of combined phosphorus atoms are contained in 27.6 g of P_4O_{10}?

17. How many grams of combined hydrogen are contained in a 92.1 g sample of ammonium acetate, $NH_4C_2H_3O_2$?

18. Determine the percentage by mass composition of the following compounds:

(a) silver oxide, Ag_2O

(b) mecuric oxide, HgO

(c) sodium sulfide, Na_2S

(d) iron (III) oxide, Fe_2O_3

(e) hydrogen peroxide, H_2O_2

(f) calcium chloride, $CaCl_2$

(g) sodium phosphate, Na_3PO_4

19. Deduce the empirical formulas for the compounds that have the following percentage-by-mass composition:

(a) 36.1% Ca and 63.9% Cl

(b) 37.2% C, 7.81% H, and 55.0% Cl

(c) 48.7% C, 13.6% H, and 37.8% N

(d) 70.0% Fe and 30.0% O

(e) 42.1% Na, 18.91% P and 39.0% O

(f) 59.3% C, 4.55% H, 23.0% N, and 13.15% O

20. Give a list of the elements that are found as gases under normal conditions.

21. Which elements are found as liquids under normal conditions?

22. Give the names and molecular formulas for the elements that occur naturally in the form of diatomic molecules.

23. Balance the following unbalanced equations:

(a) $Mg + O_2 \rightarrow MgO$

(b) $Fe + O_2 \rightarrow Fe_2O_3$

(c) $KClO_3 \rightarrow KCl + O_2$

(d) $Na + O_2 \rightarrow Na_2O$

(e) $Sb + Cl_2 \rightarrow SbCl_3$

(f) $C_2H_4 + O_2 \rightarrow CO_2 + H_2O$

(g) $Al + O_2 \rightarrow Al_2O_3$

(h) $N_2 + H_2 \rightarrow NH_3$

(i) $C + CaO \rightarrow CaC_2 + CO$

24. A 5.00-g sample of a compound of chromium and sulfur is found to contain 2.40 g of sulfur. What is the empirical formula of the compound?

25. Determine the empirical formula of a compound if a sample of the compound contains 0.200 g of calcium and 0.800 g of bromine.

26. Determine the empirical formula of a compound that is found to have the following percent-by-mass composition: 40.0% calcium, 12.0% carbon, 48.0% oxygen.

27. A penny has a mass of 3.20 g. How many copper atoms are in the penny?

28. A typical carton of salt contains 26 oz of sodium chloride, NaCl. How many moles of sodium chloride are in a carton of salt? (28.3 g = 1 oz.)

29. How many moles of iron are in one pound of iron nails? How many iron atoms?

30. A box of baking soda contains 32 oz of sodium hydrogen carbonate, $NaHCO_3$. How many moles of sodium hydrogen carbonate are in a carton? (28.3 g = 1 oz.)

31. A typical aspirin tablet contains 5.0 grains of acetylsalicylic acid $C_9H_8O_4$ (180 g/mole). How many moles of acetylsalicylic acid are in a tablet? (0.0648 g = 1 grain.)

32. If the average weight of a person is 120 lb, what mass would Avogadro's number of people be in grams? (For comparison note that the mass of the earth is about 6×10^{24} g).

33. If the volume of a sugar cube is 4.0 cm³, what is the volume of Avogadro's number of sugar cubes? How many cubic kilometers would this be?

3

Chapter 3: Atomic Structure

Objectives

The student should be able to:

1 Give a description of Rutherford's concept of a nuclear atom.
2 Define the term isotope considering nuclear composition.
3 Give a description of the Bohr concept of the atom.
4 Give a description of the quantum mechanical atom including energy level structure.
5 Write the electronic configuration of an element (atomic numbers 1 to 20) in orbital notation.
6 Sketch the outline of a periodic table and indicate the positions of the s block, p block, d block, and f block in the periodic table.
7 Indicate the positions of the representative elements, the transition elements, and the innertransition (rare earth) elements in the periodic table.
8 Indicate the positions of the alkali metals, the alkaline-earth metals, boron-aluminum group, carbon group, nitrogen group, oxygen group, and the halogens in the periodic table.

Terms to Know

Electron
Proton
Neutron
Atomic number
Energy level
Atomic orbital
Periodic law

3

Atomic Structure

3-1 The Atomic Model

Today we live in an atomic age. The atom has been found to be a tremendous source of energy. This energy has been used both destructively in atomic bombs and constructively in nuclear reactors. Chemists have found ways to make atoms combine to form various products. In this way many synthetic substances such as plastics, rubber, drugs, and medicines are manufactured for commercial use. This chapter is devoted to a discussion of atomic structure, which is of great interest to chemistry.

Elements combine with one another to form numerous compounds. However, every element will not combine with every other element; and a given element forms only certain types of compounds. Some of the elements can form more than one compound with another element. Observations of these facts led to this question: If the elements are composed of atoms, what must be the conceivable structure of atoms that will account for these observations? Elements behave as if they are made up of atoms; but their behavior indicates that the atoms of various elements must differ. We know that the atoms of elements have different relative masses. This fact alone suggests that the atoms have different structures. Furthermore, differences in combining abilities of the elements also suggest differences in atomic structure of the atoms of the elements. Much scientific effort has been expended in the last century in an attempt to develop a suitable explanation of atomic structure. An understanding of atomic structure is important for an understanding of the behavior of atoms. In this chapter, the development and details of the currently accepted model of atomic structure are discussed.

An atomic model is an attempt to describe the nature of the tiny fundamental particles of elements that cannot be directly observed but do display certain observable properties. Scientists have used measurements and observations of these properties to establish an atomic model that provides the following:

1. A mental image by which the shapes and sizes of atoms can be pictured.

2. A description of the similarities and differences of the atoms of the various elements and isotopes of elements.

3. An explanation of the ability of atoms to form compounds.

3-2   The Nuclear Atom

Ordinary atoms are electrically neutral. However, under certain circumstances atoms can become electrically charged. You have probably noticed that a comb can sometimes become electrically charged when you pass it through your hair. To observe the effect of this charge, place several grains of salt on a piece of paper, pass a comb through your dry hair a few times, and then place the comb just above the salt. Notice the attractive and repulsive forces that arise between the salt and the comb. Atoms of some elements take on electrical charges. Such charged atoms are called **ions.** Some elements form positive ions, called **cations,** and some elements form negative ions, called **anions.** Since atoms are able to form ions, it was proposed that atoms contain basic units of negative charge that could be lost to form positive ions or gained to form negative ions. These units of charge were found to be the same units of charge that characterized electricity. Such units of negative charge became known as **electrons.** Since the electron is apparently the smallest unit of electrical charge, we shall refer to its charge as -1 unit charge or quantum charge. When plus or minus charges are used in further discussion, unit charges will be intended. Experiments by J. J. Thompson and R. Millikan provided a way to calculate the mass of the electron. The mass was calculated to be 9.11×10^{-28} g or 5.86×10^{-4} amu. This is a very small mass. It would take about 1,000,000,000,000,000,000,000,000,000 (10^{27}) electrons to make a gram.

Normal atoms do not carry charges, and, thus, if they contain electrons they must also contain positive charges to neutralize the negative charges of the electrons. The question concerning the manner in which these negative and positive charges are distributed in the atom was partly answered by Lord Rutherford in 1911. His explanation was based on an important experimental observation made by H. Geiger. This observation is discussed below. Certain elements are found to disintegrate spontaneously to form other elements, charged particles, and give off energy in the process. Such spontaneous disintegration is called **radioactivity** and the elements that spontaneously disintegrate are called radioactive elements. Some radioactive elements disintegrate to form, as one of the products, a certain charged particle possessing great kinetic energy (energy of motion). As the atoms of the element disintegrate these particles are essentially ejected with great velocities. These particles have plus two charges ($+2$) and masses of about 4 amu and are called **alpha particles.** (See Section 15-3.) The production of such particles during radioactive decay suggests the presence of positively charged particles as constituent particles of atoms. Geiger used a

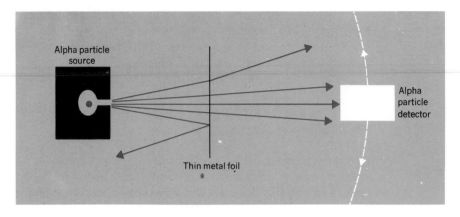

Figure 3-1   In the alpha-particle experiment, alpha particles are projected at a thin metal foil, and the detector is moved about to observe how the alpha particles are scattered.

sample of radioactive radium as a source of alpha particles, and he observed the interaction of the alpha particles with samples of other elements. He did this by subjecting thin pieces of metal, such as gold foil, to a beam of alpha particles (See Figure 3-1). Most of the alpha particles passed through the foil without deflection. Many, however, were deflected to a small degree, and a few were found to be deflected by 180° (see Figure 3-2). Rutherford interpreted these results as follows.

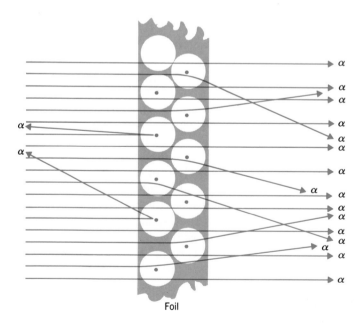

Figure 3-2   In the alpha-particle scattering experiment, most of the alpha particles (α) passed straight through the foil. Some particles were deflected at various angles, and a few were deflected (180°) back to the source.

1. The atoms appear to be made up mainly of empty space since most of the alpha particles were not deflected.

2. The alpha particles that were deflected must have come close to concentrations of positive charge.

3. Very few alpha particles were deflected by 180° so very few must have approached the concentration of positive charge head on, but the concentration of charge must be quite massive to cause a 180° deflection or a complete rebound of the alpha particles.

Based on his interpretations, Rutherford proposed a model for the structure of the atom. He suggested that the atom consists of a small, massive, positively charged nucleus surrounded by a swarm of negatively charged electrons. The term nucleus means center. This idea became known as the **nuclear model of the atom.** See Figure 3-3. From the alpha particle scattering behavior of atoms, Rutherford was able to calculate the approximate size of atoms and nuclei. He calculated the diameter of the average atom to be about 2×10^{-8} cm and the diameter of the average nucleus to be about 1×10^{-13} cm. To illustrate how small a part of the atom the nucleus is, consider that if an atom had a diameter of 100 mi the diameter of the nucleus would be about 30 in. On the other hand, most of the mass of an atom is concentrated in the nucleus. The nucleus makes up at least 99.9% of the mass of an atom.

The nuclear model of the atom provides a view of the atom in which a small positively charged nucleus is surrounded by electrons in motion. In a neutral atom (an atom having no net charge), the charge on the nucleus must be the same as the number of electrons in motion about the nucleus. Furthermore, most of the mass of the atom is found in the nucleus. A question arises: What is the nature of the nucleus? The structure of the nucleus is a topic currently being investigated by nuclear physicists. The nucleus and nuclear energy are discussed in Chapter 15. Investigations by nuclear scientists have substantiated a model of the nucleus in which the nucleus is considered to be made up of particles called **nucleons** of which there are two major types. **Protons** are nucleons that have a +1 charge and a mass of about 1 amu. Protons have charges that are equal in magnitude and opposite in sign to the charges carried by electrons. **Neutrons** are nucleons that have no charge and a mass of about 1 amu. Nuclei are made up of clusters of

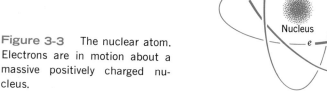

Figure 3-3   The nuclear atom. Electrons are in motion about a massive positively charged nucleus.

Table 3-1   Properties of the Fundamental Particles

Particle	Symbol	Mass	Charge (Quantum Charge)
Electron	e^-, e or $_{-1}^{0}e$	0.000549 amu	-1
Proton	p or $_1^1H$	1.007825 amu	$+1$
Neutron	n or $_0^1n$	1.008665 amu	0(neutral)

nucleons. Numerous other nuclear particles have been detected but are not fundamental to our discussion of atomic structure. Thus, the fundamental particles which make up the atoms are the electron, proton, and neutron. The relative properties of these particles are given in Table 3-1.

The difference in the masses of atoms of elements are the result of the differences in the composition of the nuclei of the atoms. Nuclei of atoms can vary in the number of protons and neutrons present. The nuclei of atoms of a given element have the same number of protons, and, therefore, the atoms must also have the same number of electrons. The number of protons in the nucleus of an atom is called the **atomic number.** It is possible to determine the atomic number of elements, and the elements can be classified according to these atomic numbers, and, as discussed in Section 3-10, elements are listed in the periodic table of elements according to increasing atomic numbers. The fact that atoms of different elements have different numbers of protons and electrons partially accounts for the differences in relative masses of the elements. The neutrons also contribute to the masses of atoms. The atoms of most elements have been found to have nuclei consisting of more than one combination of neutrons and protons. From the nuclear point of view, **isotopes** can be considered to be atoms of the same element which contain the same number of protons but different numbers of neutrons. See Figure 3-4 for an illustration of the nature of the isotopes of several elements.

Rutherford's nuclear model of the atom provided a reasonable view of the atom, but it did not explain how the negatively charged electrons could be in motion about the positively charged nucleus. Neils Bohr developed an atomic model that described such motion.

3-3   The Bohr Atom

In 1913, Neils Bohr described the atom as consisting of a very small positively charged central nucleus with electrons in motion around the nucleus in definite circular orbits. According to his model, the hydrogen atom consisted of a nucleus with a $+1$ charge around which an electron is revolving in a circular path. (See Figure 3-5.) The path that the electron follows is circular or elliptical and is always a certain fixed distance from the nucleus. Because of its motion and position, the electron possesses energy. The distance between the electron and the nucleus depends on the energy of the electron. However, Bohr assumed in his

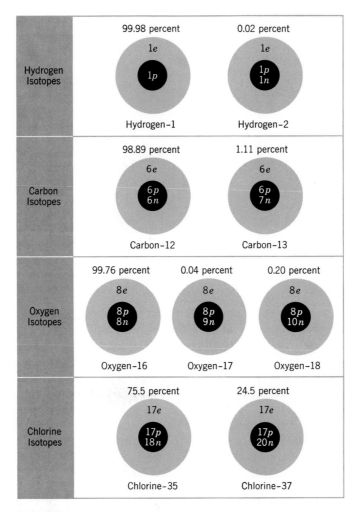

Figure 3-4   The compositions of some common isotopes found in nature. The percentages indicate the contribution of the isotopes of each element to the make-up of the naturally occurring element; e = electron, p = proton, n = neutron. From *Understanding Chemistry: From Atoms to Attitudes* by T. R. Dickson, Wiley, N.Y., 1974.

theory that the electron could only be found specific distances from the nucleus in specific orbits. In other words, he assumed that the energy of the electron was quantized in that the energy could have only certain values. The idea of **quantized energy** is new to us. As an analogy to the quantized energy of an electron, consider that when you climb a

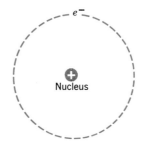

Figure 3-5   The Bohr model of the hydrogen atom. An electron is in a circular orbit about the positive nucleus.

Figure 3-6   The person on the ladder will take on only specific potential energies depending on which rung he is located. Since he can occupy only certain energy levels, his energy is quantized.

8th energy level

7th energy level

6th energy level

5th energy level

4th energy level

3rd energy level

2nd energy level

1st energy level
(ground state)

ladder you can take on only "quantized" positions as you go up as shown in Figure 3-6. That is, you cannot remain at any position between rungs on the ladder. As you climb the ladder your energy (potential energy with respect to the ground) is of a certain value (first rung) and increases by some integer multiple thereof (second rung, third rung, etc.). The hydrogen atom, according to Bohr's model, will have an electron located in an orbit that depends on the energy of the electron. The possible quantized positions of the electron are called the **energy states** or **energy levels** of the electron. (See Figure 3-7.) In normal hydrogen atoms, the electrons take on the lowest quantized energy state allowable. Atoms in which the electrons are the lowest possible energy states are called ground state atoms. Using the Bohr model of the atom, we can describe what happens when an electron in a ground state hydrogen atom gains energy from some external source such as a flame

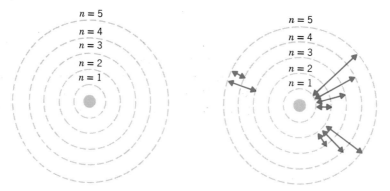

$n = 5$

$n = 4$

$n = 3$

$n = 2$

$n = 1$

$n = 5$

$n = 4$

$n = 3$

$n = 2$

$n = 1$

Figure 3-7   The energy levels in the Bohr atom are indicated as possible values of n. Five energy levels are shown in the figure. An electron may make a quantum jump from a lower energy level to any higher energy level. An excited electron can fall from a higher energy level to a lower energy level. The possible quantum jumps involving the first five energy levels are indicated by the arrows.

or electrical arc. As the electron gains the proper amount of energy, it jumps from the lowest energy level to some higher energy level. Such a jump is called a **quantum jump** during which the electron changes energy by a specific amount. (See Figure 3-7.) When an electron jumps to a higher energy level, it is said to be excited, and an atom with such electrons is called an excited atom. When an electron is greatly excited, it may actually break away from the atom leaving behind a positive ion. Thus, we consider that positive ions can be formed by the loss of electrons by atoms. Excited atoms are unstable, and the excited electrons tend to drop back to lower energy levels and finally to the ground state. As the electron drops to lower energy states, it gives off the excess energy. This is analogous to what happens to you when you drop to the ground (state) from the ladder. You give off the potential energy you stored in the form of kinetic energy or energy of motion. The energy that is given off by electrons falling to lower energy states in atoms usually takes the form of radiant energy or light. In fact, much of the light you see in a flame or the colored light given off by fireworks comes from the loss of energy of excited electrons.

So far, we have described only the hydrogen atom in terms of Bohr's model. Bohr's model does provide a good picture of the hydrogen atom, but what about atoms with more than one electron? When there is more than one electron, we view the atom as consisting of a nucleus with the electrons distributed in various orbits. However, all of the electrons cannot occupy the same orbit. In fact, only specific numbers of electrons can occupy a given orbit, and, thus, the electrons take on different energy values depending on which orbit they occupy. Actually, the Bohr theory of atomic structure did not provide a good model of atoms more complicated than hydrogen since it could not be used to explain certain experimental evidence involving complex atoms. Thus, it was necessary to develop a new theory of atomic structure. The modern theory provides a model of the atom called the quantum mechanical model.

3-4   The Quantum Mechanical Model

The development of the quantum mechanical model represented a significant change in the view of the atom compared to the Bohr model. First, it was realized that the electrons were moving at very high speeds about the nucleus and circumvented the nucleus in rather complicated three-dimensional patterns. Moreover, since the electrons could not be directly observed, their positions were best referred to in terms of the probability of occupying a position about the nucleus. As an analogy consider that when you view a spinning bicycle wheel, the spokes appear as a blur around the entire wheel. In fact, there is only a certain chance or probability that one spoke will be in a specific position at a given instant. The fact that electrons in atoms possess energy, and that this energy is quantized, was incorporated into the new atomic model. However, it was realized that motion of these energetic electrons would best be described as a wave motion. We know that energy can be possessed and transmitted by a water wave, but in an atom the electrons can be described by a continuous three-dimensional wave pattern about the nucleus. In the 1920s Ervin Schrödinger developed a mathematical equation called a wave equation that described the wave behavior of electrons in atoms. All of these ideas concerning the electrons in atoms provided a rather complex atomic model. However, the model was quite successful and met all the requirements of a model stated in Section 3-1. Moreover, the model provided an excellent mental view of the atom that has had a profound influence on the development of chemistry. In the next section this atomic view is described.

3-5   The Modern Atom

The modern theory of atomic structure was developed primarily by Erwin Schrödinger, P. A. M. Dirac, and Werner Heisenburg in the 1920s. Many other scientists were also involved in the development of this valuable theory, which is known as the **quantum mechanical** or **wave mechanical theory of atomic structure.** Today, this theory stands as one of the foundations of modern science. The actual theory involves complicated mathematics, but we shall be concerned only with the picture of the atom that the theory allows us to develop. Actually, the quantum mechanical model of the atom is somewhat similar to Bohr's model except that it allows for the description of atoms containing many electrons. According to this model, the atom still consists of a nucleus surrounded by electrons. The electrons can be located only in certain allowable energy states, which are quantized as they were in the Bohr model. However, these energy states are more complicated than those in the Bohr model. In the new model there are several possible main or principal energy states in which the electrons can be found. These main energy states are called **energy levels,** and each is made up of one or more energy states called **energy sublevels.** Furthermore, each energy sublevel consists of one or more electron energy states. These are called **orbitals** or **electron orbitals.** It is in these orbitals that the electrons are located.

Let us summarize this picture of the atom. (See Figure 3-8.) The atom consists of main energy levels, made up of one or more sublevels that contain one or more energy positions called orbitals. This is a more complicated picture of the energy positions of the electrons than we had in the Bohr model, but it provides a way to describe atoms with many electrons. This description of the atom according to energy states does not provide us with a physical picture of the atom that we can visualize. Therefore, a question still remains concerning how we can visualize an atom according to this theory. Actually, the theory does provide a good visual representation of atoms. This representation develops from the shapes of the electron orbitals.

According to the quantum mechanical model, electrons are not visualized as particles with specific positions in the atom. Instead, they are viewed with respect to the probability of being located in certain portions of the atom. Furthermore, since electrons are negatively charged and are in rapid motion within the atoms, they can be viewed as clouds of negative charge, called **electron clouds.** The shapes and densities of

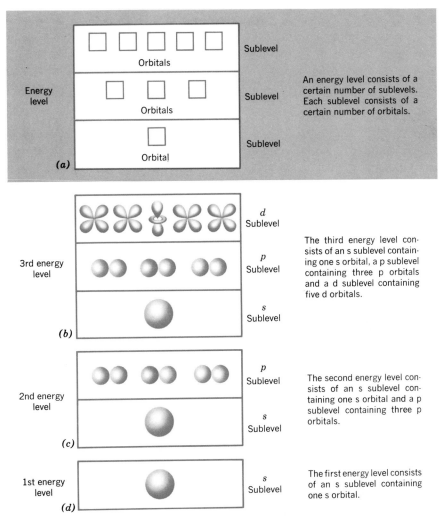

Figure 3-8   A representation of the quantum model of the atom.

these clouds correspond to the probability of finding the electrons in specific portions of the atom. Consequently, these electron clouds are sometimes called **probability volumes.** An electron of a specific energy may be viewed as being within the volume about the nucleus in which there is a specific probability of finding such an electron. This may seem to be a strange way to view electrons, but actually it provides a good picture of the atom. Since electrons are viewed as charged clouds of certain shapes, it is possible to develop a three-dimensional view of the atom. The atom can be visualized as consisting of a nucleus surrounded by electron clouds. The shape of the atom depends on the shapes of the electron clouds. The quantum mechanical theory allows us to actually determine the approximate shapes of these electron clouds or electron orbitals. (See Figure 3-9.)

Consider the hydrogen atom according to this model. The hydrogen atom in the ground state will consist of the nucleus surrounded by a spherical cloud of negative charge, which is the electron located in the

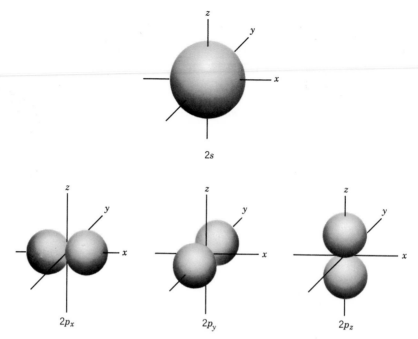

Figure 3-9   Some atomic orbital shapes.

lowest energy state. Excited hydrogen atoms can be pictured in a similar manner except that the shape and size of the electron cloud will depend on the energy state of the electrons. We can extend this picture to atoms of other elements. We can view the atom of any element as consisting of a nucleus, with a charge corresponding to the atomic number, surrounded by electron clouds. The number of electrons will be given by the atomic number. The electrons will occupy the lowest available energy states. However, several questions arise at this point. How many electrons can fit in each main energy level and in each sublevel? How many electrons can fit in each electron orbital? What are the relative energies of the various orbitals?

Once these questions are answered, it is possible to deduce the arrangement of electrons in the atoms of a given element. Since the number of electrons in the atoms is known, we just hypothetically place the electrons in the orbitals starting at the lowest energy positions and work up. Once the electronic arrangement in an atom has been deduced, it is called the **electronic structure.** Furthermore, once the electronic structure is known, we can develop a picture of the atom depending on which orbitals are occupied by electrons. However, before we do this, the questions posed above must be answered.

3-6   Electron Energy Levels

The quantum mechanical model provides answers to these questions. First, the atom has **main energy levels.** These levels are referred to by number as the first energy level (1), the second energy level (2), third energy level (3), fourth energy level (4), fifth energy level (5), and so on. The main energy levels are sometimes referred to by the letters $K, L, M, N, O,$ etc.

$$
\begin{array}{cccccc}
\text{energy level} & 1 & 2 & 3 & 4 & 5 & \ldots \\
& K & L & M & N & O & \ldots
\end{array}
$$

distance from nucleus increases $\longrightarrow$

energy increases $\longrightarrow$

Each energy level consists of a certain number of **sublevels.** The sublevels are referred to by the letters s, p, d, and f. The first energy level has one sublevel, an s sublevel. The second energy level has two sublevels, an s sublevel and a p sublevel. The third energy level has three sublevels, s, p, and d, The fourth energy level has four sublevels, s, p, d, and f.

$$
\begin{array}{ccccc}
\text{sublevel} & s & p & d & f & \ldots
\end{array}
$$

increasing energy $\longrightarrow$

These sublevels are usually denoted and distinguished by giving the main energy level number followed by the letter corresponding to the sublevel. So, we can refer to the $1s$ (pronounced one-ess) sublevel, $2p$ (pronounced two-pee) sublevel or the $3d$ (pronounced three-dee) sublevel. Let us pause and summarize the structure of the first five energy levels.

Energy level	Sublevels
1st	1s
2nd	2s, 2p
3rd	3s, 3p, 3d
4th	4s, 4p, 4d, 4f
5th	5s, 5p, 5d, 5f, . . .

Now each sublevel consists of a certain number of **electron orbitals.** An s sublevel consists of one orbital referred to as an s orbital. A p sublevel consists of three orbitals called p orbitals. A d sublevel consists of five orbitals called d orbitals. An f sublevel consists of seven orbitals called f orbitals. This is summarized below and in Table 3-2.

Sublevel	Number of Possible Orbitals
s	One s orbital
p	Three p orbitals
d	Five d orbitals
f	Seven f orbitals

Table 3-2   Sublevel Structure in the First Four Energy Levels

Energy Level	Type of Sublevel	Number of Orbitals of Given Type	Electron State Notation
First	s	One s orbital	1s electrons
Second	s	One s orbital	2s electrons
	p	Three p orbitals	2p electrons
Third	s	One s orbital	3s electrons
	p	Three p orbitals	3p electrons
	d	Five d orbitals	3d electrons
Fourth	s	One s orbital	4s electrons
	p	Three p orbitals	4p electrons
	d	Five d orbitals	4d electrons
	f	Seven f orbitals	4f electrons

The orbitals making up a sublevel correspond to the same electron energies, and they are called degenerate orbitals. As can be seen, the higher energy levels are more complex than the lower energy levels. The electrons in normal atoms tend to occupy the lowest energy positions available. The important question at this point is: How many electrons can be in a given orbital? An orbital can contain a maximum of two electrons. An orbital that contains two electrons is called a filled orbital while an orbital containing only one electron is a half-filled orbital. Given the fact that the orbital capacity is two electrons, it is possible to deduce the number of electrons that can be contained within a given sublevel. Of course, this is done by considering the number of orbitals within a sublevel. An s sublevel has one orbital and, thus, can accommodate only 2 electrons. A p sublevel with three orbitals can accommodate a maximum of 6 electrons. A d sublevel can contain a maximum of 10 electrons and an f sublevel a maximum of 14 electrons.

To deduce how the electrons are distributed in an atom of an element, all that is necessary is to distribute hypothetically the electrons of the atom in the various energy positions. This is done by starting at the lowest position and working up to a higher position as the capacity of a given level is reached.

By referring to the periodic table given in the back cover of the book, it is seen that the elements are arranged in rows (left to right) according to increasing atomic numbers. Recall that the atomic number of an element is equal to the number of electrons in the atoms of that element. The atomic numbers increase by one unit, starting with hydrogen (atomic number 1) going up to lawrencium (atomic number 103). This means that atoms of each element have one more electron than the atoms of the preceeding element in the table. Of course, the nuclei of the atoms also differ considerably; the numbers of neutrons and protons vary. To state the electronic structure of the atoms of an element, we must state how the electrons are distributed in the energy levels. Since the electrons in an unexcited atom tend to occupy the lowest available energy positions, it is important to consider the **order of increasing energy of the orbitals or sublevels.** This order is generally $1s$ $2s$ $2p$ $3s$ $3p$ $4s$ $3d$ $4p$ $5s$ $4d$ $5p$ $6s$ $4f$ $5d$ $6p$ $7s$ $5f$ $6d$. . . . This order can be easily remembered by using the mnemonic table shown in Figure 3-10. It is important to note that the order of energies shows some orbitals in higher principal energy levels having lower energies than some orbitals in lower principal energy levels (i.e., the $4s$ orbital is of lower energy than a $3d$ orbital). This situation is referred to as **energy overlap** of the energy levels and must not be interpreted as indicating that the lower energy orbital is physically closer to the nucleus than the higher energy orbital. That is, the $4s$ orbital is usually not located closer to the nucleus than a $3d$ orbital. The energy increase resulting from increasing complexity of orbitals is comparable to the energy increase resulting from a change in main energy level; this results in an energy overlap of some of the energy levels. Note the occurrence of several such energy level overlaps that are indicated in Figure 3-10.

3-7   The Electronic Configuration of the Elements

If we keep in mind the number of electrons that can be accommodated in a sublevel, we can use the above order of orbital energies to describe the electronic structure of the atoms of the elements. An s sublevel can accommodate 1 or 2 electrons. The p sublevel can accommodate from 1 to 6 electrons, the d sublevel from 1 to 10 electrons, and the f sublevel from 1 to 14 electrons. Now, it is possible to describe the electronic structure of atoms by stating the sublevel or orbital distribution of the electrons. The electrons are distributed in the lower energy sublevels first, and, when these are completely occupied, the next higher energy sublevel is used.

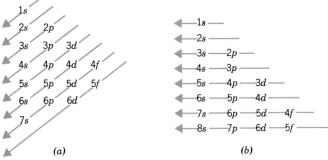

(a)   (b)

(a) The sublevels are written as shown. The direction of the arrows from upper right to lower left indicates the general energies of the energy sublevels.

(b) The sublevels are written in columns as shown. The relative energies are given in the direction of horizontal arrows from right to left.

Figure 3-10   Relative energies of sublevels.

The one electron in a hydrogen atom is located in the s sublevel of the first energy level. A shorthand method of indicating this **electronic configuration** is

$$1s^1$$

This is called **orbital notation** and indicates one electron in the s sublevel of the first energy level. The number of electrons in the sublevel is indicated by the superscript digit to the right of the sublevel notation. The two electrons in a helium atom are located in the lowest energy position that is in the s sublevel of the first energy level. The electronic configuration of helium in orbital notation is

$$1s^2$$

The next element, lithium, has atoms that have three electrons. Two of these electrons occupy the lowest energy position, the $1s$ sublevel, and the third electron must occupy the next lowest available energy position that would be the $2s$ sublevel. Thus, the electronic structure of lithium in orbital notation is

$$1s^2 2s^1$$

which would be interpreted as indicated in the previous sentence. To determine the electronic configuration for an element, all we have to do is decide how many electrons are involved and then distribute these electrons in the sublevels starting in the lower energy sublevels and working up until all of the electrons have been distributed. Beryllium atoms have four electrons so the electronic configuration of beryllium must be

$$1s^2 2s^2$$

The electronic configuration of boron is $1s^2 2s^2 2p^1$, and the electronic configuration of carbon is $1s^2 2s^2 2p^2$. By using the order of energies of the sublevels and the electron capacities of the sublevels (s-2 electron capacity, p-6 electron capacity, d-10 electron capacity, f-14 electron capacity), the electronic configuration of an element can be deduced given the atomic number.

Example 3-1   What is the electronic configuration of phosphorus, atomic number 15? Phosphorus atoms have 15 electrons. The order of energy of the sublevel is

$$1s\ 2s\ 2p\ 3s\ 3p\ \ldots$$

Distributing the 15 electrons in these sublevels so that the lower energy sublevels are filled to capacity gives the electronic configuration

$$1s^2 2s^2 2p^6 3s^2 3p^3$$

What is the electronic configuration of vanadium, atomic number 23? The order of energy sublevels is

$$1s\ 2s\ 2p\ 3s\ 3p\ 4s\ 3d\ \ldots$$

Distributing the 23 electrons gives

$$1s^2 2s^2 2p^6 3s^2 3p^6 4s^2 3d^3$$

Note that, in terms of energy, the $4s$ sublevel comes before the $3d$ sublevel. However, when writing electronic configurations, normally the distribution is written so that all of the sublevels of a given energy level are grouped together. So the configuration of vanadium should be written as

$$1s^2 2s^2 2p^6 3s^2 3p^6 3d^3 4s^2$$

Such notation emphasizes the number of outer energy level electrons. We shall soon see why this notation is more convenient than the other notation. The electronic configurations of all the elements are given in Appendix IV.

3-8   Hund's Rule

We can now express the electronic configuration for the atoms of any element in terms of the sublevel distribution of the electrons. Sometimes it is important to know how the electrons are distributed in the degenerate orbitals that comprise a sublevel. Of course, in a filled sublevel each orbital contains two electrons, but the question is how the electrons are distributed in a partly filled sublevel. When electrons are located in a given sublevel, they take positions according to **Hund's rule.** Hund's rule can be simply stated as:

If two or more electrons have energies that place them in the same sublevel, the lower energy state will be the state in which the electrons are not in the same orbital.

This means that when electrons are in a sublevel, the lower energy situation will be the state in which the electrons do not share the same orbital. This rule can be illustrated if we consider the figure below. In this figure, the squares represent three degenerate p orbitals that comprise a p sublevel and the arrows represent electrons.

p orbitals	↑↓			higher energy state
p orbitals	↑	↑		lower energy state

In the first case, the electrons are in the same orbital. This is a higher energy situation than the second case in which the electrons are in two of the orbitals. The same type of situation would apply if a third electron were located in these orbitals.

A pictorial representation of electron distribution in orbitals can be developed by considering squares to represent orbitals in which arrows may be placed to indicate electrons. The squares can be arranged so that the increasing energies of the orbitals are indicated. This is illustrated in Figure 3-11. Using this figure, the electron distribution in various atoms can be represented by placing arrows in the appropriate boxes.

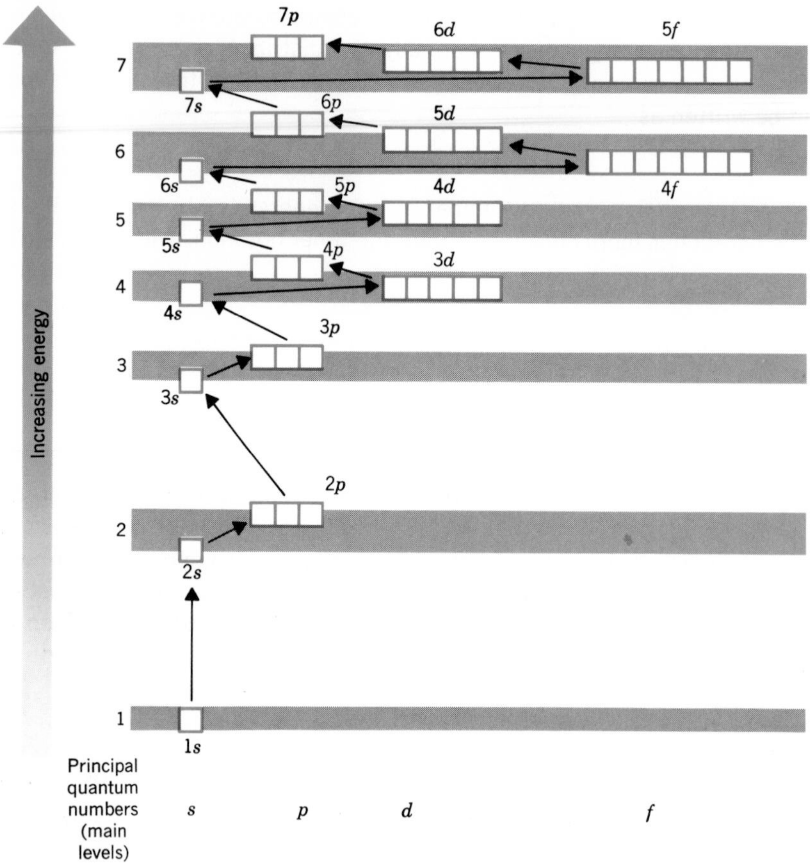

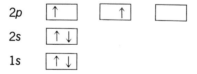

Increasing energy →

7p   6d   5f

7s   6p   5d

6s   5p   4d   4f

5s   4p   3d

4s   3p

3s

2p

2s

1s

Principal
quantum
numbers   s   p   d   f
(main
levels)

Figure 3-11   Box representation of possible electron states.

Example 3-2   What is the distribution of electrons in a carbon atom?
The atomic number of carbon is 6, so the distribution would be

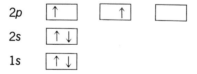

2p   [↑]   [↑]   [ ]
2s   [↑↓]
1s   [↑↓]

Hund's rule requires two unpaired 2 p electrons.

What is the electronic distribution in a nitrogen atom?
The seven electrons are distributed as follows.

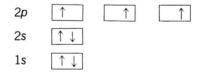

2p   [↑]   [↑]   [↑]
2s   [↑↓]
1s   [↑↓]

Example 3-3   How are the electrons distributed in an iron atom?
The 26 electrons are distributed as follows.

3d   [↑↓]   [↑]   [↑]   [↑]   [↑]

4s   [↑↓]

3p   [↑↓]   [↑↓]   [↑↓]

3s   [↑↓]

2p   [↑↓]   [↑↓]   [↑↓]

2s   [↑↓]

1s   [↑↓]

3-9 The Visualization of Atoms

It is possible to determine the general shapes of electron orbitals in atoms. These orbitals are the probability volumes for electrons in specific energy positions in atoms. Using the general shapes of orbitals, it is possible to develop a visual picture of an atom. To do this we must consider that the electron energy levels of atoms are made up of subenergy levels, and that these sublevels contain specific numbers of electron orbitals. Each orbital can contain a maximum of two electrons. An *s* sublevel contains one *s* orbital. An **s orbital** can be pictured as a spherically shaped cloud.

The sizes of *s* orbitals increase at higher electron energy levels.

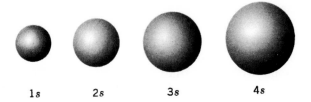

1s   2s   3s   4s

A *p* sublevel contains three possible **p orbitals.** A *p* orbital can be pictured as a dumbbell-shaped cloud.

The sizes of *p* orbitals increase at higher energy levels.

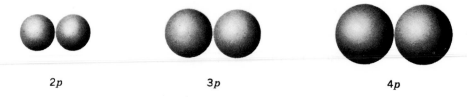

2p   3p   4p

The *p* orbitals comprising a given *p* sublevel will be oriented perpendicular (at right angles) to one another.

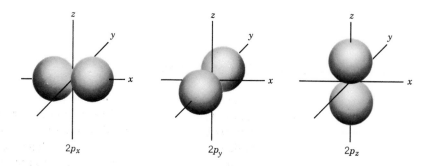

2p_x   2p_y   2p_z

A *d* orbital can be pictured as a four-lobed cloud.

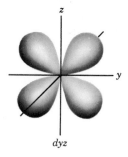

dyz

A *d* sublevel contains five possible *d* orbitals and, consequently, has a complex shape.

Atoms can be pictured as consisting of a nucleus surrounded by the electron orbital clouds in which the electrons are located. The number of orbitals occupied depends on how many electrons are in the atom. Atoms with more than two electrons can be visualized as two or more orbitals superimposed on one another about the nucleus. As a result of the superimposed orbitals, atoms can be visualized as having essentially a spherical or ball-like shape.

3-10   The Periodic Table

Certain groups of elements have been found to have very similar properties. That is, they form similar kinds of compounds with other elements. For example, the elements lithium, sodium, and potassium each form a compound with chlorine that involves one combined atom of the element for every one combined atom of chlorine. The elements oxygen, sulfur, selenium, and tellurium each form a compound with hydrogen involving two combined atoms of hydrogen to every combined atom of the element. For many years prior to an understanding of atomic structure, such similarities in elements were observed. Chemists tried to arrange the elements in a tabular fashion according to this repetition in properties. Around 1860, L. Meyer and D. I. Mendeleev developed such a table based on a repeating pattern in the observed properties of the elements. This table, called the **periodic table of the elements,** is one of the foundations of chemistry. The periodic table is based on what has become known as the **periodic law,** which is stated as follows:

The properties of the elements are a periodic function of the atomic numbers.

The law indicates that as we compare the elements we find a pattern in which the elements of higher atomic number have similar properties to elements of lower atomic number. For instance, the eleventh element, sodium, is found to be similar to the third element, lithium; the twelfth element, magnesium, is similar to the fourth element, beryllium; the thirteenth element, aluminum, is similar to the fifth element, boron; the fourteenth element, silicon, is similar to the sixth element, carbon; and so on. When the elements are arranged in columns according to this periodic repetition in properties, the periodic table results.

It was not until electronic structure was related to the properties of the elements that an explanation of the table became apparent. It was found that the elements with the same number of electrons in the outer energy level of the atoms have similar properties. That is, those elements with one outer energy level electron are similar, those with two are similar, those with three are similar, and so on.

For example, the electronic configurations of Li, Na, K, Rb, and Cs are:

Li $1s^2 2s^1$
Na $1s^2 2s^2 2p^6 3s^1$
K $1s^2 2s^2 2p^6 3s^2 3p^6 4s^1$
Rb $1s^2 2s^2 2p^6 3s^2 3p^6 3d^{10} 4s^2 4p^6 5s^1$
Cs $1s^2 2s^2 2p^6 3s^2 3p^6 3d^{10} 4s^2 4p^6 4d^{10} 5s^2 5p^6 6s^1$

These five metals have similar physical properties and form similar compounds with other elements. The electronic configurations indicate that all of these metals have one electron in the outer energy level. When the elements are tabulated according to increasing atomic number so that the elements of similar electronic configurations are arranged in columns, the periodic table is obtained. (See Figure 3-12.)

The form of the periodic table can be remembered if it is interpreted in terms of the distribution of the electrons in the various orbitals according to the order of energy of the orbitals given previously. The periodic table given in Figure 3-12 emphasizes the positions in the table corresponding to the various orbitals. This orbital interpretation of the

IA	IIA	IIIB	IVB	VB	VIB	VIIB	VIII	VIII	VIII	IB	IIB	IIIA	IVA	VA	VIA	VIIA	NOBLE GASES
H $1s^1$																H $1s^1$	He $1s^2$
Li $2s^1$	Be $2s^2$											B $2s^2 2p^1$	C $2s^2 2p^2$	N $2s^2 2p^3$	O $2s^2 2p^4$	F $2s^2 2p^5$	Ne $2s^2 2p^6$
Na $3s^1$	Mg $3s^2$											Al $3s^2 3p^1$	Si $3s^2 3p^2$	P $3s^2 3p^3$	S $3s^2 3p^4$	Cl $3s^2 3p^5$	Ar $3s^2 3p^6$
K $4s^1$	Ca $4s^2$	Sc $3d^1 4s^2$	Ti $3d^2 4s^2$	V $3d^3 4s^2$	Cr $3d^5 4s^1$	Mn $3d^5 4s^2$	Fe $3d^6 4s^2$	Co $3d^7 4s^2$	Ni $3d^8 4s^2$	Cu $3d^{10} 4s^1$	Zn $3d^{10} 4s^2$	Ga $4s^2 4p^1$	Ge $4s^2 4p^2$	As $4s^2 4p^3$	Se $4s^2 4p^4$	Br $4s^2 4p^5$	Kr $4s^2 4p^6$
Rb $5s^1$	Sr $5s^2$	Y $4d^1 5s^2$	Zr $4d^2 5s^2$	Nb $4d^4 5s^1$	Mo $4d^5 5s^1$	Tc $4d^5 5s^2$	Ru $4d^7 5s^1$	Rh $4d^8 5s^1$	Pd $4d^{10}$	Ag $4d^{10} 5s^1$	Cd $4d^{10} 5s^2$	In $5s^2 5p^1$	Sn $5s^2 5p^2$	Sb $5s^2 5p^3$	Te $5s^2 5p^4$	I $5s^2 5p^5$	Xe $5s^2 5p^6$
Cs $6s^1$	Ba $6s^2$	La $5d^1 6s^2$	Hf $5d^2 6s^2$	Ta $5d^3 6s^2$	W $5d^4 6s^2$	Re $5d^5 6s^2$	Os $5d^6 6s^2$	Ir $5d^7 6s^2$	Pt $5d^9 6s^1$	Au $5d^{10} 6s^1$	Hg $5d^{10} 6s^2$	Tl $6s^2 6p^1$	Pb $6s^2 6p^2$	Bi $6s^2 6p^3$	Po $6s^2 6p^4$	At $6s^2 6p^5$	Rn $6s^2 6p^6$
Fr $7s^1$	Ra $7s^2$	Ac $6d^1 7s^2$															

Figure 3-12 Periodic table showing the similarities in the outer electronic configurations of the groups of elements.

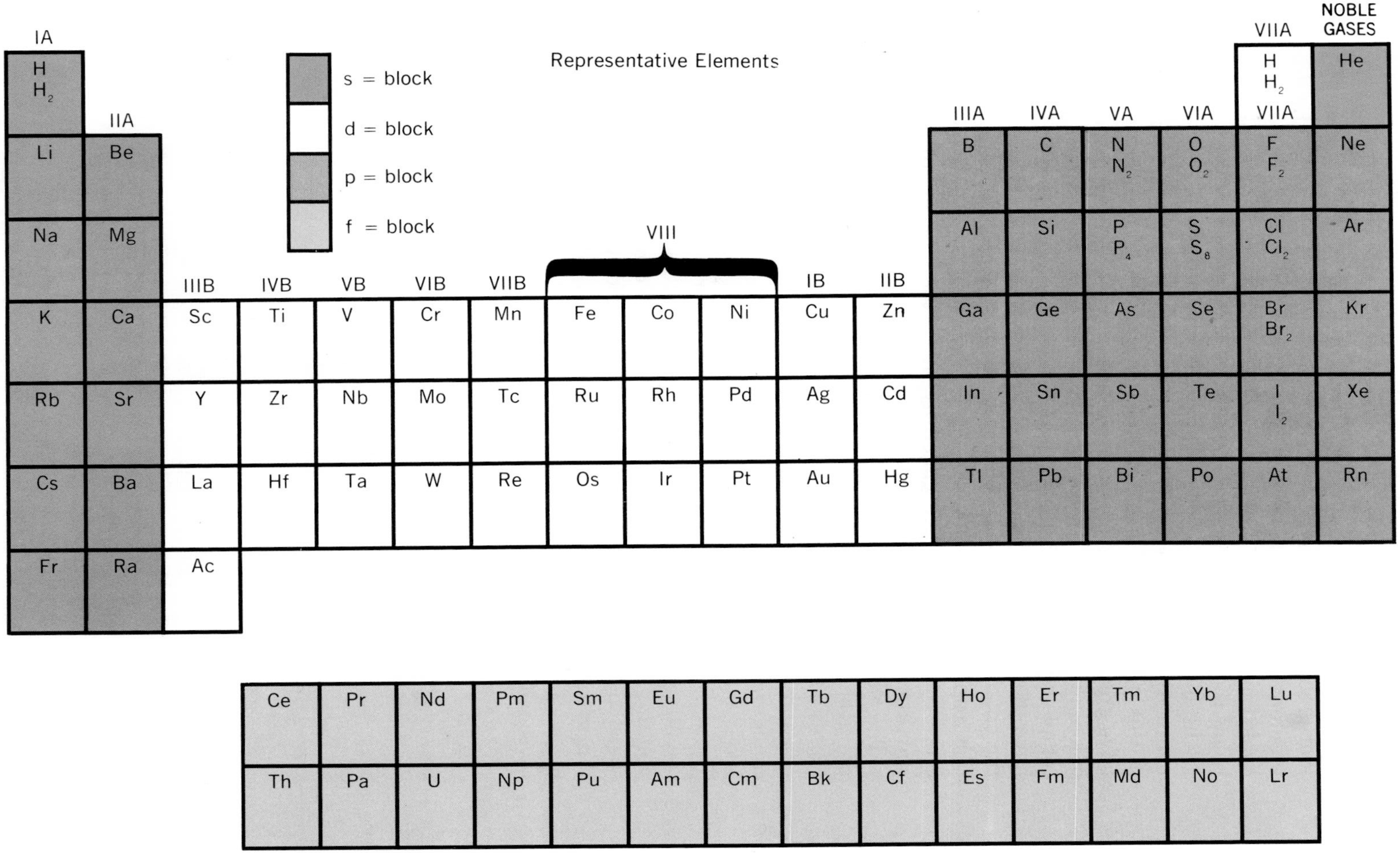

Figure 3-13 Periodic table showing the *s*, *p*, *d*, and *f* blocks.

periodic table is very useful. The elements are sometimes classified according to the part of the periodic table in which they are found. Such a classification is shown in Figure 3-13. The elements are classified into four groups; the *s*-block elements, the *p*-block elements, the *d*-block elements, and the *f*-block elements. In order to shorten the space occupied by the periodic table, the *f*-block elements are normally placed below the *d*-block elements.

The elements in the far-right column of the *p*-block elements and helium comprise a group referred to as the **noble gases**. The elements of this group are all gases under normal conditions, and they do not have great tendencies to form chemical compounds. With the exception of helium, which has the configuration of $1s^2$, all of the inert gases have completely filled *s* and *p* sublevels in the outer energy level. Thus, they have the outer energy level configuration of ns^2np^6, where *n* indicates the outer energy level. This configuration is called the octet configuration (two *s* electrons and six *p* electrons) or the noble gas configuration. Since these elements are generally inert, there seems to be a certain stability associated with such an electronic configuration.

The elements that comprise the *s* and *p* blocks, with the exception of the noble gases, are called the **representative elements**. The *s*-block representative elements have the outer energy configuration of ns^1 or ns^2, and the *p*-block representative elements have an outer energy level configuration ranging from ns^2np^1 to ns^2np^5. The term "outer energy level" means the highest numbered energy level containing electrons. The *d*-block elements have electronic configurations that correspond to the filling of the *d* sublevels. These elements are called the **transition metals**. The *f*-block elements have electronic configurations that correspond to the filling of the *f* sublevels. The *f*-block elements are called the **innertransition metals** or the **rare earth metals**. The innertransition metals involving the 4*f* sublevel are called the lanthanide series, and those which involve the 5*f* sublevel are called the actinide series.

Group number	IA Alkali metals		IIA Alkaline earth metals		IIIA Boron–aluminum group		IVA Carbon group		VA Nitrogen group		VIA Oxygen group		VIIA Halogens	
Lithium	3 Li	Beryllium	4 Be	Boron	5 B	Carbon	6 C	Nitrogen	7 N	Oxygen	8 O	Fluorine	9 F	
Sodium	11 Na	Magnesium	12 Mg	Aluminum	13 Al	Silicon	14 Si	Phosphorus	15 P	Sulfur	16 S	Chlorine	17 Cl	
Potassium	19 K	Calcium	20 Ca	Gallium	31 Ga	Germanium	32 Ge	Arsenic	33 As	Selenium	34 Se	Bromine	35 Br	
Rubidium	37 Rb	Strontium	38 Sr	Indium	49 In	Tin	50 Sn	Antimony	51 Sb	Tellurium	52 Te	Iodine	53 I	
Cesium	55 Cs	Barium	56 Ba	Thallium	81 Tl	Lead	82 Pb	Bismuth	83 Bi	Polonium	84 Po	Astatine	85 At	
Francium	87 Fr	Radium	88 Ra											

Figure 3-14   Some common groups or families of elements. From *Understanding Chemistry: From Atoms to Attitudes* by T. R. Dickson, Wiley, N.Y., 1974.

The elements that occur in a **column** of the periodic table are called **groups** or **families,** while the horizontal sequences of elements are referred to as **rows** or **periods.** The groups of representative elements have reference symbols and group names. These are given in Figure 3-14. Notice that the Roman numeral group number gives the total number of electrons in the outer energy level of each group. The names and symbols are often used to refer to groups of the representative elements.

Questions and Problems

1. Describe Rutherford's concept of a nuclear atom.

2. What is the atomic number of an element?

3. How do the various atoms of the elements differ in terms of subatomic particles?

4. Use the atomic numbers to answer the following. (See the periodic table at the back of the book.)
   (a) How many protons are in a nitrogen atom?
   (b) How many protons are in a chlorine atom?
   (c) How many electrons are in an oxygen atom?

5. What are isotopes?

6. No one has ever seen an atom, but it remains as one of the fundamental "beliefs" upon which modern science is based. What do you think of the atom, and what kind of mental picture do you have of the atom?

7. Describe the Bohr atom.

8. What is an excited atom according to the Bohr concept?

9. The energy in picoergs (pergs) of an electron in the Bohr atom is given by the expression

$$E = \frac{-21.7 \text{ pergs}}{n^2}$$

where n is the number of the energy level. When an excited electron falls to a lower energy level, light is given off with an energy corresponding to the difference between the electron energies of the levels.

$$E = \text{energy higher level} - \text{energy lower level}$$

$$E = 21.7 \left(\frac{1}{n^2} - \frac{1}{m^2} \right) \text{pergs}$$

where n is the lower energy level number, and m is the higher energy level number. Calculate the energy of light that is given off when an electron falls from the second to the first energy level. Repeat for an electron falling from the fourth to the first energy level.

10. Describe the quantum mechanical model of the atom and indicate the orbital and sublevel structure of the energy levels.

11. What is an electron (atomic) orbital?

12. Which of the following electron states do not exist? 3p, 2d, 3s, 4f, 1p, 3p, 2f.

13. What is the maximum number of electrons that can be contained in:
    (a) one electron orbital
    (b) an s sublevel
    (c) a p sublevel
    (d) a d sublevel
    (e) an f sublevel
    (f) the first energy level
    (g) the second energy level
    (h) the third energy level

14. Give the electron configuration in orbital notation for the following elements:
    (a) Na (sodium)
    (b) Si (silicon)
    (c) Cl (chlorine)
    (d) O (oxygen)
    (e) S (sulfur)
    (f) Al (aluminum)
    (g) N (nitrogen)
    (h) Ca (calcium)
    (i) Zn (zinc)
    (j) Au (gold)

15. Give the orbital distribution of electrons for the following elements (Refer to Section 3-8 and Figure 3-11).
    (a) O (oxygen)
    (b) Zn (zinc)
    (c) Mg (magnesium)
    (d) Cr (chromium)
    (e) Ne (neon)
    (f) Br (bromine)

16. Refer to Figure 3-10 and Figure 3-13 and indicate two examples of energy overlaps between energy sublevels. (e.g., the 3d sublevel overlaps the 4s sublevel.)

17. Give a statement of the periodic law.

18. Sketch an outline of the periodic table and indicate the following:
    (a) The s block, the p block, the d block, and the f block.
    (b) Label the portions of the table corresponding to the filling of the 2s sublevel, the 3d sublevel, the 4p sublevel, the 4f sublevel.

19. How many elements are included in each row of the following:
    (a) the s block
    (b) the p block
    (c) the d block
    (d) the f block

20. Sketch an outline of the periodic table and indicate the positions of the following:
  (a) the noble gases, the representative elements, the transition elements, and the innertransition elements.
  (b) the alkali metals, the alkaline-earth metals, the boron-aluminum group, the nitrogen group, the carbon group, the oxygen group, and the halogens.

21. Refer to the periodic table at the back of the book and classify the following as representative element, transition element, inner-transition element, or noble gas:
  (a) oxygen
  (b) neon
  (c) chromium
  (d) uranium
  (e) magnesium
  (f) zinc
  (g) silicon
  (h) lithium

22. Classify each of the elements in question 21 as metal, nonmetal, metalloid, or noble gas.

4

Chapter 4: Chemical Bonding

Objectives

The student should be able to:

1 Describe the ionic bond.
2 Give the electron dot symbol for a representative element.
3 Predict the likely formulas for the monoatomic ions formed by representative elements.
4 Describe the covalent bond.
5 Write the electron dot structure of a simple molecule involving representative elements given the bonding sequence.
6 Describe a multiple covalent bond.
7 Give the names and formulas of common monatomic and polyatomic ions.
8 Describe the variation of the electronegativities of the elements in the periodic table.
9 Describe a polar bond and a polar molecule.

Terms to Know

Chemical bond
Octet rule
Ionic compound
Molecule
Molecular or covalent compound
Polyatomic ion
Electronegativity
Mole

4

Chemical Bonding

4-1 The Nature of Chemical Compounds

Atoms are the fundamental chemical particles. They serve as the building blocks of matter. Some substances consist of aggregations of atoms. For instance, a piece of iron consists of a large aggregation of iron atoms. On the other hand, many substances contain atoms of various elements that have entered into atomic combinations to form compounds. When atoms combine with one another they do not retain their original electronic structures. Instead, they are transformed by the loss, gain, or mutual sharing of electrons into new kinds of chemical particles that compose compounds. These chemical particles, called ions and molecules, are discussed in this chapter.

Sodium chloride, NaCl, is a compound composed of the elements sodium and chlorine. When crystalline sodium chloride is heated to form a melt, it can be decomposed into the constituent elements by passing an electrical current through the melt. This electrolysis of sodium chloride is represented by the equation

$$2NaCl \xrightarrow{\text{electrolysis}} 2Na + Cl_2$$

When metallic sodium and chlorine gas are intermixed they react to form sodium chloride. This is the reverse of the above reaction.

$$2Na + Cl_2 \rightarrow 2NaCl$$

Water can also be decomposed into the constituent elements by electrolysis.

$$2H_2O \xrightarrow{\text{electrolysis}} 2H_2 + O_2$$

Furthermore, when hydrogen gas and oxygen gas are mixed and ignited they react to form water.

$$2H_2 + O_2 \rightarrow 2H_2O$$

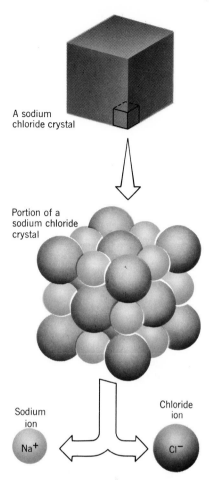

Figure 4-1   The constituent particles of sodium chloride are sodium ions and chloride ions.

Sodium chloride and water are typical examples of the compounds that can be decomposed to their elements or compounded from the elements. However, specific properties of these two compounds reveal that there is a distinct difference in the way in which the sodium and chlorine combine in sodium chloride and the way in which hydrogen and oxygen combine in water. As illustrated in Figure 4-1 if we consider the fundamental particles that structure sodium chloride we find that these particles are not ordinary atoms but instead atoms that have deficient or excess electrons and, thus, carry positive and negative charges. Such charged atoms are called ions, and sodium chloride is composed of positive sodium ions and negative chloride ions. Now let us consider water as illustrated in Figure 4-2. The fundamental particles that compose water are not independent atoms but instead aggregations of two hydrogen atoms and one oxygen atom held in association by mutual sharing of electrons. Such aggregations of atoms are called molecules, and water is composed of water molecules.

The compounds discussed above are examples of the two types of

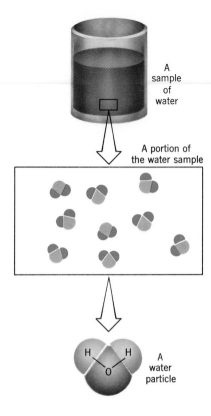

Figure 4-2   The characteristic particle of water is an H_2O molecule.

chemical combinations that are involved in compound formation. Such chemical combinations involve the formation of chemical bonds that are discussed in this chapter.

4-2   Electron Dot Symbols

When atoms of elements form compounds, the outer energy level electrons are usually the only electrons involved. These outer energy level electrons are called **valence electrons.** Rather than write the electronic configurations for elements in order to indicate the number of valence electrons, a method for the symbolic representation of such electrons has been developed. To indicate the valence electrons of an element, the electron dot symbol is used. The **electron dot symbol** is written by giving the usual symbol of the element around which the outer energy level (valence) electrons are indicated as dots. For example, the electron dot symbol for hydrogen is H·. These electron dot symbols are used mainly for the representative elements, and for them the following general pattern can be followed. The electron dot symbol gives the symbol of the element surrounded by a number of dots corresponding to the number of valence electrons. The number of valence electrons for a representative element is given by the Roman numeral group number in the periodic table. When writing a dot symbol, imagine a square

Table 4-1   Electron Dot Symbols of Some Representative Elements

IA	IIA	IIIA	IVA	VA	VIA	VIIA	
Li·	·Be·	·Ḃ·	·Ċ·	·N̈·	:Ö·	:F̈·	H·
Na·	·Mg·	·Al·	·Si·	·P̈·	:S̈·	:C̈l·	
K·	·Ca·		·Ge·	·Äs·	:Se·	:Br·	
Rb·	·Sr·				:Te·	:Ï·	

around the symbol of the element and put a dot on each side until all the valence electrons are used. Double up electrons only when necessary. For example, the electron dot symbol for sodium (Group IA) is

$$Na·$$

The electron dot symbol for carbon (Group IVA) is

$$·\overset{\cdot}{\underset{\cdot}{C}}·$$

The electron dot symbol for oxygen (Group VIA) is

$$:\overset{\cdot}{O}·\quad \text{or}\quad ·\overset{\cdot}{\underset{\cdot\cdot}{O}}:$$

The electron dot symbol for fluorine (Group VIIA) is

$$:\overset{\cdot\cdot}{\underset{\cdot\cdot}{F}}·\quad \text{or}\quad ·\overset{\cdot\cdot}{\underset{\cdot\cdot}{F}}:$$

The side used for pairing dots is not important.

The dot symbols for most of the representative elements are given in Table 4-1. Note carefully that the dot symbols for all the elements in a group are the same since the number of valence electrons is the same. Although these dot symbols are simple representations of the distributions of valence electrons, they are extremely useful in the discussion of the chemical bonding between atoms when elements form compounds.

4-3   Chemical Bonds

The forces that hold atoms in combination with one another in compounds are called **chemical bonds.** The ability of an atom to form chemical bonds is related to the distribution of electrons in the atom. Specifically, the combining ability of an atom depends on the number of outer energy level electrons or valence electrons. Around 1920, W. Kossel and G. Lewis noted that the representative elements tend to enter into chemical combinations involving the loss, gain, or sharing of electrons. In fact, based on their study of numerous compounds, the **octet rule** was proposed. This rule states that atoms tend to lose, gain, or share electrons so that they attain a total of eight (octet) outer energy level electrons. The noble gases, which form very few compounds, have an outer energy level of eight electrons in the uncombined state. An octet of outer energy level electrons can be shown as eight dots around the elemental symbol. For example, the octet of neon is shown as

$$: \overset{\displaystyle ..}{\underset{\displaystyle ..}{Ne}} :$$

Another view of the Kossel-Lewis rule is that atoms of the representative elements enter into combinations in which they lose, gain or share valence electrons to attain the same out energy level configuration as the nearest noble gas. Those elements with few valence electrons tend to lose those electrons leaving the noble gas configuration. Refer to the periodic table on the back cover of the book and note that the Group IA element sodium tends to lose one electron to attain the same configuration as neon. Those elements with numerous valence electrons tend to gain or share electrons to attain the nearest noble gas configuration. The Group VIA element oxygen tends to gain or share two electrons to attain the same configuration as neon. The Group VIIA element chlorine tends to gain or share one electron to attain the same configuration as argon, the nearest noble gas. Hydrogen often shares one electron to attain the same configuration as helium, the nearest noble gas. Even though, there are exceptions to this rule it is observed in a great many cases and is a most useful concept.

There are two fundamental kinds of chemical bonds. The ionic bond results when electrons are transferred from one atom to another. The covalent bond results when two atoms share electrons.

4-4   Ionic Bonding and Ion Formation

When two elements are mixed, the constituent atoms may interact to form a chemical compound. Such an interaction may result in the outer energy level of electrons in the atoms of one element moving to the outer energy level of the atoms of another element. Such a transfer of electrons allows both kinds of atoms to attain the octet of outer energy level electrons. Atoms that lose electrons have an octet in the next lowest energy level, and those that gain electrons gain enough electrons to complete an octet in the outer energy level. When an atom loses electrons this means that negative charges are removed (each electron carries a single negative charge), leaving behind a positively charged particle consisting of a nucleus and the remaining electrons. Such a charged particle is called a **cation** or a **positive ion** (see Figure 4-3). Atoms that gain electrons take on the negative charges of the electrons. Such a negatively charged particle is also called an ion but this time an **anion** or **negative ion** (see Figure 4-3).

Which elements form positive ions and which form negative ions? The elements with few (one, two, or three) outer energy level electrons tend to lose electrons. This includes nearly all the metals. The elements with many outer energy level electrons (five, six, or seven) tend to gain electrons. This includes many of the nonmetals. The Group IA elements (see the periodic table) tend to lose one electron, and the Group VIIA elements tend to gain one electron when they react with the atoms of other elements. For example, the sodium atoms lose electrons to the chlorine

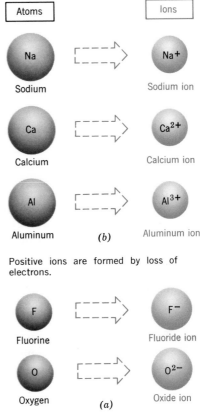

Figure 4-3   Positive and negative ions. From *Understanding Chemistry: From Atoms to Attitudes* by T. R. Dickson, Wiley, N.Y., 1974.

Positive ions are formed by loss of electrons.

Negative ions are formed by gain of electrons.

atoms to form sodium ions and chloride ions. This process can be represented as

$$\text{Na·} + \text{·}\ddot{\underset{..}{\text{Cl}}}\text{:} \rightarrow \text{Na}^+ + \text{Cl}^-$$

Such a transfer of electrons produces positive sodium ions and negative chloride ions. By transfer of electrons, the chlorine atoms attain an octet of outer level electrons, and the sodium atoms retain an octet in the inner energy level. When such negative and positive ions are produced, they aggregate as a result of the attraction between oppositely charged particles. The force of attraction between oppositely charged ions which is strong enough to hold the ions in an aggregation is called an **ionic bond.** That is, when sodium metal reacts with chlorine gas, the numerous ions that are produced aggregate together by ionic bonds to form a three-dimensional stack of ions that usually takes the form of a crystalline solid. Sprinkle a few grains of salt on a piece of paper. Look closely at this crystalline solid. Imagine that such crystals are composed of numerous sodium ions and chloride ions stacked together in space.

It is important to realize that ions are a different kind of chemical particle compared to atoms. They have different properties from the original atoms. We sprinkle salt on our food and eat it. In fact, we need

Table 4-2   Ions Formed by Group IA, IIA, and IIIA Metals

Na· → Na$^+$ sodium sodium ion		IA	One electron lost
·Ca· → Ca^{2+} calcium calcium ion		IIA	Two electrons lost
·Äl· → Al^{3+} aluminum ion		IIIA	Three electrons lost

it to keep our bodies functioning correctly. No one would dare eat sodium or chlorine. They are highly toxic and dangerous. Chemists and professional people sometimes use terminology that may be misleading to a beginner. For instance, a doctor may refer to a low-sodium diet, or a nutritionist may claim that we need a certain amount of calcium in our diets. These professionals and chemists know that they are referring to sodium ion and calcium ion and not sodium or calcium metal. As you learn more chemistry you will become more aware of this terminology.

An ion is represented by a formula using the symbol of the element from which the ion was formed with the charge of the ion as a superscript following the element symbol. A sodium ion is represented as Na$^+$ (read "N-A-plus"), and the chloride ion is represented as Cl$^-$ (read "C-L-minus"). The sodium ion has a single positive charge, and the chloride

Table 4-3

Common Group IA, IIA, and IIIA Ions

Li$^+$	Lithium ion	Mg^{2+}	Magnesium ion	Al^{3+}	Aluminum ion
Na$^+$	Sodium ion	Ca^{2+}	Calcium ion		
K$^+$	Potassium ion	Sr^{2+}	Strontium ion		
		Ba^{2+}	Barium ion		

Common Transition and Group IVA Metal Ions

Cr^{3+}	Chromium(III) ion (chromic ion)	Pb^{2+}	Lead(II) ion (plumbous ion)
Fe^{2+}	Iron(II) ion (ferrous ion)	Sn^{2+}	Tin(II) ion (stannous ion)
Fe^{3+}	Iron(III) ion (ferric ion)	Sn^{4+}	Tin(IV) ion (stannic ion)
Cu$^+$	Copper(I) ion (cuprous ion)		
Cu^{2+}	Copper(II) ion (cupric ion)		
Zn^{2+}	Zinc ion		
Cd^{2+}	Cadmium ion		
Hg^{2+}	Mercury(II) ion (mercuric ion)		

Common Group VIA and VIIA Negative Ions

O^{2-}	Oxide ion	F$^-$	Fluoride ion
S^{2-}	Sulfide ion	Cl$^-$	Chloride ion
		Br$^-$	Bromide ion
		I$^-$	Iodide ion

Table 4-4   Ions Formed by Group VIA and VIIA Nonmetals

$:\overset{..}{\underset{.}{O}}\cdot \rightarrow$ O^{2-} Oxide ion	VIA	Two electrons gained
$:\overset{..}{\underset{..}{F}}\cdot \rightarrow$ F^- Fluoride ion	VIIA	One electron gained

ion has a single negative charge. The formula of the calcium ion, Ca^{2+} (read "C-A-two-plus"), indicates that the calcium ion carries a double positive charge.

For any of the representative metals (Groups IA, IIA, and IIIA metals), the number of electrons lost in forming positive ions is given by the group number. (See Table 4-2). Many of the transition metals (B groups) and Groups IVA and VA metals form ions by the loss of one, two, three, and sometimes more electrons. (See Table 4-3.)

Groups VIIA, VIA, and VA nonmetals tend to gain electrons to form negative ions. The number of electrons gained is given by the group number subtracted from eight, as shown in Table 4-4.

Generally, when a metal is mixed with a nonmetal, the transfer of electrons occurs, resulting in the formation of ions which ionically bond to form an aggregate of ions. Such a compound between a metal and a nonmetal is called an **ionic compound.** Thus, we can say that binary or two-element compounds involving a metal and a nonmetal are ionic compounds and contain metal ions and nonmetal ions. For every atom that loses electrons, other atoms gain them. Even though ionic compounds are composed of ions, they are neutral. That is, they have the same amount of positive and negative charge. The formula of the compound reflects this. Consider the compound between calcium (Group IIA) and chlorine (Group VIIA). Calcium forms Ca^{2+} ions, and chlorine forms chloride ions, Cl^-. Thus, two chloride ions are formed for every calcium ion that forms.

$$\cdot Ca\cdot + :\overset{..}{\underset{..}{Cl}}\cdot + :\overset{..}{\underset{..}{Cl}}\cdot \rightarrow Ca^{2+}(Cl^-)_2$$

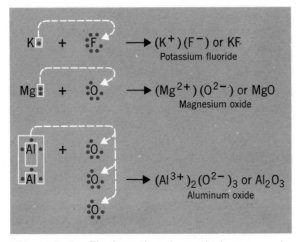

Figure 4-4   The formation of some ionic compounds.

Table 4-5   Names and Formulas of Some Common Ions

Positive Ions		Negative Ions	
Ammonium ion	NH_4^+	Acetate ion	$C_2H_3O_2^-$
Copper(I) ion	Cu^+	Bromide ion	Br^-
(cuprous ion)			
Hydronium ion	H_3O^+	Chlorate ion	ClO_3^-
Potassium ion	K^+	Chloride ion	Cl^-
Silver ion	Ag^+	Cyanide ion	CN^-
Sodium ion	Na^+	Fluoride ion	F^-
Barium ion	Ba^{2+}	Hydrogen carbonate ion	HCO_3^-
Calcium ion	Ca^{2+}	(bicarbonate ion)	
Cobalt(II) ion	Co^{2+}	Hydrogen sulfate ion	HSO_4^-
Copper(II) ion	Cu^{2+}	Hydroxide ion	OH^-
(cupric ion)			
Iron(II) ion	Fe^{2+}	Iodide ion	I^-
(ferrous ion)			
Lead(II) ion	Pb^{2+}	Nitrate ion	NO_3^-
Magnesium ion	Mg^{2+}	Nitrite ion	NO_2^-
Manganese(II) ion	Mn^{2+}	Permanganate ion	MnO_4^-
Mercury(I) ion	Hg_2^{2+}	Carbonate ion	CO_3^{2-}
(mercurous ion)			
Mercury(II) ion	Hg^{2+}	Chromate ion	CrO_4^{2-}
(mercuric ion)			
Nickel(II) ion	Ni^{2+}	Dichromate ion	$Cr_2O_7^{2-}$
Tin(II) ion	Sn^{2+}	Oxalate ion	$C_2O_4^{2-}$
Zinc ion	Zn^{2+}	Oxide ion	O^{2-}
Aluminum ion	Al^{3+}	Sulfate ion	SO_4^{2-}
Chromium(III) ion	Cr^{3+}	Sulfide ion	S^{2-}
Iron(III) ion	Fe^{3+}	Sulfite ion	SO_3^{2-}
(ferric ion)			
		Phosphate ion	PO_4^{3-}

However, we usually leave out the charges and the parentheses (when not needed) in writing the formula.

$$CaCl_2$$

A few more examples are given in Figure 4-4. A list of ions formed by the common metals and nonmetals is given in Tables 4-3 and 4-5.

4-5   **The Covalent Bond**

The elements with many outer level electrons can form compounds in which the atoms attain an octet (or duo in the case of hydrogen) outer energy level electrons by the sharing of electron pairs between atoms. The nonmetals are the elements that commonly enter into such combinations. For example, the formation of fluorine gas from fluorine atoms can be represented as

$$:\ddot{F}\cdot + \cdot\ddot{F}: \rightarrow :\ddot{F}:\ddot{F}:$$

The fluorine atoms mutually share a pair of electrons. This results in an octet of electrons about each fluorine if the shared pair is considered to be associated with both atoms. The mutual sharing of electron pairs by atoms results in the atoms being bonded to one another. The force of attraction arising when two atoms share electrons is called a **covalent bond.** The covalent bond is a very common and important type of chemical bond.

Since no transfer of electrons is involved, the atoms that share electrons form stable aggregates that can be considered to be chemical species or chemical particles. Such a chemical species formed from two or more covalently bonded atoms is called a **molecule.** The molecule is a very common chemical species. Compounds in which the atoms are combined in molecules are called **molecular** or **covalent compounds.**

The number of covalent bonds normally formed by an atom of a nonmetal is given by the group number subtracted from 8. This can also be seen by noting the electron dot symbol of an element. The elements tend to form as many covalent bonds as there are single (unpaired) electrons in the dot symbol. Oxygen (Group VIA) tends to form two covalent bonds:

$$:\ddot{O}\cdot$$

Hydrogen tends to form one:

$$H \cdot$$

The compound formed between oxygen and hydrogen is water, which involves molecules containing two atoms of hydrogen covalently bonded to an oxygen atom.

$$:\ddot{O}\cdot \ \cdot H \rightarrow :\ddot{O}:H$$
$$\ \ H \qquad\qquad \ddot{H}$$

Such a molecule is represented by a structural formula,

$$:\overset{..}{O}{-}H$$
$$\ \ |$$
$$\ \ H$$

where the lines represent electron pairs as covalent bonds, or more commonly by the molecular formula H_2O. The formula H_2O indicates that water consists of molecules in which two atoms of hydrogen are bonded to an oxygen atom. Look at a glass of water. Imagine that such a sample of water consists of a vast (about 10^{25}) collection of H_2O molecules. Covalent compounds may involve one element, two elements, or three or more elements. Often, covalent compounds involve a few atoms bonded together to form simple molecules. However, molecules can consist of several atoms, tens of atoms, hundreds of atoms, or even thousands of

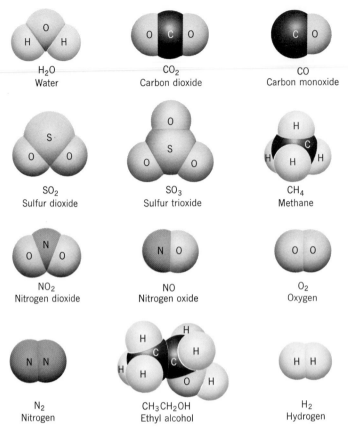

Figure 4-5   Pictorial representations of some common molecules. From *Understanding Chemistry: From Atoms to Attitudes* by T. R. Dickson, Wiley, N.Y., 1974.

covalently bonded atoms. For example, the paper in this page is composed mainly of the compound cellulose, which consists of molecules made up of thousands of covalently bonded carbon, hydrogen, and oxygen atoms. Some covalent compounds are solids, some are liquids, and some are gases. Figure 4-5 illustrates the structures of a few different molecules.

You might wonder how it is possible for atoms to share electrons. A simple way to picture the formation of a covalent bond is to consider the bond formation as involving the valence electron orbitals of the atoms concerned. Recall that an orbital is filled when it contains two electrons. Orbitals containing one electron are said to be half-filled. As two atoms become covalently bonded, the half-filled orbitals overlap, so that the electron pair can be shared. In other words, a covalent bond can be considered to result from the overlap of half-filled atomic orbitals. The aggregate of atoms linked by such overlapped orbitals is a molecule. The formation of a molecule of hydrogen fluoride is illustrated in Figure 4-6. In this case, the overlap of the $1s$ orbital of hydrogen and the $2p$ orbital of fluorine constitutes the covalent bond. The overlap of the half-filled orbitals makes it possible for the atoms to mutually share the electrons.

Seven of the nonmetals have such a great tendency to form covalent bonds in the uncombined state (not combined with other elements) that

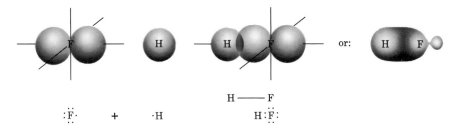

$$H \text{———} F$$

$$:\ddot{F}\cdot \quad + \quad \cdot H \qquad\qquad H:\ddot{F}:$$

Figure 4-6   A bond that is symmetrical about the bonding axis may also form from overlap between a *p* orbital and an *s* orbital, as illustrated here for hydrogen fluoride, H—F. Note that the maximum electron density does not fall halfway between the two nuclei. This maximum is shifted toward the fluorine end of the bond, and the electron density in the immediate vicinity of the nucleus of hydrogen is *less* than it was in the hydrogen atom. Similarly, the total electron density in the vicinity of the nucleus of fluorine is greater than in the fluorine atom. The H—F molecule is therefore polar as described in the text. (From *Principles of Physical, Organic and Biological Chemistry* by John Holum, Wiley, N.Y., 1969.)

they occur in the form of molecules in which two atoms are covalently bonded. Uncombined oxygen is found in the form of such molecules and, thus, is represented by the formula O_2. Such a molecule is called a **diatomic molecule,** and elements that occur in this form are called **diatomic molecular elements.** This means that when we have a sample of such an element, the element can be represented by the formula of the diatomic molecule. However, this does not mean that these elements remain in the form of diatomic molecules when they are combined with other elements. Do you remember which elements exist in the form of diatomic molecules? (See Section 2-11.)

4-6   Electron Dot Structures

A simple way to represent molecules involving covalent bonds is to use the electron dot symbols of the elements involved, so that the shared pairs of electrons are indicated. Such an arrangement of electron dot symbols is called a **Lewis electron dot structure.** For hydrogen, which can be represented by the Lewis electron dot structure,

$$H:H$$

the two electrons between the two combined hydrogen atoms indicate the mutual possession of the electrons by the hydrogens. In other words, the electron pair between the two hydrogen atoms represents the covalent bond between the atoms. The Lewis electron dot symbol for diatomic chlorine is

$$:\ddot{Cl}:\ddot{Cl}:$$

Each combined chlorine atom has the configuration of the nearest inert gas, which is argon. These electron dot structures are not meant to convey any information concerning the shapes of the electron clouds or to indicate that electrons are dots but merely to serve as a convenient representation of molecules. Often the shared pairs of electrons are represented by lines connecting the atoms. For instance, the dot structure of chlorine is often written as

$$:\ddot{C}l—\ddot{C}l:$$

When such dot structures are written, the line should be interpreted as a shared pair of electrons or a covalent bond.

Since the nonmetals tend to react with one another to form covalent compounds in which the noble gas configuration is attained by sharing electron pairs, we can often deduce the dot structure of the molecules of such compounds by arranging the electron dot symbols of the elements in order to satisfy the octet rule. That is, combined atoms of each element should have an octet of electrons around them in the electron dot structures. Of course, hydrogen should have just one shared pair of electrons associated with it in the dot structure. The covalent compound water, H_2O, consists of molecules that involve two combined hydrogen atoms and one combined oxygen atom. The electron dot symbol for hydrogen is H· and that for oxygen is $:\dot{O}·$. Arrangement of two hydrogen symbols and one oxygen symbol to satisfy the octet rule for oxygen gives the electron dot structure for water:

$$:\overset{..}{O}—H$$
$$|$$
$$H$$

The covalent compound ammonia has the formula NH_3 and, since the electron dot symbol for nitrogen is $·\overset{..}{N}·$ and that for hydrogen is H·, the electron dot structure of ammonia is

$$H—\overset{..}{N}—H$$
$$|$$
$$H$$

The formula for a molecule of methane is CH_4 and, since the electron dot symbol of carbon is $·\dot{C}·$ and that for hydrogen is H·, the electron dot structure of methane is

$$\begin{matrix} & H & & & H & \\ & \overset{..}{} & & & | & \\ H:&C&:H & \text{or} & H—C—H \\ & \overset{..}{H} & & & | & \\ & & & & H & \end{matrix}$$

Example 4-1 What is the electron dot structure of hydrogen chloride, HCl?
The electron dot symbol of hydrogen is H· and that for chlorine is $·\overset{..}{C}l:$, so a possible electron dot structure could be

$$H:\overset{..}{\underset{..}{C}l}: \quad \text{or} \quad H—\overset{..}{\underset{..}{C}l}:$$

Example 4-2 What is the electron dot structure for chloroform, $CHCl_3$?
The electron dot symbol for iodine is $:\overset{..}{C}l·$ and that of carbon is $·C·$, so a possible dot structure for chloroform would be

$$\overset{\text{H}}{\underset{\ddot{\underset{\cdot\cdot}{\text{Cl}}}}{\vdots}}\overset{\cdot\cdot}{\underset{\cdot\cdot}{\text{Cl}}}\text{:}\overset{\cdot\cdot}{\underset{\cdot\cdot}{\text{C}}}\text{:}\overset{\cdot\cdot}{\underset{\cdot\cdot}{\text{Cl}}}\text{:} \qquad \text{or} \qquad$$

(See image of electron dot and structural formulas.)

4-7 Multiple Bonds

When we attempt to write the electron dot structures of some compounds, we find that, in order to obey the octet rule, it is necessary to move electron pairs around so that more than one pair is shared by some atoms. For instance, the electron dot symbol for nitrogen is $\cdot\ddot{\text{N}}\cdot$, but in order to write the electron dot structure for a molecule of diatomic nitrogen, it is necessary to move the electrons in the electron dot symbols of the two nitrogen atoms around so that the following electron dot structure is obtained:

$$:\text{N}::\text{N}: \qquad \text{or} \qquad :\text{N}\equiv\text{N}:$$

Since there is a total of 10 outer energy level or valence electrons (five from each nitrogen) involved in diatomic nitrogen, it is necessary to have the two combined nitrogen atoms share three pairs of electrons so that the octet rule is satisfied. An electron dot structure of carbon dioxide, CO_2, can be represented as

$$\overset{\cdot}{\underset{\cdot\cdot}{\text{O}}}:\ :\text{C}:\ :\overset{\cdot}{\underset{\cdot}{\text{O}}} \qquad\qquad \overset{\cdot\cdot}{\underset{\cdot}{\text{O}}}=\text{C}=\overset{\cdot\cdot}{\underset{\cdot}{\text{O}}}$$

Since there are 16 valence electrons (6 from each oxygen and 4 from the carbon), it is necessary to place two pairs of electrons between the carbon and each oxygen in order to satisfy the octet rule. The sharing of more than one pair of electrons by atoms is entirely possible and is referred to as **multiple covalent bonding.** The sharing of two pairs of electrons between two atoms is called a **double bond,** and the sharing of **three pairs** of electrons between two atoms is called a triple bond. The sharing of more than three pairs of electrons between two atoms does not occur.

In order to deduce a possible electron dot structure for molecules in which multiple bonding occurs, it is necessary to decide which atoms are bonded to one another and then to write down the symbols of these atoms according to the bonding sequence. Which atoms are bonded to one another is sometimes obvious, but, in some cases, more information in addition to the formula is needed to establish the bonding sequence. Next, the total number of valence electrons is determined from the electronic structures or group numbers of the atoms present in the compound. Finally, these valence electrons are distributed in pairs so that each atom (except hydrogen) has an octet of electrons.

Example 4-3 What is the electron dot structure of formaldehyde, H_2CO, in which the carbon is bonded to the hydrogens and the oxygen? The bonding sequence gives the arrangement of symbols shown on the next page.

H

C   O

H

There are 12 valence electrons available (2 from the hydrogen, 4 from the carbon, and 6 from the oxygen). Distribution of these electrons to satisfy the octet rule gives as a possible electron dot structure

$$
\begin{array}{c}
\text{H} \\
| \\
\text{C}\!=\!\ddot{\text{O}}\!: \\
| \\
\text{H}
\end{array}
$$

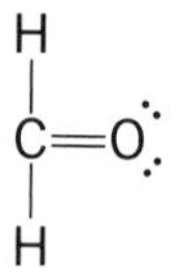

Example 4-4   What is the electron dot structure of acetylene, C_2H_2, in which a hydrogen is bonded to each carbon and the carbons are bonded?
The bonding sequence is

H   C   C   H

Since there are 10 valence electrons available (2 from the hydrogens and 8 from the carbons) a possible electron dot structure is

H—C≡C—H

For some compounds it is not possible to write electron dot structures in which the octet rule is satisfied. These are **exceptions** to the octet rule. For example, the electron dot structure for boron trichloride

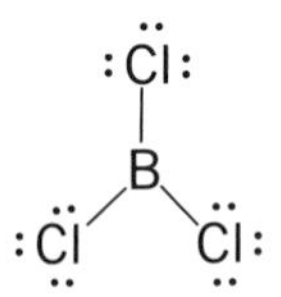

shows boron with 6 electrons rather than 8. Phosphorus, in the compound phosphorus pentachloride,

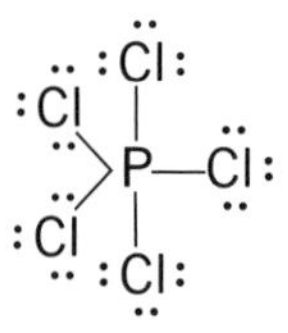

$$
\begin{array}{c}
\qquad\quad :\ddot{\text{Cl}}: \\
:\ddot{\text{Cl}} \quad | \\
\qquad\!\!\diagup\!\!\text{P}\!-\!\ddot{\text{Cl}}: \\
:\ddot{\text{Cl}} \quad | \\
\qquad\quad :\ddot{\text{Cl}}:
\end{array}
$$

has five bonding pairs of electrons.

4-8   Polyatomic Ions

In the natural environment the elements are most often found as compounds. A few elements such as the noble gases, sulfur, gold, and copper are often found in the uncombined state. Such elements are just aggregations of atoms. However, the atoms of most elements are found

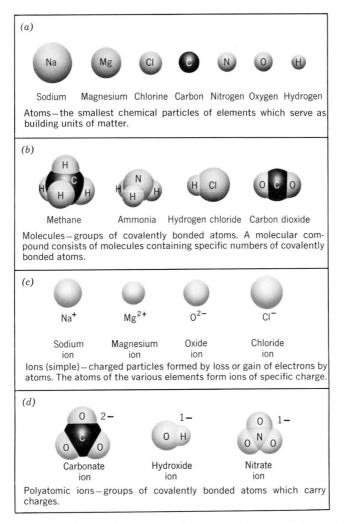

Figure 4-7   The four kinds of chemical species or particles that serve as building units of matter. From *Understanding Chemistry: From Atoms to Attitudes* by T. R. Dickson, Wiley, N.Y., 1974.

combined as ions in ionic compounds or covalently bonded to other atoms in the molecules of molecular compounds.

Positive and negative ions and molecules are chemical particles of much importance since it is in these forms that combined atoms are found. Thus, it can be said that the three important chemical particles or chemical species are atoms, ions, and molecules. (See Figure 4-7.) The ions we have referred to are those that result when atoms lose or gain electrons. These are monoatomic or simple ions. When we investigate the combined elements in the natural environment and certain manufactured chemicals, another kind of ion is found to occur. These ions are composed of two or more covalently bonded atoms that carry an electrical charge as a unit. Such an ion is a **polyatomic ion.** For instance, a piece of limestone or marble is composed of the calcium ion, Ca^{2+}, in combination with the polyatomic ion carbonate ion, CO_3^{2-} (read "C-O-three-two-minus"). The carbonate ion is an ion composed of three oxygen atoms covalently bonded to a carbon atom. The four-atom particle carries a double negative charge. Negative polyatomic ions occur in combination with metallic ions in many common ionic compounds. In such an ionic compound, the amount of positive charge and negative charge must be the same, and the net charge is thus zero. The formula of the ionic compound involving calcium ion and carbonate ion is

$$Ca^{2+}CO_3^{2-}$$

We usually write the formula without the charges as

$$CaCO_3$$

Table 4-5 includes several common polyatomic ions. Only one of the common polyatomic ions is a positive ion (ammonium ion, NH_4^+), and it is found in combination with negative ions. For instance, ammonium ion can combine with chloride ion, Cl^-, to form ammonium chloride, NH_4Cl, or it can combine with sulfate ion, SO_4^{2-}, to form ammonium sulfate:

$$NH_4^+ \qquad SO_4^{2-}$$
two of these ions
needed for each of these ions,
resulting in the formula
$$(NH_4)_2SO_4$$

Note that, since the charge of ammonium ion is 1+ and the charge of a sulfate ion is 2−, two ammonium ions are needed for every sulfate ion. To denote two ammonium ions, the formula for ammonium ion is enclosed in parentheses with a subscript of two, $(NH_4)_2SO_4$ (read "N-H-four-taken twice-S-O-four"). Figure 4-8 gives a few more examples of such ionic compounds.

4-9   The Shapes of Molecules

The properties and behavior of molecules are related to the shapes of molecules. The shape of a molecule refers to the spatial distribution of the atoms within the molecule. A simple way to visualize the shape of a molecule is to consider the spatial arrangement of the orbitals which are involved in the covalent bond formation. Refer to Section 3-9 to note the shapes of orbitals. The formation of a water molecule is illustrated in Figure 4-9. From this illustration we can see that a water molecule could be visualized as an angular molecule resulting from the overlap of the $1s$ orbitals of the hydrogen atoms with two perpendicular $2p$ orbitals of the oxygen atom. Similarly, as illustrated in Figure 4-10, an ammonia molecule can be visualized as having a shape resulting from the overlap of the $1s$ orbitals of the hydrogen atoms with the three perpendicular $2p$ orbitals of the nitrogen atom. The shapes of many other molecules can be visualized in a similar manner. See Section 4-12.

Since water is a very important and common compound, let us con-

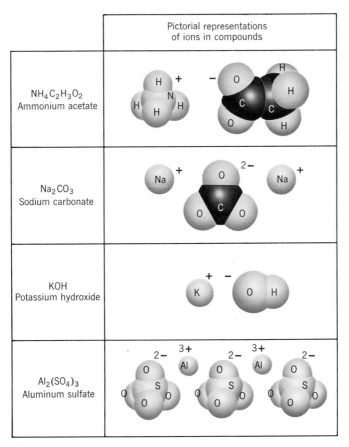

Figure 4-8   Some Ionic compounds including polyatomic ions. From *Understanding Chemistry: From Atoms to Attitudes* by T. R. Dickson, Wiley, N.Y., 1974.

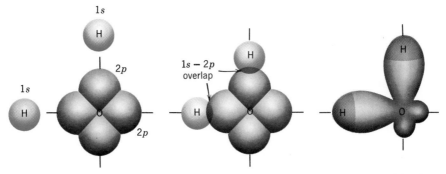

Figure 4-9   The covalent bond formation in water can be idealized as resulting from the overlap of the 1s orbitals of the hydrogen atoms with two perpendicular 2p orbitals of the oxygen atom. Once the overlap occurs, the electron density is increased between the hydrogens and the oxygen. This concentration of electron density constitutes the covalent bonds and produces the angular water molecule. (The other atomic orbitals of oxygen, which are not used in the formation of the bonds, are not shown in the figure.)

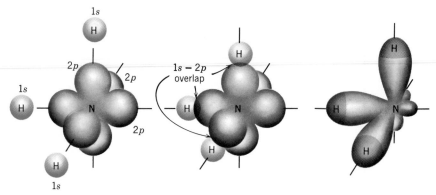

Figure 4-10   The covalent bond formation in ammonia can be idealized as resulting from the overlap of the 1s orbitals of the hydrogen atoms with the three perpendicular 2p orbitals of the nitrogen atom. Once the overlap occurs, the electron density is increased between the hydrogens and nitrogen. This concentration of electron density constitutes the covalent bonds and produces an ammonia molecule. (The other atomic orbitals of nitrogen, which are not used in the formation of bonds, are not shown in the figure.)

sider the shape of water molecules in detail. A water molecule can be visualized as

Two smaller hydrogens are sharing electrons with an oxygen. One distinct characteristic of a water molecule is the angular shape. The hydrogens protrude from one side of the oxygen at an angle of 105° to each other. As we shall see in the next section and in Section 10-5, the shape of water influences its properties.

4-10   Polar Molecules and Electronegativity

Molecules have no overall charge since the total number of electrons equals the total number of protons present in the various nuclei involved in the molecule. However, in some molecules a region of negative and positive charge may be associated with certain parts of the molecule. In a neutral atom, the electrons surround the nucleus and usually the average center of negative charge coincides with the position of the nucleus. In molecules, the average positions of some of the electrons change as a result of the sharing of electron pairs. A hydrogen molecule can be pictured as shown in Figure 4-11. The bonding electron cloud occupies a position between the two positive nuclei. Since the two nuclei are the same, they equally share the electron pair. This results in the average center of positive and negative charges coinciding as shown in

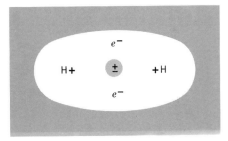

Figure 4-11   Centers of positive and negative charge coincide in a hydrogen molecule.

Figure 4-11. When a molecule involves different types of atoms, the nuclei do not generally share electron pairs equally. That is, certain atoms attract the shared electron pairs to a greater extent than other atoms as described in the next paragraph. For example, in a molecule of hydrogen fluoride, HF, the fluorine attracts the bonding electron pair more extensively than the hydrogen. This can be interpreted as meaning that there is a higher probability of finding the electron pair in the vicinity of the fluorine than in the vicinity of the hydrogen. In such a molecule, the average centers of positive and negative charge do not coincide. (See Figure 4-12.) There is an uneven distribution of charge; when this occurs in a covalent bond, it is referred to as a **polar bond.** The extent of the polarity depends on the distance between the centers of unlike charge. The greater the separation of charge, the greater the polarity of the bond. Such polar bonds are often represented as

$$\overset{\delta+ \quad \delta-}{\text{H} - \text{F}}$$

where the δ (delta) indicates a small amount. In other words, a small amount of positive charge is associated with the hydrogen and a small amount of negative charge is associated with the fluorine. The presence of these delta charges indicates a polar bond. A molecule, such as hydrogen fluoride, in which there is a net separation of positive and negative charge centers, is called a **polar molecule** and is said to possess a dipole moment.

The polarity of a bond depends on whether or not an atom involved in

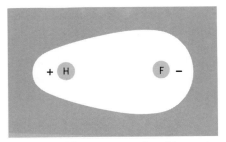

Figure 4-12   Centers of positive and negative charge do not coincide in a hydrogen fluoride molecule.

Table 4-6   Electronegativities of the Elements (as devised by Linus Pauling)

Li	Be												B	C	N	O	F
1.0	1.5					H							2.0	2.5	3.0	3.5	4.0
						2.2											
Na	Mg												Al	Si	P	S	Cl
0.9	1.2												1.5	1.8	2.1	2.5	3.0
K	Ca	Sc	Ti	V	Cr	Mn	Fe	Co	Ni	Cu	Zn	Ga	Ge	As	Se	Br
0.8	1.0	1.3	1.4	1.6	1.6	1.5	1.8	1.8	1.8	1.9	1.6	1.6	1.8	2.0	2.4	2.8
Rb	Sr	Y	Zr	Nb	Mo	Tc	Ru	Rh	Pd	Ag	Cd	In	Sn	Sb	Te	I
0.8	1.0	1.2	1.4	1.6	1.8	1.9	2.2	2.2	2.2	1.9	1.7	1.7	1.8	1.9	2.1	2.5
Cs	Ba	La-Lu	Hf	Ta	W	Re	Os	Ir	Pt	Au	Hg	Tl	Pb	Bi	Po	At
0.7	0.9	1.1	1.3	1.5	1.7	1.9	2.2	2.2	2.2	2.4	1.9	1.8	1.9	1.9	2.0	2.2
Fr	Ra	Ac	Th	Pa	U	Np	Pu									
0.7	0.9	1.1	1.3	1.5	1.7	1.2	1.2									

the bond attracts the shared electron pairs to a greater extent than the other atom involved. The tendency of an atom of a given element to attract electrons in a covalent bond is called the **electronegativity.** A relative measure of electronegativities of the atoms of the elements has been devised by observing the bond-forming behavior of the elements. A relative scale of electronegativity, suggested by Linus Pauling, establishes the electronegativity of fluorine, which has the greatest electronegativity, as 4.0. The electronegativities of atoms of the other elements can be expressed relative to fluorine. Table 4-6 gives a periodic table in which electronegativities for the elements are listed. These electronegativities apply to bonded atoms of the elements. Notice that the electronegativities of the elements within a given group of representative elements tend to decrease from the top to the bottom of the group. Furthermore, notice that within a given row of representative elements the electronegativities decrease from right to left. The transition elements generally have electronegativities that are less than the nonmetals. The elements in the upper right of the periodic table have the greatest electronegativities. Fluorine has the largest electronegativity (4.0) while oxygen has the next largest (3.5). The greater the electronegativity of an element, the greater the tendency for the atoms of that element to attract electron pairs in covalent bonds. The metals generally have low electronegativities. Because of this, they are said to be **electropositive,** which is the opposite of electronegative.

When atoms of two different elements are sharing electrons in a covalent bond, the electrons will be attracted toward the atom having the higher electronegativity. Thus, we can predict whether or not a bond is polar by comparing the electronegativities of the atoms involved. In water, H_2O, we would predict that the bonds in water are polar, since oxygen is more electronegative than hydrogen. We can represent these polar bonds as

$$\delta^-O\delta^-$$
$$\delta^+H \qquad H\delta^+$$

Since water has these polar bonds, is it a polar molecule? Whether or not it is a polar molecule depends on whether or not the average centers of positive and negative charge are separated. We know that a water molecule is angular in shape. This results in the oxygen end of the molecule taking on the slight negative charge and the hydrogen end taking on the slight positive charge. The angular shape results in a net separation of charge, and, thus, water molecules are polar molecules.

The molecules of many other compounds are polar and many are non-polar. A few typical examples are given in Table 4-7.

4-11   Covalent and Ionic Compounds

Actually, most substances are of such a nature that they are neither purely ionic or purely covalent. We sometimes refer to such substances as partially ionic or partially covalent. The ionic character of a covalent bond is related to the polarity of the bond. The more polar the bond, the greater its ionic character. In fact, the pure ionic bond can be considered to be completely polarized. The transition of compounds from purely covalent to purely ionic can be represented as a transition of the degree of polarity of the bonds involved as illustrated in Figure 4-13. Even though most compounds are neither purely covalent nor purely ionic, it is convenient to classify most compounds as ionic or covalent. Those compounds that are considered to be composed of discrete molecules are covalent while those compounds that are apparently composed of ions are considered to be ionic. Compounds composed of metals combined with nonmetals or polyatomic ions are generally considered to be ionic compounds. Compounds composed of nonmetals only are generally covalent compounds.

4-12   The Visualization of Molecules

As was mentioned previously, molecules involve the sharing of electron pairs between atoms facilitated by the overlap of atomic orbitals. A convenient way to visualize molecules is to consider the sharing of electrons as an intersection or overlap of the spherical or balllike atoms of the elements involved. For example, a molecule of ammonia, NH_3, can be visualized as the overlap of three hydrogens with one nitrogen.

Table 4-7   Some Polar and Nonpolar Molecules

Compound		Polarity of Molecule
Nonpolar Chlorine	Cl — Cl	(Nonpolar bond)
Sulfur trioxide	O∖S∕O over O	(The molecule has polar bonds but the symmetry of the molecule makes it nonpolar.)
Carbon tetrachloride	Cl—C—Cl (with Cl above and Cl below)	(The molecule has polar bonds but the symmetry of the molecule makes it nonpolar.)
Polar Hydrogen chloride	H — Cl	⊕ H — Cl ⊖ (Polar bonds with unsymmetrical molecule.)
Ammonia	H—N—H over H	⊖ N, H—N—H, H ⊕ (Polar bonds with unsymmetrical molecule.)
Water	H∖O∕H	⊖ O, H H, ⊕ (Polar bonds with unsymmetrical molecule.)
Ethyl alcohol		⊕ structure; (Polar C — O and O — H bonds with unsymmetrical molecule.)

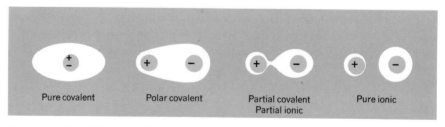

Pure covalent   Polar covalent   Partial covalent / Partial ionic   Pure ionic

Figure 4-13   The transition from purely covalent to completely ionic compounds.

Another way to visualize molecules that is convenient but less realistic is to view the bonded atoms as balls bonded by sticks. The ball and stick view of ammonia is

A third method sometimes used for the visualization of the bond sequence and shapes in molecules is to show the element symbols connected by lines representing bonds. Ammonia viewed in this manner is

$$H \diagup \overset{\displaystyle N}{\underset{\displaystyle H}{|}} \diagdown H$$

The visualizations of a few typical molecules are given in Figure 4-5.

4-13   The Mole Revisited

Now that we are familiar with the way in which atoms enter into combination we know that the smallest characteristic or elementary units of substances can be atoms, molecules, or ions. These three units are the chemical species that structure matter. We have also learned that the formula of a substance often suggests its nature. That is, if a substance is represented by an element symbol, the elementary units are atoms. A substance made up of combinations of nonmetals consists of molecules as elementary units. A substance that contains a metal in combination with nonmetals usually is made up of ions as elementary units.

Considering the various possible chemical species it is possible to state a general definition of the mole so that all species are included. This definition is as follows:

Mole. The amount of a substance in grams that contains as many elementary units (chemical species) as there are carbon atoms in exactly 12 g of carbon-12. The elementary unit must be specified and may be an atom, molecule, ion, electron, etc., or any specified group of such entities.

Since the number of particles comprising a mole is Avogadro's number or

$$\left(\frac{6.02 \times 10^{23} \text{ particles}}{1 \text{ mole}} \right)$$

an alternate view for the mole of a substance is the mass of 6.02×10^{23} elementary units. A mole of atoms is the mass of Avogadro's number of atoms. A mole of ions is the mass of Avogadro's number of ions. A mole of molecules is the mass of Avogadro's number of molecules. Recall that as long as we know the formula for a chemical species we can express the molar mass of the species by adding up the contribution of each element in the species using a table of atomic weights.

Based on the definition of the mole, the number of grams per mole of any substance can be referred to as long as the species is specified. Thus, when the molar mass of a substance is stated, the species formula must be indicated. For example,

$$\left(\frac{16.00\ g}{1\ mole\ O}\right)$$ Oxygen atoms are indicated.

$$\left(\frac{32.0\ g}{1\ mole\ O_2}\right)$$ Oxygen molecules are indicated.

$$\left(\frac{16.00\ g}{1\ mole\ O^{2-}}\right)$$ Oxide ions are indicated.

$$\left(\frac{74.6\ g}{1\ mole\ KCl}\right)$$ Potassium chloride formula units are indicated. A formula unit is the number of positive and negative ions indicated by the formula of the compound.

Recall that the molar masses of chemical species can be used to determine the number of moles in a specific mass or the mass of a specific number of moles.

Example 4-5   If a sample of aluminum oxide was known to contain 64.0 g of oxide ion, O^{2-}, how many moles of oxide ion is this? The molar mass of oxide ion obtained from the atomic weight of oxygen can be used to convert the grams of oxide ion to moles.

$$64.0\ g\left(\frac{1\ mole\ O^{2-}}{16.00\ g}\right) = 4.00\ moles\ O^{2-}$$

Example 4-6   What is the mass in grams of 0.500 moles of potassium chloride, KCl? The molar mass of potassium chloride found from the atomic weights of potassium and chlorine is used to convert the moles to grams.

$$0.500\ moles\ KCl\left(\frac{74.6\ g}{1\ mole\ KCl}\right) = 37.3\ g$$

Questions and Problems

1. What is a chemical bond?

2. Give a statement of the octet rule.

3. What is a monoatomic or simple ion?

4. Give the formulas of the positive monoatomic ions formed by the following elements (refer to the periodic table).
   (a) lithium
   (b) barium
   (c) aluminum

   (d) potassium
   (e) calcium
   (f) sodium

5. Give the formulas of the negative monoatomic ions formed by the following elements (refer to the periodic table).
   (a) sulfur
   (b) fluorine
   (c) nitrogen
   (d) oxygen
   (e) phosphorus
   (f) iodine

6. The mass of an electron is about 0.0006 amu. Using the atomic weight of oxygen, determine the mass in amu of a typical oxide ion, O^{2-} (two extra electrons). Using the mass of an oxide ion express the mass in grams of one mole of oxide ions to four digits. How does the molar mass of oxide ion compare to the molar mass of oxygen atoms (to four digits). Using the atomic weight of sodium, determine the mass in amu of a typical sodium ion, Na^+ (one less electron). Using the mass of a sodium ion express the mass in grams of one mole of sodium ions to four digits. How does the molar mass of sodium ion compare to the molar mass of sodium atoms (to four digits)?

7. Describe the ionic bond.

8. Describe the covalent bond.

9. Define the following terms:
   (a) ionic compound
   (b) polyatomic ion
   (c) molecule
   (d) molecular compound

10. Give the names or formulas for the following ions:
   (a) aluminum ion   (i) Mg^{2+}
   (b) bromide ion   (j) Hg^{2+}
   (c) hydrogen carbonate ion   (k) NO_3^-
   (d) chromate ion   (l) PO_4^{3-}
   (e) copper (I) ion   (m) K^+
   (f) iron (II) ion   (n) SO_4^{2-}
   (g) hydronium ion   (o) SO_3^{2-}
   (h) OH^-

11. Give electron dot structures for the following molecules:
   (a) iodine, I_2
   (b) dichlorine oxide, Cl_2O (two Cl bonded to one O)
   (c) carbon tetrachloride, CCl_4 (four Cl bonded to C)
   (d) phosphorus tribromide, PBr_3 (three Br bonded to P)
   (e) hydrogen sulfide, H_2S (two H bonded to S)
   (f) diiodomethane, CH_2I_2 (two I and two H bonded to C)
   (g) hydrazine, N_2H_4 (two H bonded to each N)
   (h) hydrogen peroxide, H_2O_2 (one H bonded to each O)
   (i) ethane, C_2H_6 (3H bonded to each C)

12. What is a multiple covalent bond?

13. Give the electron dot structures for the following molecules involving multiple bonds:
   (a) carbon monoxide, CO
   (b) carbon disulfide, CS_2 (two S bonded to C)
   (c) phosgene, $COCl_2$ (two Cl and one O bonded to C)
   (d) tetrafluoroethene, C_2F_4 (two F bonded to each C)
   (e) vinyl chloride, C_2H_3Cl (two H bonded to one C and one H and one Cl bonded to the other C)

14. Sketch a water molecule and describe the shape.

15. Define electronegativity.

16. Sketch an outline of the main portion (s, p, d block) of the periodic table and indicate with arrows how the electronegativities of the elements vary from left to right in the table and from the bottom to top of a group.

17. List the four most electronegative elements in descending order. Which group contains the elements of lowest electronegativity?

18. What is a polar bond?

19. Describe a polar molecule and give an example.

20. What is the difference between an ionic compound and a molecular compound? Which elements tend to form ionic compounds? Which elements tend to form molecular compounds?

21. The following is a list of the 18 top (according to tonnage) manufactured chemicals in the United States. Indicate which of these chemicals are ionic compounds and which are molecular compounds.
   (a) sulfuric acid H_2SO_4
   (b) ammonia NH_3
   (c) oxygen O_2
   (d) sodium hydroxide   NaOH
   (e) chlorine Cl_2
   (f) ethylene C_2H_4
   (g) sodium carbonate Na_2CO_3
   (h) nitric acid HNO_3
   (i) ammonium nitrate NH_4NO_3
   (j) phosphoric acid H_3PO_4
   (k) nitrogen N_2
   (l) benzene C_6H_6
   (m) propylene C_3H_6
   (n) ethylene dichloride $C_2H_4Cl_2$
   (o) urea CH_4N_2O
   (p) polyethlene $(CH_2)_n$
   (q) toluene C_7H_8
   (r) ammonium sulfate $(NH_4)_2SO_4$

22. Describe the three types of chemical species which make up matter.

23. Give a general definition of the mole.

24. Determine the number of moles of chloride ion in 250 g of sodium chloride, NaCl. (Hint: find the number of moles of sodium chloride and note the moles of chloride ion per mole of sodium chloride.)

25. Determine the number of grams of hydroxide ion in 240 g of sodium

hydroxide. (Hint: find the moles of hydroxide ion, see question 24, and then use the molar mass of hydroxide ion to find the mass.)

26. If one drop of water is 0.050 ml, and the density of water is 1.00 g/ml, determine the number of water molecules in one drop of water.

27. Cotton is composed of cellulose that exists in the form of very large molecules having an average molar mass of 500,000 g. Assuming that a cotton ball is pure cellulose, determine the approximate number of cellulose molecules in a 0.34-g cotton ball.

28. Benzene has an empirical formula of CH. However, the molar mass of benzene is 78 g. What is the molecular formula of benzene? (Hint: since the empirical formula gives the simplest molar ratio of the elements, the actual molar ratio can be found by dividing the molar mass of the compound by the molar mass of the empirical formula.)

29. Propylene has an empirical formula of CH_2. However, the molar mass of propylene is 42 g. What is the molecular formula of propylene? (Hint: see question 28.)

5

Chapter 5: Periodic Properties of the Elements

Objectives

The student should be able to:

1 Describe how the ionization energy of the elements varies within the periodic table.
2 Describe how the sizes of atoms vary in the periodic table.
3 State the common oxidation numbers for hydrogen and oxygen.
4 Deduce the oxidation number of an element in a compound or polyatomic ion given the formula of the ion and the oxidation numbers of the other elements present.
5 List the expected oxidation numbers for each group of representative elements.
6 Give the formula, predicted on the basis of expected oxidation numbers, for a compound involving two elements.
7 Describe the refining of iron and aluminum.
8 List two metals that are potential environmental threats.

Terms to Know

Periodic law
Ionization energy
Angstrom unit
Oxidation number
Refining

5

Periodic Properties of the Elements

5-1 The Periodic Table

In the 1800s, numerous elements were discovered and added to the list of known elements. During this period, chemists were successfully determining relative atomic weights of the elements. As the list of elements grew, and the properties of these elements were studied, it became apparent that the properties of some elements were quite similar and some repeating pattern of properties seemed to exist among the elements.

The periodic table as we know it today was first proposed by a Russian chemist, Dmitri I. Mendeleev in 1869 and independently by a German chemist, Lother Meyer, in 1870. Mendeleev's original table is shown in Figure 5-1. This table was based on studies of the similarities in the physical and chemical properties of the elements known at that time. Such studies led Mendeleev to conclude that "elements arranged according to the size of their atomic weights show clear periodic properties . . . the size of the atomic weight determines the nature of the elements." His conclusions were a statement of the periodic law on which the form of the periodic table is based. A statement of the periodic law is that the properties of the elements are a periodic function of the atomic weights of the elements. Later, based on the discovery of isotopes and measureable atomic numbers, it became apparent that a better statement of the **periodic law** is:

The properties of the elements are a periodic function of the atomic numbers of the elements.

Mendeleev studied the elements for years and pondered their similarities. It is thought that he finally developed the periodic law and the

Figure 5-1   Periodic Table   D. I. Mendeleev — 1869

			Ti	Zr	
			V	Nb	Ta
			Cr	Mo	W
			Mn	Rh	Pt
			Fe	Ru	Ir
			Ni, Co	Pd	Os
H			Cu	Ag	Hg
	Be	Mg	Zn	Cd	
	B	Al		Ur	Au
	C	Si		Sn	
	N	P	As	Sb	Bi
	O	S	Se	Te	
	F	Cl	Br	I	
Li	Na	K	Rb	Cs	Tl
		Ca	Sr	Ba	Pb
			Ce		
	Er	La			
	Yt	Di			
	In	Th			

periodic table while listening to a string quartet playing a piece of music with a repetitious or periodic theme. The crowning success of Mendeleev's arrangement of the elements was that he was able to predict the existence and properties of some of the undiscovered elements that would fit into his table. For example, he predicted that the missing element that he called "eka-silicon" would have properties intermediate between those of silicon and tin. He predicted the new element would have an atomic weight of about 72, and a density of about 5.5 g/cm^3. He also predicted that it would form a dioxide with oxygen and a tetrachloride with chlorine. In 1886, just such an element named germanium, Ge, was discovered. As Mendeleev had predicted, germanium has an atomic weight of 72.59 and a density of 5.32 g/cm^3. It forms the compounds germanium dioxide, GeO_2 and germanium tetrachloride, $GeCl_4$.

Since the time Mendeleev established the periodic table, numerous other elements have been discovered and fit into the table. Figure 5-2 shows a periodic table with the dates of discovery of the elements indicated. Today, the periodic table is the foundation for the study of chemistry.

Knowledge of the electronic configuration of the elements provides an explanation for the observed periodic properties of the elements. Elements that have similarities in properties also have similarities in electronic configuration. This generalization that provides an alternate view of the **periodic law** can be stated as follows.

The properties of elements are directly related to the electronic configurations of the atoms of the elements; and the elements that have similar outer energy level electronic configurations have similar properties.

1	2	3	4	5	6	7	8	9	10	11	12	13	14	15	16	17	18
1 H 1766																	2 He 1868
3 Li 1817	4 Be 1828											5 B 1808	6 C Ancient	7 N 1772	8 O 1774	9 F 1886	10 Ne 1898
11 Na 1807	12 Mg 1808											13 Al 1827	14 Si 1824	15 P 1669	16 S Ancient	17 Cl 1774	18 Ar 1894
19 K 1807	20 Ca 1808	21 Sc 1879	22 Ti 1791	23 V 1801	24 Cr 1797	25 Mn 1774	26 Fe Ancient	27 Co 1735	28 Ni 1751	29 Cu Ancient	30 Zn 1746	31 Ga 1875	32 Ge 1886	33 As 1250	34 Se 1817	35 Br 1826	36 Kr 1898
37 Rb 1861	38 Sr 1808	39 Y 1843	40 Zr 1789	41 Nb 1801	42 Mo 1778	43 Tc 1937	44 Ru 1844	45 Rh 1804	46 Pd 1803	47 Ag Ancient	48 Cd 1817	49 In 1863	50 Sn Ancient	51 Sb Ancient	52 Te 1782	53 I 1811	54 Xe 1898
55 Cs 1860	56 Ba 1808	57 La 1839	72 Hf 1923	73 Ta 1802	74 W 1783	75 Re 1925	76 Os 1803	77 Ir 1803	78 Pt 1735	79 Au Ancient	80 Hg Ancient	81 Tl 1861	82 Pb Ancient	83 Bi 1753	84 Po 1898	85 At 1940	86 Rn 1900
87 Fr 1939	88 Ra 1898	89 Ac 1899															

58 Ce 1803	59 Pr 1885	60 Nd 1885	61 Pm 1945	62 Sm 1879	63 Eu 1896	64 Gd 1880	65 Tb 1843	66 Dy 1886	67 Ho 1879	68 Er 1843	69 Tm 1879	70 Yb 1907	71 Lu 1907
90 Th 1828	91 Pa 1917	92 U 1789	93 Np 1940	94 Pu 1940	95 Am 1944	96 Cm 1944	97 Bk 1950	98 Cf 1950	99 Es 1952	100 Fm 1953	101 Md 1955	102 No 1958	103 Lr 1961

Figure 5-2   The discovery dates of the elements. The shaded elements were discovered before 1869. (The existence of flourine, F, was suspected in 1869 but had not been isolated.)

For example, the electronic configurations of Li, Na, K, Rb, and Cs are:

Li $1s^2 2s^1$
Na $1s^2 2s^2 2p^6 3s^1$
K $1s^2 2s^2 2p^6 3s^2 3p^6 4s^1$
Rb $1s^2 2s^2 2p^6 3s^2 3p^6 3d^{10} 4s^2 4p^6 5s^1$
Cs $1s^2 2s^2 2p^6 3s^2 3p^6 3d^{10} 4s^2 4p^6 4d^{10} 5s^2 5p^6 6s^1$

These five metals have similar properties and form similar compounds with other elements. The electronic configurations indicate that all of these metals have one electron in the outer energy level. When the elements are tabulated according to increasing atomic number so that the elements of similar electronic configurations are arranged in columns, the periodic table is obtained. The periodic table takes on a specific form that corresponds to filling of the various electron energy sublevels in the atoms.

5-2   Ionization Energy

The term periodic refers to the fact that when we consider the elements according to the increasing atomic number it is noted that elements of higher atomic number have properties that are similar to those of lower atomic number. This pattern is repeated several times. For instance, the eleventh element, sodium, is similar to the third element, lithium, the twelfth element, magnesium, is similar to the fourth element, beryllium, and so on, up to the eighteenth element, argon. The pattern repeats at the nineteenth element, potassium, which is similar to the eleventh element, sodium. Let us consider some properties that show this periodic behavior.

The amount of energy required to remove one electron from a neutral gaseous atom can be measured experimentally. This energy is called the first **ionization energy** and is usually expressed in units of kilocalories per mole. That is, it will take a certain number of kilocalories (kcal) of energy to remove the electrons from a mole of atoms. The ionization energy can be measured by subjecting atoms of an element in the vapor state to a beam of free electrons. If the beam of electrons is of the correct energy, the bombarding electrons may interact with the outer energy level electrons of the atom resulting in the formation of a singly charged ion. (See Figure 5-3.) This ionization process can be generally represented as

$$\text{energy} + \text{E} \rightarrow \text{E}^+ + \text{e}^-$$

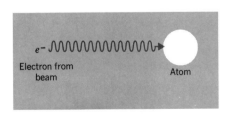
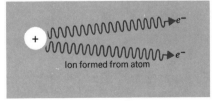

Figure 5-3   The ionization of an atom by an electron beam of the proper energy.

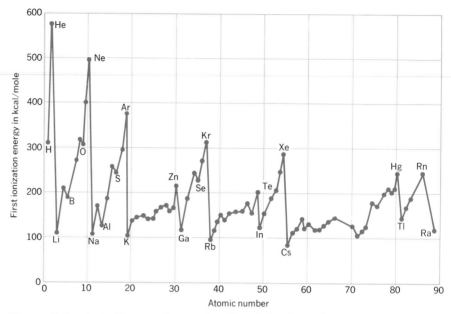

Figure 5-4   Ionization energies as a function of atomic numbers.

where E represents a specific element and e⁻ represents the electron that has been removed to form the E^+ ion. The ionization energy is determined by observing the minimum energy of the electron beam that results in the formation of the singly charged ion. The ionization energies of the elements are related to the atomic structures. A plot of these ionization energies versus atomic number is given in Figure 5-4. This plot illustrates the periodic behavior of this property of the elements quite well. Note that the ionization energies of the elements tend to decrease from the top to the bottom of a given group and to increase from left to right within a given row. This is to be expected if we consider the factors that affect the ionization energies. Higher energy level electrons are generally further from the nucleus than lower energy level electrons. Furthermore, the outer energy level electrons are shielded from the positive nucleus by the inner core of the lower energy

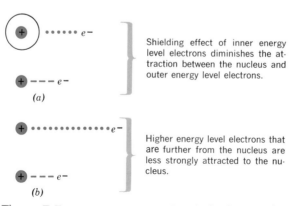

Figure 5-5   Two factors that affect ionization energies.

level electrons. This **shielding effect** tends to decrease the attraction of the nucleus for the outer energy level electrons. This is illustrated in Figure 5-5. These two factors, the increased distance of outer energy level electrons from the nucleus and the shielding effect, can account for the general decrease in the ionization energies of the elements within a group. That is, as we move down a given group, the outer energy level electrons in the atoms of the elements will be further from the nucleus, and the attraction by the nucleus will be diminished. These two effects result in the electrons being less tightly bound to the atoms and, thus, more easily removed. The increase in ionization energy from left to right within a given row of the table can be generally considered to result from the increase in nuclear charge without an increase in shielding. The increase in nuclear charge as we go from left to right in a given row is usually sufficient to cause an increase in the attraction between the nucleus and the outer energy level electrons. This results in these electrons being more tightly bound to the atoms and, thus, it would require more energy to produce ionization.

5-3   Atomic and Ionic Sizes

It is possible to measure experimentally some atomic and ionic sizes. Sometimes it is possible to indirectly calculate approximate **atomic** and **ionic sizes** using experimental information concerning compounds. If

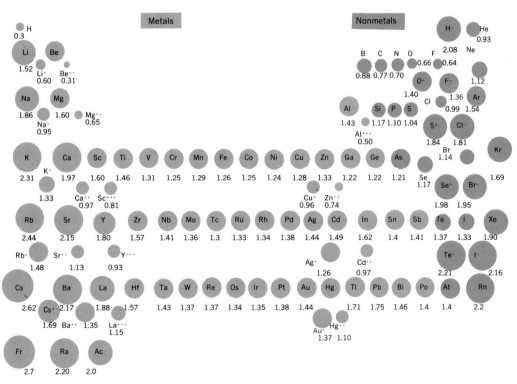

Figure 5-6   Periodic table showing the atomic sizes in terms of atomic radii. The radii of some monoatomic ions are also given. (All radii are expressed in Angstrom units.) (Adapted from Cambell in the *Journal of Chemical Education*.)

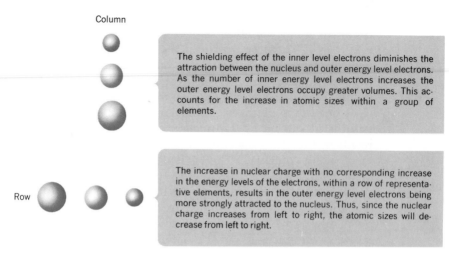

Column

The shielding effect of the inner level electrons diminishes the attraction between the nucleus and outer energy level electrons. As the number of inner energy level electrons increases the outer energy level electrons occupy greater volumes. This accounts for the increase in atomic sizes within a group of elements.

Row

The increase in nuclear charge with no corresponding increase in the energy levels of the electrons, within a row of representative elements, results in the outer energy level electrons being more strongly attracted to the nucleus. Thus, since the nuclear charge increases from left to right, the atomic sizes will decrease from left to right.

Figure 5-7   Effect of number of electrons, shielding, and nuclear charges on atomic size.

we assume atoms and monoatomic ions to be essentially spherical in shape, we can express sizes in terms of atomic and ionic radii. These radii are usually expressed in Angstrom units. The **Angstrom unit** is a unit of length used to describe atomic, molecular, and ion sizes. The relation between the Angstrom and the centimeter is

$$1 \text{ Angstrom (A)} = 10^{-8} \text{ cm}$$

The Angstrom is used only for conveniencce. Just as we express our height in units of feet and inches rather than miles (how tall are you in units of miles?), we express atomic sizes in units of the Angstrom. Figure 5-6 gives a periodic table in which the atomic radii of the elements are indicated along with the ionic radii of some monoatomic ions. The periodic relationship between sizes are those that we would expect, since the factors affecting the sizes are essentially the same as those discussed above for ionization energies. The increase in atomic sizes with a given group is because higher energy level electrons are located further from the nucleus. Of course, the shielding effect that decreases the attraction between outer energy level electrons and the nucleus would result in these electrons occupying greater volumes about the nucleus. The decrease in atomic sizes from left to right within a given row would be because of the increase in nuclear charge that would tend to result in a greater attraction for the outer energy level electrons. These factors are illustrated in Figure 5-7. It is interesting to note the differences in the sizes of atoms and their corresponding ions. The positive ions result from the loss of the outer energy level electrons, and, thus, we would expect these ions to be smaller than the parent atoms. On the other hand, the negative ions are formed by the gain of outer energy level electrons which results in these ions being larger than the parent atoms. The relative sizes of some ions and parent atoms are given in Figure 5-8.

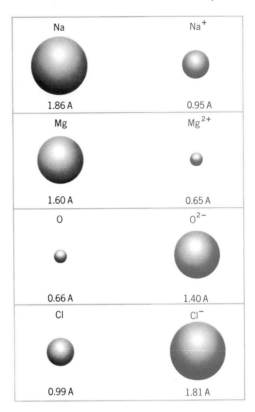

Na

Na$^+$

1.86 A

0.95 A

Mg

Mg^{2+}

1.60 A

0.65 A

O

O^{2-}

0.66 A

1.40 A

Cl

Cl$^-$

0.99 A

1.81 A

Figure 5-8   The relative sizes of some atoms and ions.

5-4   Some Other Properties

The physical states and forms of the elements were discussed in Chapter 2. As was pointed out in this chapter, most elements are solids, two are liquids, and 11 are gases under normal conditions. Figure 5-9 summarizes the physical states, the chemical forms, and the colors of the elements under normal conditions. Two of the metals, copper and gold. have unique colors. The noble gases occur as monoatomic gases under normal conditions and form very few compounds. The nonmetals occur in various forms in which the atoms are involved in covalent bonds. Hydrogen, nitrogen, oxygen, fluorine, and chlorine occur as gases consisting of covalent diatomic molecules (H_2, N_2, O_2, F_2, Cl_2). Bromine occurs as a liquid consisting of covalent diatomic molecules, Br_2. Iodine occurs as a solid made up of diatomic covalent molecules, I_2. The other nonmetals occur as molecular solids of more complex forms in which there is extensive covalent bonding. Among the important nonmetals that have distinct colors, fluorine is a pale yellow gas, chlorine is a greenish-yellow gas, bromine is a brownish-red liquid, iodine is a deep purple solid, and sulfur is a pale yellow solid. Gaseous hydrogen, nitrogen, and oxygen are colorless. The colors of the other elements are given in Figure 5-9.

Gas (dark shading) · Liquid (light shading) · Solid (unshaded)

3 dim covalent = three dimensional network of covalently bonded atoms

IA	IIA	IIIB	IVB	VB	VIB	VIIB	VIII	VIII	VIII	IB	IIB	IIIA	IVA	VA	VIA	VIIA	NOBLE GASES
H H_2 Colorless																H H_2 Colorless	He Colorless
Li metal	Be metal											B Colorless or brown	C diamond graphite 3-dim covalent	N N_2 Colorless	O O_2 Colorless	F F_2 pale yellow	Ne Colorless
Na metal	Mg metal											Al metal	Si 3-dim covalent blue-silver	P P_4 white or violet	S S_8 rings yellow	Cl Cl_2 green-yellow	Ar Colorless
K metal	Ca metal	Sc metal	Ti metal	V metal	Cr metal	Mn metal	Fe metal	Co metal	Ni metal	Cu reddish metal	Zn metal	Ga metal	Ge 3-dim covalent	As grey or yellow	Se chain molecules grey	Br Br_2 red-brown	Kr Colorless
Rb metal	Sr metal	Y metal	Zr metal	Nb metal	Mo metal	Tc metal	Ru metal	Rh metal	Pd metal	Ag metal	Cd metal	In metal	Sn gray or metallic	Sb metal	Te chain molecules grey	I I_2 purple	Xe Colorless
Cs metal	Ba metal	La metal	Hf metal	Ta metal	W metal	Re metal	Os metal	Ir metal	Pt metal	Au yellow metal	Hg metal	Tl metal	Pb metal	Bi metal	Po metal	At ?	Rn Colorless
Fr metal	Ra metal	Ac metal															

The metals range in color from dull gray to bright silver

Ce metal	Pr metal	Nd metal	Pm metal	Sm metal	Eu metal	Gd metal	Tb metal	Dy metal	Ho metal	Er metal	Tm metal	Yb metal	Lu metal
Th metal	Pa metal	U metal	Np metal	Pu metal	Am metal	Cm metal	Bk metal	Cf metal	Es metal	Fm metal	Md metal	No metal	Lr metal

Figure 5-9 The states, forms, and colors of the elements.

5-5   Oxidation Numbers

As we know, the atoms of elements enter into chemical combinations by sharing or transfer of electrons. Moreover, the kinds of ions and the number of covalent bonds formed by atoms of a particular element depend on the number of outer energy level electrons. We would expect the elements with similar electronic configurations to possess similar chemical properties. This, indeed, is observed, since the various groups of elements tend to form compounds and have chemical properties that are quite similar. This does not mean that all of the elements within a group have precisely the same properties. In fact, many elements have quite unique and interesting chemical properties that set them apart from the others.

We can make certain generalizations concerning the chemical behavior of certain groups of elements; but the behavior of certain elements is of such importance that a specific study of them is required. For example, organic chemistry is a branch of chemistry that is basically devoted to the study of compounds involving carbon called organic compounds. Nonorganic compounds are called inorganic compounds.

Literally millions of chemical compounds exist in nature or have been synthetically produced in the laboratory. Some compounds can be made in the laboratory by carrying out appropriate chemical reactions designed to produce the compound. The study of the compounds of the elements has led to the establishment of certain generalizations concerning the ability of chemical elements to form compounds. That is, certain patterns involving the abilities of elements to form compounds have been observed. These patterns are often useful for predicting the nature of the compounds that are formed between specific elements. Let us consider the combining ability of oxygen as an example. Oxygen is widely distributed in the environment both as uncombined molecular oxygen and in combination with a variety of other elements. In fact, oxygen can enter into chemical combination with most of the other elements in the periodic table. Some examples of such oxygen-containing compounds, called oxides, are given in Table 5-1. Oxygen has six outer level electrons as shown by the electron dot symbol

$$:\ddot{O}\cdot$$

Oxygen forms the oxide ion by gain of two electrons.

$$O^{2-}$$

Thus, oxygen is said to display an **ionic valence** of 2−. Oxygen commonly forms two covalent bonds as in water.

Table 5-1   Some Typical Oxides of the Representative Elements

H_2O						
Li_2O	BeO	B_2O_3	CO_2	N_2O_5		OF_2
Na_2O	MgO	Al_2O_3	SiO_2	P_4O_{10}	SO_3	Cl_2O
K_2O	CaO	Ga_2O_3	GeO_2	As_4O_{10}	SeO_3	Br_2O
Rb_2O	SrO		SnO_2	Sb_4O_{10}	TeO_3	
Cs_2O	BaO					

$$H\!-\!\ddot{O}:$$
$$|$$
$$H$$

Thus, oxygen is said to display a **covalence** of 2.

It has been observed that elements tend to form ions of specific charges and to form a specific number of covalent bonds. To generalize these observations, a number (or sometimes two or more) is assigned to each element that relates to the state (or states) of combination an element may assume. These numbers, called **oxidation numbers,** have been established by noting the various formulas of compounds and polyatomic ions of the elements. For example, as described above, the element oxygen is found to readily form the oxide ion, O^{2-}, rather than O^{1-} or O^{3-}, and to form two covalent bonds. Thus, we say that oxygen can have an oxidation number of -2. This number helps us predict the formulas of the compounds and ions that oxygen could form. The oxidation numbers of the elements are established by following certain rules. The oxidation numbers can be considered to be numbers related to the possible states of combination of the elements. They indicate the ability of an element to form compounds having certain formulas. As we shall see, the concept of the oxidation number is quite useful and convenient when dealing with the formulas of compounds.

The oxidation number or oxidation state of an element is a positive or negative number assigned according to the following rules.

1. The oxidation number of an element in the natural form uncombined with other elements is zero, 0. For example, the oxidation state of hydrogen in H_2 is 0, the oxidation state of chlorine in Cl_2 is 0, and the oxidation state of sodium in metallic sodium, Na, is 0.

2. The oxidation number of an element in a monoatomic ion is given by the charge of the ion. For example, the oxidation number of sodium in Na^+ is $+1$, and the oxidation number of sulfur in S^{2-} is -2.

3. Hydrogen normally has an oxidation number of $+1$ in its compounds. ·

4. Oxygen in the combined state normally has an oxidation number of -2.

5. In a given compound, the algebraic sum of the oxidation numbers of the elements present must equal zero. This means that the sum of the oxidation numbers of the elements in positive oxidation states must be numerically equal to the sum of the oxidation numbers of the elements in negative oxidation states, but, of course, the two sums will be of opposite signs. For example, in water, H_2O, the two hydrogens are in the $+1$ oxidation state, and the oxygen is in the -2 oxidation state. The sum of the positive oxidation numbers corresponds to the sum of the negative oxidation numbers, and the algebraic sum equals zero.

$$H_2O$$
$$2(+1) + (-2) = +2 - 2 = 0$$

6. In a polyatomic ion, the algebraic sum of the oxidation numbers of the elements present must equal the charge of the ion. For example, in the hydroxide ion, OH^-, the hydrogen is in the $+1$ oxidation state, and the oxygen is in the -2 oxidation state. The algebraic sum of the oxidation numbers equals the charge of the ion.

$$OH^-$$
$$(-2) + (+1) = -1$$

7. The more electronegative elements usually have negative oxidation numbers in the combined state; and the less electronegative elements usually have positive oxidation numbers in the combined state. Thus, metals normally take on positive oxidation numbers since they generally have low electronegativities. The more electronegative non-metals take on negative oxidation numbers.

8. The oxidation number of an element in a compound or a polyatomic ion involving several combined elements can be determined by assigning reasonable oxidation numbers to the other elements and then applying rule 5 or 6.

Example 5-1   What is the oxidation number of sulfur in hydrogen sulfate ion, HSO_4^-? The oxidation number of sulfur in hydrogen sulfate ion, HSO_4^-, is determined by assigning a $+1$ oxidation number to hydrogen and a -2 oxidation number to oxygen. The oxidation number of sulfur must be of the specific value that results in the overall ionic charge of -1. So, applying rule 6, we have

$$\begin{array}{cccc} H & S & O_4^- & \text{Charge} \\ (+1) + & X & + 4(-2) = & -1 \\ (+1) + & X & + (-8) = & -1 \\ (+1) + (+6) + & (-8) = & -1 \end{array}$$ The oxidation number of sulfur is $+6$.

Example 5-2   What is the oxidation number of nitrogen in nitrous acid, HNO_2? Assigning an oxidation number of $+1$ to hydrogen and -2 to oxygen and using rule 5 gives

$$\begin{array}{ccc} H & N & O_2 \\ (+1) + & X & + 2(-2) = 0 \\ (+1) + & X & + (-4) = 0 \\ (+1) + (+3) + & (-4) = 0 \end{array}$$ A $+3$ oxidation number for nitrogen satisfies rule 5.

Example 5-3   What is the oxidation number of chromium in dichromate ion, $Cr_2O_7^{2-}$? Assigning an oxidation number of -2 to oxygen and using rule 6 gives

$$\begin{array}{ccc} Cr_2 & O_7^{2-} & \text{Charge} \\ X & + 7(-2) & = (-2) \\ X & + (-14) & = (-2) \\ (+12) + (-14) & = (-2) \\ 2(+6) + 7(-2) & = (-2) \end{array}$$

Since the ion has a -2 charge, and it contains two combined chromium atoms, the sum of the positive oxidation numbers is $+12$, which gives an oxidation number of $+6$ for each chromium.

5-6   Oxidation Numbers of the Elements

Since the oxidation numbers are related to the ability of the elements to form compounds, it is reasonable that the likely oxidation numbers are related to the electronic configurations of the elements. We know that the representative elements tend to lose, gain, and share electrons, so that the noble gas configuration is obtained. This tendency is useful for predicting the likely oxidation numbers of the elements. Additional aids in predicting oxidation numbers are the tendency of elements to lose or share electrons as pairs, if possible and the tendency of elements lose or share p electrons first. By considering these factors and the electronic configurations or electron dot structures of the groups of representative elements, it is possible to make certain predictions concerning likely oxidation numbers of these elements. These predictions are often quite good but do not always include all of the observed oxidation numbers for certain elements and the observed differences in oxidation numbers within a given group. Nevertheless, these predictions can be quite useful. Let us consider the expected oxidation numbers for the various groups of representative elements.

Alkali Metals-Group IA. From the electron dot structure of M·, we would predict an oxidation number of $+1$, which results when elements of this group lose the outer energy level electron to form the 1+ monoatomic ions. These elements rarely form covalent bonds, and the 1+ mono-atomic ions of all of the elements of the group are readily formed.

Alkaline Earth Metals-Group IIA. The electron dot structure of ·M· indicates a likely oxidation number of $+2$. This oxidation state is the most commonly observed state of these elements, and the 2+ monoatomic ions of these elements (except beryllium) are readily formed.

Boron-Aluminum Group-Group IIIA. From the outer energy level configuration ns^2np^1 or electron dot structure of ·Ṁ· , we would predict two oxidation states of $+1$ (loss or sharing of one electron) or $+3$ (loss or sharing of all three electrons). The $+3$ oxidation state of these elements is the most common state, and the $+1$ state is rare.

Carbon Group-Group IVA. The outer energy level configuration of ns^2np^2 or dot structure of ·Ë· indicates the likely oxidation states of $+2$ and $+4$. These oxidation numbers are indeed those most often observed. The most common members of the groups (C and Si) tend to form covalent compounds in which the $+4$ state is most often observed.

Nitrogen Group-Group VA. The positive oxidation states of $+3$ and $+5$ would be predicted from the electron dot structure of ·Ë·. Furthermore, some of these elements are electronegative enough to take on negative oxidation numbers. We would predict a likely negative oxidation number of -3 which would correspond to the gain of three electrons to give the ns^2np^6 configuration. The -3 oxidation state is common for the more electronegative members of the group, and the $+3$ and $+5$ oxidation states are observed for all members. The -3, $+3$, and $+5$ states are common for nitrogen, but this element is also found in the -2, -1, $+1$ and $+2$ and $+4$ oxidation states.

Oxygen Group-Group VIA. The likely negative oxidation state of these elements is -2 as would be predicted from the outer energy level configuration of ns^2np^4 or dot structure of $:\ddot{\overset{..}{E}}\cdot$. The likely positive oxidation states would be $+2$, $+4$, and $+6$. The -2 state is favored by the more electronegative members of the group, and the positive oxidation states are favored by the others. Oxygen, of course, is most often found in the -2 state.

Halogen Group-Group VIIA. This group, with the outer energy level configuration of ns^2np^5 and dot structure of $:\ddot{\overset{..}{E}}\cdot$, will likely be found in the oxidation states of -1, $+7$, $+5$, $+3$, and $+1$. The -1 state is common to all of the halogens and is found in the monoatomic halide ions; fluoride ion, F^-, chloride ion, Cl^-, bromide ion, Br^-, and iodide ion, I^-. Fluorine is always found in the -1 state except in F_2.

The observed oxidation numbers of the representative elements, as given in the periodic table inside the back cover, generally fit the predictions. Notice the pattern concerning the predicted oxidation numbers of the groups of representative elements. The highest positive oxidation number is given by the group number. The lower positive oxidation numbers differ from the group number by increments of two. The negative oxidation numbers for groups VA, VIA, and VIIA are found by subtracting 8 from the group number.

It is more difficult to predict the likely oxidation states of the transition elements. The common oxidation numbers of some of the important transition metals are of interest to us. Chromium is found in the $+3$ and $+6$ oxidation states. Iron favors the $+2$ and $+3$ states. Cobalt, nickel, and copper favor the $+2$ states. However, cobalt and nickel do occur in the $+3$ state. Combined zinc and cadmium are always found in the $+2$ state, and combined silver is found in the $+1$ state. Mercury occurs in the $+2$ state and also in the $+1$ state in the form of mercury(I) ion, Hg_2^{2+}. The oxidation states of these transition metals are shown in the periodic table inside the back cover of the book.

5-7   Formulas From Oxidation Numbers

The oxidation numbers of elements can be used to predict the formulas of binary compounds involving two elements. Of course, the oxidation numbers indicate the kinds of ions and numbers of covalent bonds formed by elements, but we can use oxidation numbers to predict formulas without worrying whether they are ionic or covalent compounds. When nonmetals form binary compounds with metals, the metals have positive oxidation numbers, and the nonmetals have negative oxidation numbers. The nonmetals generally take on the expected negative oxidation number corresponding to the group (8 subtracted from the group number). The representative metals take on the common positive oxidation numbers. Transition metals that have more than one possible oxidation number can generally form more than one binary compound with a given nonmetal. We can predict the formulas of binary compounds involving metals and nonmetals by using the expected oxidation numbers. These predictions usually give the correct formulas for the compounds, but such a prediction does not necessarily mean that the compounds can be readily formed from the elements. Let us consider the prediction of the formulas of some binary compounds. When writing these formulas, we give the symbol of the metal with an appropriate subscript followed by the symbol of the nonmetal with an appropriate subscript. We expect the +1 oxidation state for the Group IA elements and a −2 oxidation state for oxygen. So, the predicted formula for a binary compound involving potassium and oxygen would be

$$\overset{+1}{K} \quad \overset{-2}{O}$$

K_2O \qquad (The total sum of oxygen numbers should equal zero.)

The compound would be predicted to contain two combined atoms of potassium for each combined oxygen atom. The formula predicted for the compound between magnesium (Group IIA) and oxygen is

$$\overset{+2}{Mg} \quad \overset{-2}{O}$$

MgO \qquad (One magnesium and one oxygen gives an oxidation number total of zero.)

The predicted formula for the compounds involving iron and oxygen are determined by considering that the two likely oxidation states of iron are +2 and +3 and the likely oxidation state of oxygen is −2. Thus, two compounds are possible.

$$\overset{+2}{Fe} \quad \overset{-2}{O} \qquad\qquad \overset{+3}{Fe} \quad \overset{-2}{O}$$

$$FeO \qquad\qquad\qquad Fe_2O_3$$

Example 5-4   What is the expected formula for a calcium and nitrogen compound? Using the likely oxidation numbers of +2 for calcium and −3 for nitrogen, the formula is

$$\overset{+2}{Ca} \quad \overset{-3}{N}$$

$$Ca_3N_2$$

Example 5-5   What are the expected formulas for chromium and sulfur compounds? Using the likely oxidation numbers of +3 and +6 for chromium and −2 for sulfur, the formulas are

$$\overset{+3}{Cr} \quad \overset{-2}{S}$$

$$Cr_2S_3$$

and

$$\overset{+6}{Cr} \quad \overset{-2}{S}$$

CrS_3 \qquad (Three sulfurs for each chromium gives an oxidation number sum of zero. Note that the formula is not Cr_2S_6 but simply CrS_3.)

Many of the nonmetals form binary compounds with one another. We can predict the formula for many of these by using the expected oxidation numbers of the nonmetals. The predicted formulas of such compounds are often the correct ones, but in some cases it is only possible to predict the simplest formula. The actual formula may be a multiple of this. When writing the predicted formulas, we usually write the symbol of the less electronegative element first, followed by the symbol of the more electronegative element. (See Section 6-4.) Normally, the symbol of the element with the positive oxidation number comes first in the formula, followed by the symbol of the element with the negative oxidation number. The predicted formula for the binary compound involving sulfur and hydrogen would be deduced using a +1 for hydrogen and a −2 for sulfur.

$$\overset{+1}{H} \quad \overset{-2}{S}$$
$$H_2S$$

The formulas of the two likely binary compounds of nitrogen and oxygen are deduced using the +3 and +5 oxidation numbers of nitrogen and the −2 oxidation number of oxygen.

$$\overset{+3}{N} \quad \overset{-2}{O}$$
$$N_2O_3$$

and

$$\overset{+5}{N} \quad \overset{-2}{O}$$
$$N_2O_5$$

Example 5-6   What is the formula of the binary compound of hydrogen and chlorine? Using a +1 oxidation number for hydrogen and a −1 for chlorine, the formula is

$$\overset{+1}{H} \quad \overset{-1}{Cl}$$
$$HCl$$

Example 5-7   What are the expected formulas for the compounds of sulfur and oxygen? Using the likely oxidation numbers of +2, +4 and +6 for sulfur and −2 for oxygen, the expected formulas are

$$\overset{+2}{S} \quad \overset{-2}{O}$$
$$SO$$

and

$$\overset{+4}{S} \quad \overset{-2}{O}$$
$$SO_2$$

and

$$\overset{+6}{S} \quad \overset{-2}{O}$$
$$SO_3$$

5-8   Mining and Refining

In an industrial country like the United States, the most important non-fuel resources are iron ore, aluminum ore, copper ore, and minerals used in fertilizers. In addition, large amounts of cement, gravel, stone, and asphalt are used. To support the industrial society of the United States, around 5 billion tons of materials and ore are extracted from the earth each year. Once an ore deposit is located, much energy is expended in the mining operations and the transportation of the ore. To make these operations of physical extraction and transportation worthwhile, the ores must contain enough of the desirable compound or element. Some low-grade ore deposits are not usable due to the economics of the mining process.

The materials extracted from the earth have to be processed for use. The physical processes of screening, separating, washing, and grinding produce wastes (some are called mine tailings) and dust (fine solid particles that become suspended in the air). Often the desirable element in an ore is in an undesirable chemical form and must be converted to a desirable one. For instance, aluminum in aluminum ore is in the form of aluminum ion, Al^{3+}, combined with oxide ion, O^{2-}. To be useful, the aluminum ion has to be converted to aluminum metal, Al. In such cases the ore needs to be processed to extract the desirable element. Chemical or electrical processing of an ore to extract an element is called the **refining** of the ore. The refining process normally involves a chemical reaction or a series of chemical reactions which produce the desired form of the element or compound sought from the ore. These chemical reactions usually require large amounts of energy and the cooling and washing processes involved utilize large amounts of water. Let us consider the refining and processing of some important ores and materials.

Iron and Steel. Iron ores are usually a mixture of iron oxides and silicate rocks. The principle oxides are Fe_2O_3 (hematite ore) and Fe_3O_4 (magnetite ore). These iron oxides are converted to iron by allowing the ore to react with carbon (in a form called coke), limestone, and air in a **blast furnace** at temperatures exceeding 1300°C. The blast furnace is illustrated in Figure 5-10. The main chemical reactions that take place in the refining process are:

$$2C + O_2 \rightarrow 2CO$$
$$3CO + Fe_2O_3 \rightarrow 3CO_2 + 2Fe \text{ (molten iron)}$$

The molten iron is tapped from the blast furnace and cast into ingots called pig iron or cast iron. Most of this iron is used to manufacture steels. **Steels** are alloys of iron: iron mixed with small amounts of carbon and other metals such as manganese, nickel, chromium, vanadium, and/or tungsten. Stainless steel is iron containing some carbon, 15 to 20% chromium and about 10% nickel. Steel is manufactured by melting pig iron and mixing in the alloying metals. The preparation of steel is carried out by use of devices called Bessemer converters, open-hearth furnaces, or electric-arc furnaces.

Steel is used widely in manufacturing to make pipes, wires, beams,

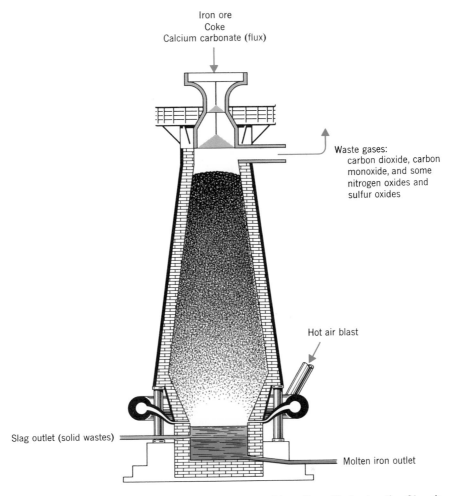

Iron ore
Coke
Calcium carbonate (flux)

Waste gases:
carbon dioxide, carbon
monoxide, and some
nitrogen oxides and
sulfur oxides

Hot air blast

Slag outlet (solid wastes)

Molten iron outlet

Figure 5-10   Blast furnace used in the refining of iron. From *Understanding Chemistry: From Atoms to Attitudes* by T. R. Dickson, Wiley, N.Y., 1974.

rails, steel sheets, and bars. The uses of iron are illustrated in Figure 5-11. In the United States, around 10 tons of steel are in use for each person. Most of this steel is in structural materials and in motor vehicles. Nearly one-half ton of steel per person is added to the environment each year, but around one-third ton per person is lost as a result of junking, dumping trash, and corrosion each year. The presence of acids in the smoggy air of heavily urbanized, industrialized areas hastens the corrosion of exposed iron. Of this discarded steel, about 40% is recycled to the steel furnaces and the rest is scrapped or buried. The buried and scrapped iron may never be utilized again. In other words, we throw away nearly 200 kg of steel per person each year.

Aluminum. Large amounts of aluminum occur in the form of aluminum compounds in the lithosphere. Aluminum is the most abundant metal in the crust of the earth. Combined aluminum is found in clays and rocks. **Bauxite** is **aluminum ore** consisting of a mixture of aluminum com-

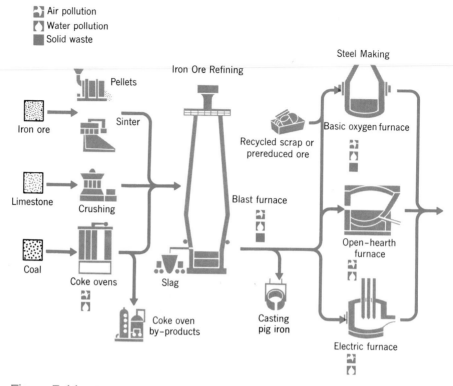

■ Air pollution
■ Water pollution
■ Solid waste

Steel Making

Iron Ore Refining

Pellets

Iron ore

Sinter

Limestone   Crushing

Coal

Coke ovens

Coke oven
by-products

Slag

Blast furnace

Casting
pig iron

Recycled scrap or
prereduced ore

Basic oxygen furnace

Open-hearth
furnace

Electric furnace

Figure 5-11   The uses of iron and the potential sources of pollution in the refining of iron ore and the manufacture and processing of steel. From "Pollution Control in the Steel Industry," by Henry C. Bramer, *Environmental Science and Technology* **5**, (10), 1004 (Oct. 1971). From *Understanding Chemistry: From Atoms to Attitudes* by T. R. Dickson, Wiley, N.Y., 1974.

pounds and some iron oxides. Aluminum oxide, Al_2O_3, can be extracted from the bauxite. The refining of aluminum oxide to aluminum is more difficult than the production of iron from iron ore. The problem is that aluminum metal cannot be produced from the oxide by a simple chemical process such as the reaction with carbon as is done in the refining of iron. To refine aluminum, a process called electrolysis, in which electric current converts the aluminum ion, Al^{3+}, to aluminum metal, is used. All aluminum is made electrolytically by the **Hall process,** named after Charles M. Hall, who first developed the process in 1886. In this process, (See Figure 5-12) the aluminum oxide is dissolved in a large vat of the molten (around 1000°C) mineral cryolite, Na_3AlF_6. When the aluminum oxide dissolves in the cryolite it forms the aluminum ion, Al^{3+}, and oxide ion, O^{2-}. Immersed in the vat are large pieces of carbon that serve as the anode or positive electrode. The carbon lining of the iron vat serves as the cathode or negative electrode. When electrical current flows, the aluminum ion is attracted to the cathode where it is converted to molten aluminum metal by gaining electrons. The oxide ion is attracted to the anode where it loses electrons and combines with carbon to form carbon dioxide. The production of aluminum by the Hall process requires large amounts of electricity and water for cooling and washing. Consequently, large aluminum plants are usually

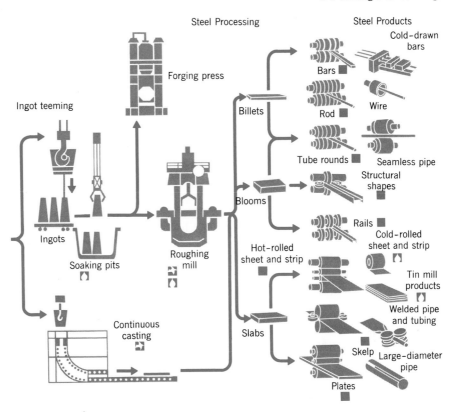

located near convenient sources of water in regions where electricity from hydroelectric sources is available.

In the United States around 5 million tons of aluminum are produced annually. Aluminum is a lightweight, nontoxic metal that can be easily formed or cast into a variety of shapes ranging from structural beams to

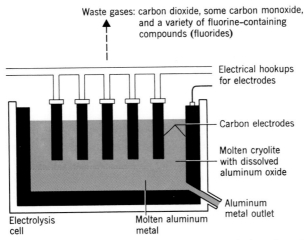

Figure 5-12   A typical Hall process electrolysis cell used in the refining of aluminum ore. From *Understanding Chemistry: From Atoms to Attitudes* by T. R. Dickson, Wiley, N.Y., 1974.

sheets and foils. It is corrosion resistant, has a high heat conductivity and is a good conductor of electricity. Aluminum alloyed with small amounts of other metals is used for such items as kitchen utensils, building decorations and structural components, aircraft construction, electrical transmission lines, mirrors, automobile parts, foils, and food and drink cans. Aluminum used in buildings and other structural components may remain in use for long periods of time. Much of the aluminum we use is thrown away in short periods of time. Such aluminum is widely distributed in junked cars and refuse disposal dumps. Since aluminum is corrosion resistant, junked aluminum oxidizes back to aluminum oxide very slowly. Some of the junk aluminum we dispose of today (i.e., aluminum cans) may be around for hundreds of years as metallic aluminum. The refining of 1 ton of aluminum requires five times the energy needed to refine 1 ton of steel. Consequently, it is an energetically expensive metal. The recycling of aluminum is possible and some effort has been directed to recycling aluminum in cans. Unfortunately, only about 20% of the annual production of aluminum returns to be used again. Over 400,000 tons of aluminum are used each year to make cans, can tops, and lids.

5-9   Toxic Metals in the Environment

Most of the elements of the earth are widely distributed in the lithosphere, hydrosphere, and atmosphere. We are exposed to these elements when we eat food, drink water, and breathe air. Some of the elements are beneficial and essential to our life processes, and some are very toxic. Fortunately, most of the toxic elements that are found in food, water and air are present in extremely small amounts. Nevertheless, there are trace amounts of many different elements entering our bodies each day. Since these trace amounts result from the natural distribution of elements, they are unavoidable and are referred to as background levels. Sometimes the endeavors of humans and, on occasion, a geological accident introduce undesirable amounts of certain toxic elements into the environment. Most notorious of these are the heavy metals mercury (Hg), lead (Pb), and cadmium (Cd). Note the position of these metals in the periodic table. However, other elements of concern are arsenic (As), beryllium (Be), antimony (Sb), vanadium (V), and nickel (Ni). The **heavy metals** are of greatest concern since they are used in great amounts and distributed widely in our industrial society.

Mercury. This element is very useful since it is the only metal that is a liquid at normal temperatures and has a high electrical conductivity. Unfortunately, mercury compounds are poisonous to all living systems. Most mercury is obtained from ore containing cinnabar, HgS, by heating in air. The heating converts the HgS to mercury metal and sulfur dioxide.

$$HgS + O_2 \rightarrow Hg + SO_2$$

Over 3000 tons of mercury are used in the United States each year. Mercury is used in a variety of industrial processes and in various products such as paints, fungicides, electrical apparatuses, and thermome-

Table 5-2   Mercury Uses in the United States (percentage of Total Tonnage Used, 1969)

Use	Percentage of Total Tonnage Used
Chlor-alkali industry	26.0
Electrical apparatus (including batteries)	22.9
Paint (fungicides)	12.2
Scientific instrumentation	6.5
Catalysts	3.7
Dental preparations (amalgams)	3.5
Agriculture (pesticides and fungicides)	3.4
General laboratory uses	2.1
Pharmaceuticals	0.9
Paper and pulp processing (fungicides)	0.7
Industrial amalgamation	0.2
Miscellaneous	17.9

From *Understanding Chemistry: From Atoms to Attitudes* by T. R. Dickson, Wiley, N.Y., 1974.

ters. Table 5-2 lists the common uses of mercury in the United States. The greatest single use of mercury is in the electrolytic manufacture of chlorine and sodium hydroxide. The second greatest use of mercury is in electrical apparatuses such as mercury vapor lamps, electronic tubes, and switches and mercury batteries.

Mercury enters the environment in the elemental form as a loss from industrial processes and scrapped equipment and in the form of mercury compounds from industrial and agricultural endeavors. Mercury also enters the environment from unexpected sources. Significant amounts of mercury are found in sewage resulting from use of small amounts of mercury-containing chemicals, pharmaceuticals, and paints by large numbers of people. Another mercury source appears to be the combustion of coal, oil, and gasoline. These fossil fuels contain small amounts of mercury. However, since fossil fuels are used in such large amounts, significant quantities of mercury enter the atmosphere by fossil fuel combustion. It is estimated that around 3000 tons of mercury enter the atmosphere from this source each year.

Mercury is incorporated into the food we eat by agricultural uses and the water supply. Mercury is toxic in the metallic form and also in the combined form. The two general forms of combined mercury are called inorganic mercury and organic mercury. Within the environment the various forms of mercury are interconverted. In fact, it appears that metallic and inorganic mercury are converted to methylated mercury by biological processes occuring in water in which mercury wastes are found. Furthermore, **methylated mercury (dimethyl mercury,** $Hg(CH_3)_2$, and **methyl mercury ion,** $HgCH_3^+$) are absorbed in the tissue of living organisms. Once absorbed, these forms of mercury can remain in an organism for long periods. As one animal eats another, the mercury can be incorporated into the food chain. This can result in the biological concentration of mercury within the food chain. For example, algae ingest mercury from the water, a small fish eats the algae, a larger fish eats the small fish. This biological concentration results in animals at the top of the food chain having more than normal (above background) amounts of mercury incorporated in the tissue. Plants can incorporate mercury compounds from the soil and mercury-containing seed coatings. Mercury enters our bodies through the plants and animals we eat and the water we drink and, thus, becomes incorporated into our body tissue. The U.S. Food and Drug Administration (FDA) has established maximum concentrations of 0.5 parts per million (ppm) of mercury in fish and 0.005 parts per million of mercury in water. The FDA periodically checks samples of food for mercury content. In 1971, the FDA discovered that commercial swordfish was highly contaminated with mercury, and all sales were banned. In the same year, certain canned-tuna samples were found to have an intolerable mercury content. As a result, large numbers of cans were recalled from retail stores.

Too much mercury in our systems can produce mercury poisoning, which can be deadly or cause permanent brain damage. In Japan, illness, deaths, and birth defects have been directly attributed to mercury-containing seafoods. Around 100 persons living in the Minimata Bay area of Japan became afflicted with a mysterious illness, and many died. It was found that the victims' main diet was mercury-contaminated fish from the bay. The mercury had been dumped into the bay in the waste water of a plastics plant. The Japanese government soon imposed strict requirements on mercury disposal. The Swedes, who also rely on a seafood diet, have strictly limited the use of mercury and mercury compounds in their country. The symptoms of acute mercury poisoning are loss of appetite, numbness of extremities, metallic taste, diarrhea, vision problems, lack of coordination, speech and hearing difficulties, and mental instability. The damage to the body by mercury poisoning is usually permanent. Not much is known about the effects of small amounts of mercury in the body since the symptoms are not specific and cannot be distinguished from other diseases.

Since mercury is used in such great amounts and can be converted to highly toxic forms that are biologically concentrated in the environment, it is necessary that we exercise greater control over the intended or unintended discharge of mercury. Alternatives to industrial discharges should be enforced. The use of mercury compounds in agriculture should be restricted or eliminated if possible. Finally, the content of mercury in the food, water, and air should be closely watched.

Lead. This metal, heavy like mercury, is a toxic metal that accumulates in the body as it is inhaled from the air or ingested from food and water. Lead has been used for centuries to fabricate devices used by humans. Lead is mentioned in the Old Testament of the Bible. The Romans used lead for water pipes and for cooking utensils. Actually, lead is somewhat of a rare element in the lithosphere. The percentage of lead in the crust of the earth is about $2 \times 10^{-5}\%$. However, lead-ore deposits consisting of the ore galena, PbS, are used as sources of lead. Lead has a low melting point and is quite a soft metal. This malleability allows lead to be cast

Table 5-3   Uses of Lead in the United States (percentage of total tonnage used, 1969)

Use	Percentage of Total Tonnage Used
Storage batteries	43.1
Metal products (ammunition, cable covering, pipes and plumbing uses, solder, and type metal for printing)	26.2
Lead alkyl (lead tetraethyl and lead tetramethyl antiknock compounds)	20.0
Paint pigments	7.5
Miscellaneous (chemicals and pottery glazes)	3.2

From *Understanding Chemistry: From Atoms to Attitudes* by T. R. Dickson, John Wiley and Sons, 1974.

and formed easily. Lead can be mixed with other metals to form useful alloys. Nearly 1.5 million tons of lead are used in the United States each year. About one half of this is recycled lead, and the rest is virgin lead. As shown in Table 5-3, lead is used in a variety of products including storage batteries, gasoline antiknock chemicals, paint pigments, and ceramic glazes. Most lead is used in storage batteries, and most of this is salvaged so that the lead can be reused. The environmental threat comes from lead that is used in chemicals. Two to three hundred thousand tons of lead are used each year to manufacture lead alkyl compounds (lead tetraethyl and lead tetramethyl) used as antiknock additives in gasoline. This lead becomes widely distributed in the environment as it is discharged in automobile exhaust. Much of the lead compounds given off in automobile exhaust are dispersed into the atmosphere as gaseous substances or tiny bits of solids called particulates. This atmospheric lead can be inhaled by people, settle out on plants and soil, or be absorbed by water. The lead from gasoline is supplemented by lead compounds produced in the burning of coal and oil and the manufacturing of lead and lead products.

Investigations show that the amounts of lead in the environment have increased greatly since the beginning of the Industrial Revolution and the amounts are currently increasing at a rapid rate. This is illustrated in Figure 5-13. The major toxic effect of lead appears to be that it interferes with the production of red blood cells and other bodily functions. Lead poisoning is a possibility if the concentration of lead in the blood is 0.8 ppm for adults or 0.4 ppm for children. Currently, the lead concentration in the blood of a typical American is around 0.2 ppm. The symptoms of mild lead poisoning are loss of appetite, fatigue, headaches, and anemia. Lead poisoning is especially dangerous and prevalent in children. Severe lead poisoning can cause permanent brain damage in children.

Another source of lead in the environment is from lead-containing paint pigments. Such pigments are not supposed to be used in interior

Figure 5-13   The increase in the lead content of the environment. Lead content of arctic snow over a period of years. The snow fall in some arctic regions is slight, and the snow remains packed in layers for centuries. These snow layers can be dated and serve as good sources of evidence of the lead content of the atmosphere in years past. As the graph indicates, the atmospheric lead content has apparently increased steadily over the last 200 years. Further, a very significant increase has occurred since 1950, corresponding to the wide use of lead alkyls in gasoline and the increase in the automobile population. The graph is based on average data collected by C. Patterson of the California Institute of Technology, *Scientist and Citizen* 66 (Apr. 1968). From Understanding Chemistry: From Atoms to Attitudes by T. R. Dickson, Wiley, N.Y., 1974.

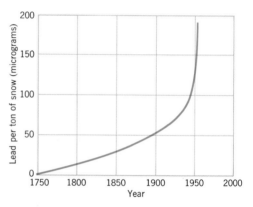

paints today but were used before World War II. In fact, U.S. Food and Drug Administration regulations call for a limit of 0.06% lead in paint pigments. Tragically, many older buildings, such as those found in ghettos, contain layers of paint with high lead content. Children suffering from malnutrition tend to nibble on bits of dirt and paint. In this way significant amounts of lead are ingested by children. It is estimated that about 400,000 children in the United States are afflicted with various degrees of lead poisoning each year. Some 200 of these children die, and around 800 suffer from permanent brain damage.

Still another source of lead in the diet comes from the lead compounds used in some glazes and ceramics. Food in contact with such glazed ceramics can sometimes extract lead in significant amounts. This does not occur with all ceramic glazes. However, it is good practice not to store carbonated beverages, fruit juices, wines, any food containing vinegar, sauerkraut, or fruit products in ceramic vessels. Furthermore, avoid using ceramic vessels with cracked, chipped, or faulty glazes.

Since lead is a cumulative poison, it is important to continuously investigate the amounts of lead that are entering the environment. This is especially true in the United States that uses nearly 60% of the lead consumed in the entire world. Of this 60%, about one half will never be used again and will become distributed in the environment. The greatest waste of lead is in the use of lead alkyls as gasoline additives. About 20% of the world consumption of lead is lost as nonrecoverable lead alkyls. In any case, the content of lead in food, water, air, and the bodies of Americans should be closely watched.

Questions and Problems

1. Give a statement of the periodic law.

2. Sketch an outline of the main portion of the periodic table (s, p, d blocks) and indicate with arrows how the ionization energies of the elements varies from left to right in rows and from top to bottom of the groups.

3. Explain how the sizes of the atoms vary with the rows and groups of the periodic table. In which part of the table are the elements with the smallest atoms found?

4. Why is a negative ion larger than the atom from which it is formed? Why is a positive ion smaller than the atom from which it is formed?

5. What is an Angstrom unit?

6. A penny is 1.0 mm thick. Assuming that the copper atoms are stacked on top of one another, calculate how many copper atoms make up the thickness of a penny. The diameter of a copper atom is 2.56 Angstrom units.

7. An approximation of the size of an iron atom can be found by dividing the molar mass by the density to find the volume per mole of atoms. This can be divided by Avogadro's number to find the volume per atom. Assuming an atom is spherical, the radius of an atom is found by solving for r in the expression: volume $= 4/3\pi r^3$. Find the approximate radius of an iron atom (density $= 7.9$ g/cm^3, molar mass $= 55.8$ g/mole, 1 A$^3 = 10^{24}$ cm^3). Considering that the atoms in iron are packed as spheres, why is the actual size of an iron atom smaller than the approximated size?

8. The metals of greatest density are those with the smallest atoms and the largest atomic weights. Refer to Figure 5-6 and find the region of the periodic table that contains the metals of greatest density. Which metal do you think has the greatest density?

9. Refer to Figure 5-4 and explain why, within each row of elements, the Group IA element has the lowest ionization energy and the noble gas has the highest ionization energy.

10. List all of the elements that normally occur as gases.

11. List all of the elements that normally occur in the form of diatomic molecules.

12. List all of the elements that you know have unique colors and indicate the color of each of these elements.

13. What is an oxidation number?

14. Make a list of the expected oxidation numbers for each group of representative elements.

15. Determine the oxidation number of the element other than hydrogen or oxygen in each of the following (use the expected oxidation numbers for hydrogen and oxygen):

(a) N_2O

(b) HNO_3

(c) Al_2O_3

(d) NH_3

(e) N_2H_4

(f) $HClO_3$

(g) HCl

(h) H_2SO_3

(i) MnO_4^-

(j) HCO_3^-

(k) CrO_4^{2-}

(l) SO_4^{2-}

(m) PO_4^{3-}

(n) $C_2H_3O_2^-$

16. Give the formulas for the compounds formed between the following metals and nonmetals. Use the expected positive oxidation numbers for the metals and the expected negative oxidation number for the nonmetals.
    (a) compound of lithium and chlorine
    (b) compound of barium and bromine
    (c) compound of calcium and oxygen
    (d) compound of magnesium and arsenic
    (e) compound of aluminum and sulfur
    (f) compounds of iron and bromine
    (g) compounds of copper and oxygen
    (h) compounds of mercury and chlorine

17. Give the formulas for the compounds formed between the following non-metals. Use the expected positive oxidation numbers for the first element given and the expected negative oxidation number of the second element.
    (a) compound of hydrogen and sulfur
    (b) compound of hydrogen and bromine
    (c) compounds of phosphorus and chlorine
    (d) compounds of tellurium and oxygen
    (e) compounds of silicon and oxygen
    (f) compound of hydrogen and nitrogen (nitrogen is written first in the formula.)
    (g) compounds of sulfur and fluorine
    (h) compounds of carbon and sulfur

18. What is an ore?

19. Describe the process involved in the refining of iron ore.

20. Describe the Hall process used to refine aluminum.

21. What is meant by the term "heavy metals"?

22. What are the major uses and potential danger of mercury?

23. What are the major uses and potential danger of lead?

6

Chapter 6: Nomenclature of Chemical Compounds

Objectives

The student should be able to:

1 Name an ionic compound given the formula or give the formula from the name.

2 Name a compound of a metal by the Stock method given the formula or give the formula from the name.

3 Name a binary nonmetal-nonmetal compound by the prefix method given the formula or give the formula from the name.

4 Name an acid given the formula or give the formula from the name.

5 Name an oxyacid given the formula or give the formula from the name.

6 Name an oxyanion given the formula or give the formula from the name.

7 Name a hydrate given the formula or give the formula from the name.

6

Nomenclature of Chemical Compounds

6-1   Chemical Nomenclature

The ability to name chemical compounds and to deduce the chemical composition from the name is very important to your study of chemistry. Systems of **nomenclature** have been developed, so that the naming of compounds can be accomplished easily. Names derived according to some system of nomenclature are called **systematic names.** The names of many chemical compounds developed historically before any systems of nomenclature had been developed. These names generally are not logical and usually do not convey any information concerning the structure of the compound. Such names are called **common** or **trivial names.** Some of these are so familiar that they are almost always used. For example, the name water for H_2O is a common name that is invariably used. Table 6-1 lists the formulas of some important compounds that have common names that are accepted. Some compounds have common names that are used on occasion but are not used as technical names. These are often encountered in nontechnical literature, but the systematic name is preferred in a technical context. Table 6-2 lists the formulas, common names, and systematic names for some typical substances.

We have considered two kinds of chemical compound: ionic and molecular. Recall that compounds of metals and nonmetals and metals and polyatomic ions are ionic and compounds of nonmetals with nonmetals are molecular. There are specific methods used to name ionic compounds and different methods for molecular compounds.

6-2   Nomenclature of Ionic Compounds

Ionic compounds are, of course, composed of positive and negative ions. The names of ionic compounds utilize the names of the ions involved.

Table 6-1   Some Compounds with Officially Accepted Common Names

Compound Formula	Name
CH_4	Methane
SiH_4	Silane
GeH_4	Germane
SnH_4	Stannane
C_2H_6	Ethane
C_3H_8	Propane
C_6H_6	Benzene
NH_3	Ammonia
PH_3	Phosphine
ArH_3	Arsine
SbH_3	Stibine
H_2O	Water
H_2S	Hydrogen sulfide
H_2Se	Hydrogen selenide
H_2Te	Hydrogen Telluride

Table 6-2   Trivial and Systematic Names for Some Common Compounds

Formula	Trivial Name	Systematic Name
Al_2O_3	Alumina	Aluminum oxide
$NaHCO_3$	Baking soda	Sodium hydrogen carbonate
$Na_2B_4O_7 \cdot 10H_2O$	Borax	Sodium tetraborate 10-water
$CaCO_3$	Calcite or marble	Calcium carbonate
$KHC_4H_4O_6$	Cream of tartar	Potassium hydrogen tartrate
$MgSO_4 \cdot 7H_2O$	Epsom salt	Magnesium sulfate 7-water
$CaSO_4 \cdot 2H_2O$	Gypsum	Calcium sulfate 2-water
C_2H_5OH	Grain alcohol, alcohol	Ethyl alcohol or ethanol
$Na_2S_2O_3$	Hypo	Sodium thiosulfate
N_2O	Laughing gas, nitrous oxide	Dinitrogen oxide
PbO	Litharge	Lead(II) oxide or plumbous oxide
CaO	Lime	Calcium oxide
$NaOH$	Lye	Sodium hydroxide
$2CaSO_4 \cdot H_2O$	Plaster of paris	2-calcium sulfate 1-water
K_2CO_3	Potash	Potassium carbonate
NH_4Cl	Sal ammoniac	Ammonium chloride
$NaNO_3$	Saltpeter	Sodium nitrate
$Ca(OH)_2$	Slaked lime	Calcium hydroxide
$C_{12}H_{22}O_{11}$	Sugar	Sucrose or α-D-glucopyranosyl β-D-fructofuranoside
$NaCl$	Salt	Sodium chloride
$Na_2CO_3 \cdot 10H_2O$	Washing soda	Sodium carbonate 10-water
CH_3OH	Wood alcohol	Methyl alcohol or methanol

Positive monoatomic ions are named by using the name of the element followed by the word ion (i.e., Na^+ sodium ion). Negative monoatomic ions are named by using the root of the name of the element with an -ide ending followed by the word ion (i.e., O^{2-} oxide ion). The names of some polyatomic ions are trivial, and some are named according to specific rules. A tabulation of some typical ions appears inside the front cover of the book.

Before considering the naming of ionic compounds, let us consider the formulas of some ionic compounds derived from combinations of ions. An ionic compound will contain the correct number of positive ions and negative ions so that the total positive charge equals the total negative charge. This dictates the formula used to represent ionic compounds. For instance, the formula for the compound containing sodium ion, Na^+, and fluoride ion, F^-, is

$$Na^+ \qquad F^-$$
NaF (one sodium ion for each fluoride ion)

The formula for the compound containing potassium ion, K^+, and sulfide ion, S^{2-}, is

$$K^+ \diagdown \quad S^{2-}$$
K_2S (two potassium ions for each sulfide ion)

The formula for the compound containing aluminum ion, Al^{3+}, and oxide ion, O^{2-}, is

$$Al^{3+} \diagdown \quad O^{2-}$$
Al_2O_3 (two aluminum ions to three oxide ions)

Formulas for metal-polyatomic ion compounds are deduced in the same fashion. The formula for the compound containing sodium ion and nitrate ion, NO_3^-, is

$$Na^+ \qquad NO_3^-$$
$NaNO_3$

The formula for the compound of calcium ion, Ca^{2+}, and hydroxide ion, OH^- is

$$Ca^{2+} \diagdown \quad OH^-$$
$Ca(OH)_2$ (Two hydroxide ions are needed for each calcium ion. This is denoted by enclosing the hydroxide in parentheses and using a subscript.)

The formula for the compound of magnesium, Mg^{2+}, and phosphate ion, PO_4^{3-}, is shown on the next page

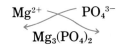

$$Mg_3(PO_4)_2$$

An ionic compound can be named by deducing which ions make up the compound and then using the name of the positive ion followed by the name of the negative ion. Some examples are:

NaF	sodium fluoride
K_2S	potassium sulfide
Al_2O_3	aluminum oxide
$NaNO_3$	sodium nitrate
$Ca(OH)_2$	calcium hydroxide
$Mg_3(PO_4)_2$	magnesium phosphate

Example 6-1   Name the compound having the formula $MgCl_2$.
The compound is made up of magnesium ions and chloride ions and is named magnesium chloride.

Example 6-2   What is the name of $Al(NO_3)_3$?
The compound consists of aluminum ions and nitrate ions. Thus, it is named aluminum nitrate.

Example 6-3   What is the name of $(NH_4)_2CO_3$?
The constituent ions are ammonium ion and carbonate ion that would result in the name ammonium carbonate.

6-3   Stock and -ous -ic Nomenclature

Compounds involving metals that have more than one possible oxidation state are named in a manner that distinguishes one oxidation state from another. For example, the name iron chloride is not meaningful, since iron has the two possible oxidation states of +2 and +3. The names of compounds of this type must indicate the oxidation state of the metal. The best method for accomplishing this is called the **Stock system** of nomenclature. This method is very similar to the one discussed above except that the name of the metal is followed by a set of parentheses containing its oxidation number expressed in Roman numerals. Thus the name of the compound consists of the name of the metal followed by the oxidation number in parentheses, which is, in turn, followed by the name of the negative ion. As an example, consider the two possible binary compounds involving iron and chlorine. Since iron occurs in either a +2 or +3 oxidation state and chlorine in a −1 state, the formulas for these two compounds would be $FeCl_2$ and $FeCl_3$. The names of these compounds according to the Stock method would be

$FeCl_2$ iron(II) chloride

and

$FeCl_3$ iron(III) chloride

These names are read iron-two-chloride and iron-three-chloride. The Roman numeral indicates the oxidation number of the iron and, since the oxidation number of chlorine would be -1, the formulas are easily deduced from the names. Keep in mind that the Roman numerals do not refer to the second element but only to the first. The formulas for iron(II) sulfide and iron(III) sulfide would be FeS and Fe_2S_3, respectively, since the oxidation number of sulfur would be -2. The Stock method gives the oxidation number of the first element and, in order to deduce the formula from the name or to obtain the name from the formula, we must know the oxidation numbers of the second element.

Example 6-4   Name the following compounds according to the Stock method of nomenclature.

(a) UCl_5 uranium(V) chloride   ⎧We know chlorine has a -1 oxidation number.
(b) UCl_4 uranium(IV) chloride   ⎨From this we obtain the oxidation number of
(c) UCl_3 uranium(III) chloride ⎩uranium.

(d) MnO_2 manganese(IV) oxide   ⎧We know oxygen has a -2 oxidation number.
(e) Mn_2O_7 manganese(VII) oxide   ⎨From this we deduce the oxidation number of
                                      ⎩manganese.

When dealing with compounds involving mercury in the $+1$ oxidation state, keep in mind that in this state mercury occurs in compounds as Hg_2^{2+}.* For example, the formula for mercury(I) chloride is Hg_2Cl_2, and the name of the compound Hg_2O is mercury(I) oxide.

A different method is occasionally used to name compounds that involve metals that can occur in at least two different oxidation states. This method uses the root of the name of the metal with an -ous ending for the lower oxidation state and an -ic ending for the higher oxidation state of the metal. For example, the names of $FeCl_2$ and $FeCl_3$ are ferrous chloride and ferric chloride, respectively. The root ferr- is derived from the Latin name for iron, ferrum. Table 6-3 lists some metals for which this ous-ic method can be used. Note that in some cases the Latin roots are used and in some cases the normal root is retained. This method is not recommended because it is necessary to know the oxidation states of the metals involved and, sometimes, the Latin names for the metals. That is, the metal name root with an -ous or an -ic ending does not convey the oxidation state of the metal, so it is necessary to know which oxidation state corresponds to each ending. The -ous-ic method should be avoided if possible but, unfortunately, it is in wide use and is found in some of the older literature.

Example 6-5   Give the formulas and names of two compounds involving iron and oxygen.

* Mercury in the $+1$ oxidation state is unique among the metals in that it occurs as an ion, Hg_2^{2+}, involving two Hg^+ ions joined by a covalent bond. The Hg^+ does not exist as a species.

Table 6-3   The ous-ic Names of Some Metals

Element	Symbol	Oxidation Number	Name
Chromium	Cr	+2	Chromous
		+3	Chromic
Manganese	Mn	+2	Manganous
		+3	Manganic
Iron	Fe	+2	Ferrous
		+3	Ferric
Cobalt	Co	+2	Cobaltous
		+3	Cobaltic
Copper	Cu	+1	Cuprous
		+2	Cupric
Mercury	Hg	+1	Mercurous
		+2	Mercuric
Tin	Sn	+2	Stannous
		+4	Stannic

The +3 oxidation state of iron forms the oxide Fe_2O_3, which is named iron(III) oxide or ferric oxide.

The +2 oxidation state of iron forms the oxide FeO, which is named iron(II) oxide or ferrous oxide.

6-4   Naming Nonmetal-Nonmetal Compounds

Since the nonmetals usually occur in several possible oxidation states, there is often more than one binary compound involving two different nonmetals in combination. For instance, the common positive oxidation numbers of phosphorus are +3 and +5. Thus, the two likely phosphorus-chlorine compounds are

$$\begin{array}{cc} +3 & -1 \\ P & Cl \end{array}$$
$$PCl_3$$

$$\begin{array}{cc} +5 & -1 \\ P & Cl \end{array}$$
$$PCl_5$$

Table 6-4   Roots of the Names of Some Common Elements

Element	Root	Element	Root
Hydrogen	hydr-	Oxygen	ox-
Boron	bor-	Sulfur	sulf- or sulfur-
Carbon	carb-	Selenium	selen-
Silicon	silic-	Tellurium	tellur-
Nitrogen	nitr-	Fluorine	fluor-
Phosphorus	phosph-	Chlorine	chlor-
Arsenic	arsen-	Bromine	brom-
Antimony	antimon-	Iodine	iod-

Binary nonmetal-nonmetal and metalloid-nonmetal compounds are named by a method that involves stating the name of the first element followed by the root of the name of the second element with an -ide ending. Table 6-4 lists the roots used for the nonmetals. Each part of the name is preceded by a Greek or Latin prefix indicating the number of combined atoms of that element in the compounds. The prefixes used are:

mono	one
di	two
tri	three
tetra	four
penta	five
hexa	six
hepta	seven
octa	eight
nona	nine
deca	ten

Which nonmetal is named first, and which is given the -ide ending? The preferred order used for naming the nonmetal-nonmetal or metalloid-nonmetal compounds is

B, Ge, Si, C, Sb, As, P, N, H, Te, Se, S, I, Br, Cl, O, and F

Notice that except for hydrogen and oxygen, this order is from bottom to top of each group and from left to right in the periodic table. Naming compounds by this method provides a way in which the formula of the compound is definitely defined by the name. For example, the compound BF_3 would be named boron trifluoride. This name indicates that the compound formula involves one boron in combination with three fluorines. Some additional examples are:

N_2O	dinitrogen oxide
N_2O_5	dinitrogen pentoxide (the "a" of penta is omitted for proper pronunciation)
P_4O_{10}	tetraphosphorus decoxide (the "a" of deca is omitted for proper pronunciation)
OF_2	oxygen difluoride (the mono prefix is usually omitted)
P_4S_7	tetraphosphorus heptasulfide
PCl_3	phosphorus trichloride
PCl_5	phosphorus pentachloride

6-5   Nomenclature of Acids

The binary compounds involving hydrogen and the VIA and VIIA elements are generally named according to the nonmetal-nonmetal nomenclature rules. For example, HCl is hydrogen chloride, and H_2S is hydrogen sulfide or dihydrogen sulfide. When the binary hydrogen compounds of these groups are dissolved in water, the resulting solutions display specific properties called acidic properties. Some common properties of acidic solutions are they taste sour, dissolve some metals, can burn the skin, and turn blue litmus paper red. (See Section

12-6.) These water solutions are called **acids** and are given specific names. This is a somewhat confusing situation since we normally give names only to pure compounds. However, since many of these acid solutions are very important in chemistry, special names are assigned to them. These names are formed by using the prefix hydro- with the root of the name of the element other than hydrogen followed by an -ic ending. This part of the name is then followed by the word acid. The general form for these names is

hydro(root)ic acid

Some examples of this general pattern are given below.

Formula of pure compound	Name of pure compound	Name of water solution
HCl	hydrogen chloride	hydrochloric acid
HI	hydrogen iodide	hydroiodic acid
H_2S	hydrogen sulfide	hydrosulfuric acid

The water solutions of acids are sometimes denoted by giving the formula of the compound followed by a parenthetical aq to indicate an aqueous or water solution (i.e., HCl(aq), H_2S(aq)).

Many of the nonmetals form ternary compounds that involve hydrogen and oxygen. Most of these compounds and the corresponding water solutions of these compounds also have acidic properties. (See Section 12-6.) These compounds are called **oxyacids.** The nomenclature of the oxyacids is somewhat confusing because the same name is usually used to refer to both the pure oxyacid and water solutions of the oxyacid. Furthermore, some of the oxyacids do not exist in the pure state but only in water solution. In such cases, the name only refers to the water solution. The nomenclature and formulas of the oxyacids can be fit to a certain pattern. The formulas of the common oxyacids of the halogens (Group VIIA) and nitrogen fit the general form

HXO_y

where X represents any of the halogens (except fluorine, which does not form any oxyacids) or nitrogen, and the y subscript is 4, 3, 2, or 1 for the halogens and 3 or 2 for nitrogen. The nomenclature rules for these oxyacids can be illustrated using the chlorine oxyacids. The chlorine oxyacids and the corresponding names are as follows.

Molecular formula	Name	Oxidation number of chlorine
$HClO_4$	*per*chlor*ic* acid	+7
$HClO_3$	chlor*ic* acid	+5
$HClO_2$	chlor*ous* acid	+3
HClO	*hypo*chlor*ous* acid	+1

Each acid name uses the root of the element name with a prefix and/or ending followed by the word acid. The name of the oxyacids involving the +7 oxidation state has the per- prefix and the -ic ending. The name of the acid that involves the +5 oxidation number has the -ic ending. The name of the oxyacid corresponding to the +3 oxidation state has the -ous ending. The name of the acid in which the element is in the +1 oxidation state has the hypo- prefix and the -ous ending.

Example 6-6   Nitrogen forms one oxyacid containing nitrogen in the $+3$ oxidation state and another containing nitrogen in the $+5$ state. Give the molecular formulas and the names of these two acids. According to the pattern the oxyacids of nitrogen should be

HNO_3 nitric acid
HNO_2 nitrous acid

The formulas of the common oxyacids of the group VIA elements fit the pattern

$$H_2EO_y$$

where E represents any of the elements in the group, except oxygen, and the y subscript can be 3 or 4. For example, sulfur forms the two oxyacids shown below.

Molecular formula	Name	Oxidation number of sulfur
H_2SO_4	sulfuric acid	$+6$
H_2SO_3	sulfurous acid	$+4$

The name of the oxyacid involving the $+6$ oxidation state has the root of the element name with an -ic ending, and the name of the oxyacid involving the $+4$ state has the root of the element name with the -ous ending. The per-ic and hyp-ous names are not needed since there are only two oxidation states.

The formulas of the common oxyacids of phosphorus and arsenic (Group VA) fit the general pattern

$$H_3EO_y$$

where E represents phosphorus or arsenic, and y is 3 or 4. For example, phosphorus forms the common oxyacids shown below.

Molecular formula	Name	Oxidation number of phosphorus
H_3PO_4	phosphoric acid	$+5$
H_3PO_3	phosphorous acid	$+3$

Example 6-7   Give the names and molecular formulas for the oxyacids of selenium (Group VIA) and arsenic (Group VA). According to the pattern these oxyacids are

H_2SeO_4 selenic acid
H_2SeO_3 selenous acid
H_3AsO_4 arsenic acid
H_3AsO_3 arsenous acid

6-6   Nomenclature of Oxyanions

Oxygen and many of the nonmetals occur in the form of polyatomic ions called **oxyanions.** In fact, the oxyanions are related to the oxyacids in the sense that they can be formed by loss of hydrogen ions (H^+) by the oxyacids. For instance, the sulfate ion, SO_4^{2-}, which is an oxyanion of sulfur, can be formed by the loss of two hydrogen ions from sulfuric acid ($H_2SO_4 \rightarrow SO_4^{2-} + 2H^+$). Table 6-5 lists some common oxyanions. Oxyanions are named in a manner related to the oxyacids. The formulas of

Table 6-5   The Common Oxyanions

Group IVA	Group VA	Group VIA	Group VIIA
CO_3^{2-} Carbonate ion	NO_3^- Nitrate ion		
	NO_2^- Nitrite ion		
	PO_4^{3-} Phosphate ion	SO_4^{2-} Sulfate ion	ClO_4^- Perchlorate ion
		SO_3^{2-} Sulfite ion	ClO_3^- Chlorate ion
			ClO_2^- Chlorite ion
			ClO^- Hypochlorite ion

these ions follow the same pattern as the formulas of the oxyacids except that hydrogen is not present, and the ions carry charges that correspond to the absence of hydrogen (one minus charge for every hydrogen that is absent). The names of the oxyanions can be deduced by changing the ending of the name of the corresponding oxyacid and adding the word ion. The -ic ending of the acid name is changed to an -ate and the -ous ending of the acid name is changed to an -ite. Thus, in order to name a given oxyanion, we merely change the ending of the name of the corresponding oxyacid. For example:

$HClO_4$	perchlor*ic* acid	ClO_4^-	perchlor*ate* ion
$HClO_3$	chlor*ic* acid	ClO_3^-	chlor*ate* ion
$HClO_2$	chlor*ous* acid	ClO_2^-	chlor*ite* ion
$HClO$	hypochlor*ous* acid	ClO^-	hypochlor*ite* ion

Example 6-8   What are the names for the two oxyanions of nitrogen: NO_3^- and NO_2^-? Considering the two related oxyacids of nitrogen, the names of the oxyanions are

HNO_3	nit*ric* acid	NO_3^-	nit*rate* ion
HNO_2	nit*rous* acid	NO_2^-	nit*rite* ion

Many of the acids and oxyacids that involve more than one hydrogen can form polyatomic ions that result from the loss of one or more hydrogens (protons, H^+) from the parent acid. The charges of these ions correspond to the absence of the hydrogens. These ions are named in the same manner as the acid anions except that the anion name is preceeded by the word hydrogen that has a prefix indicating the number of combined hydrogens that are associated with the ion. The prefix mono is usually omitted.

Example 6-9   Name the following ions.

(a) $H_2PO_4^-$	dihydrogen phosphate ion	(H_3PO_4 less one H^+)
(b) HSO_4^-	hydrogen sulfate ion	(H_2SO_4 less one H^+)
(c) HSe^-	hydrogen selenide ion	(H_2Se less one H^+)
(d) HCO_3^-	hydrogen carbonate ion	

6-7   Nomenclature of Hydrates

Water is a compound that is widely distributed in nature. It is the major component of the natural waters of the earth, and it is intimately involved in the cells and fluids of animals and plants. It is even found incorporated in the minerals and rocks of the earth. In fact, some complex compounds found in the earth actually contain water in combination with an ionic solid such compounds in association with water are called **hydrates.** They are compounds containing a specific number of moles of water per mole of the original solid (parent compound). Since they have definite compositions, hydrates are actually compounds. For example, a certain hydrate of calcium chloride contains three moles of water per mole of the parent compound. The formula of this hydrate is $CaCl_2 \cdot 3H_2O$. Normally, when we write the formula of a hydrate, the formula of the parent compound is preceded by a number indicating the number of moles of this compound associated with a mole of the hydrate. This part of the formula is followed by the formula of water, which is preceded by a number indicating the number of moles of water associated with a mole of the hydrate. The general formula of a hydrate would be

$$yA \cdot xH_2O$$

where A is the formula of the parent compound sometimes called the anhydride or anhydrous salt, y is a number indicating the number of moles of A per mole of hydrate, and x is a number indicating the number of moles of water per mole of hydrate. If x or y is absent, a 1 is understood. Notice the dot ($\cdot$) between the formula of the anhydride and the formula of water. Hydrates are named by stating the name of the anhydride preceded by a number corresponding to y above. This part of the name is followed by a number corresponding to x above that precedes the word water. Alternately, the word hydrate is used in place of the word water and, often, a Greek prefix is used in place of the number x.

Example 6-10   Name the following hydrates. (Some substances can form more than one hydrate.)

(a) $CuSO_4 \cdot 5H_2O$	copper(II) sulfate 5-water, copper(II) sulfate 5-hydrate, copper(II) sulfate pentahydrate, cupric sulfate pentahydrate
(b) $CaCl_2 \cdot 6H_2O$	calcium chloride 6-water, calcium chloride 6-hydrate, calcium chloride hexahydrate
(c) $CaCl_2 \cdot 3H_2O$	calcium chloride 3-water, calcium chloride 3-hydrate, calcium chloride trihydrate
(d) $2CdCl_2 \cdot 5H_2O$	2-cadmium chloride 5-water, 2-cadmium chloride 5-hydrate

The water involved in hydrates is called **water of hydration.** Since water of hydration is contained in hydrates as discrete water molecules, it is often possible to remove the water or **dehydrate** by heating. Heating a hydrate produces water and the **anhydrous** (without water) solid compound:

$$CaCl_2 \cdot 2H_2O \xrightarrow{\text{heat}} CaCl_2 + 2H_2O$$
$$\text{anhydrous calcium chloride}$$

The **percentage of water by mass** in a hydrate can often be found by dehydrating a known amount of the hydrate and then determining the mass of the resulting anhydrous solid. The mass of water is found by subtracting the mass of the anhydrous solid from the mass of the original sample of hydrate:

$$\text{mass } H_2O = \text{mass hydrate} - \text{mass anhydrous form}$$

Then the percentage of water can be found by dividing the mass of water by the mass of the original sample of hydrate and multiplying by 100:

$$\text{percent } H_2O = \left(\frac{\text{mass } H_2O}{\text{mass hydrate}}\right) 100$$

Some anhydrous solids and other substances can combine with water in the atmosphere. When these substances are exposed to the atmosphere, they absorb water. A substance displaying such a property is said to be **hygroscopic.** Some substances are so hygroscopic they absorb enough water from the atmosphere to form a solution. Such substances are said to be **deliquescent.** For example, a pellet of solid sodium hydroxide, when exposed to the air, will soon absorb enough water to form a solution. Such deliquescent compounds should not be unnecessarily exposed to the atmosphere. Some hygroscopic substances, such as calcium chloride and silica gel, are used as drying agents to absorb the moisture in a confined area. Such substances are called **desiccants.** For example, anhydrous calcium chloride can be used as a desiccant in a container called a desiccator used to keep other chemicals in a dry atmosphere. A few hydrates actually spontaneously lose water of hydration when exposed to the atmosphere. Such a process is known as **efflorescence.** For instance, a sample of $Na_2SO_4 \cdot 10 H_2O$ exposed to the atmosphere will soon lose the water of hydration.

6-8   Names and Formulas

We have considered the nomenclature of several common kinds of compounds. There are other kinds of compounds that are named by specific nomenclature rules, but the nomenclature covered in this chapter applies to those compounds that are important to our discussion of chemistry.

To name a compound given the formula, the compound should be classified into one of the following groups:

metal-nonmetal
metal-polyatomic ion
nonmetal-nonmetal
acid or oxyacid
hydrate

Once classified, the compound can be named using the appropriate nomenclature method. For instance, the compound CaS is classified as a metal-nonmetal compound and, this, is an ionic compound named

calcium sulfide

The compound $Zn(OH)_2$ is classified as a metal-polyatomic ion compound and, thus, is named

zinc hydroxide

The compound Cl_2O_7 is classified as a nonmetal-nonmetal compound and, thus, is named

dichlorine heptoxide

Example 6-11   Classify and name the following compounds:

H_2SO_3	oxyacid	sulfurous acid
$Ca(NO_3)_2$	metal-polyatomic ion	calcium nitrate
K_2O	metal-nonmetal	potassium oxide
PBr_3	nonmetal-nonmetal	phosphorus tribromide
$MgSO_4 \cdot 7H_2O$	hydrate	magnesium sulfate heptahydrate

It is important to consider how a formula can be read when it is encountered in readings. Formulas can be read in two ways. First, it is possible to read a formula by just reading the element symbols as letters and the subscripts as numbers. For example, consider the pronouncing of the following formulas:

H_2SO_3	H-two-S-O-three
$Ca(NO_3)_2$	C-A-N-O-three-taken twice
K_2O	K-two-O
PBr_3	P-B-R-three
$MgSO_4 \cdot 7H_2O$	M-G-S-O-four-dot-seven-H-two-O

The second approach to reading formulas is to translate the formula to the name as it is read. Admittedly, this approach requires some experience on the part of the reader.

Another aspect of nomenclature involves the translation of a name to a formula. To accomplish this, the names and formulas of ions must be known, and the formulas of oxyacids must be known. The name of a compound usually gives the components that must be put together to give the formula. For instance, what is the formula of magnesium bromide? The compound consists of magnesium ion, Mg^{2+}, and bromide ion, Br^-. Thus, the formula is shown on the next page

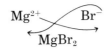

What is the formula of iron(III) sulfate? The compound consists of the iron(III) ion and the sulfate ion, so the formula is

What is the formula of sulfur hexafluoride? Binary nonmetal compound formulas are easily written since the name gives the elements and the subscripts. The formula is SF_6.

What is the formula of nitric acid? This is one of the oxyacids of nitrogen having the nitrogen in the +5 oxidation state. The formula is HNO_3. What is the formula of calcium sulfate dihydrate? This hydrate consists of calcium ion, Ca^{2+}, sulfate ion, SO_4^{2-}, and two waters. Thus, the formula is

$$CaSO_4 \cdot 2H_2O$$

Thus, the formula is $Fe_2(SO_4)_3$.

Questions and Problems

Use a periodic table as an aid in formula writing.

1. Classify each of the following compounds as metal-nonmetal, metal-polyatomic ion, nonmetal-nonmetal, or acid and name each by the appropriate nomenclature method. For example:

   CO_2 nonmetal-nonmetal   carbon dioxide

   (a) Cl_2O
   (b) $AgNO_3$
   (c) Al_2O_3
   (d) $Ba(OH)_2$
   (e) $Co(NO_3)_2$
   (f) $SnCl_2$
   (g) $K_2Cr_2O_7$
   (h) Be_3N_2
   (i) H_3PO_4
   (j) $CuSO_4$
   (k) $HgCl_2$
   (l) $CrBr_3$

(m) $ZnSO_3$
(n) $Ca(CN)_2$
(o) $NaIO_3$
(p) PCl_5
(q) $HF(aq)$
(r) SnF_4
(s) H_2SO_4
(t) N_2O_5
(u) $BaCO_3$
(v) SF_6
(w) HNO_3
(x) Fe_2S_3
(y) MnS
(z) I_2O_7
(aa) CCl_4
(bb) H_2S
(cc) PtP_2
(dd) PCl_3
(ee) $CdSe$
(ff) $GeCl_4$

2. Give the names of the following ions:

(a) Al^{3+}
(b) Br^-
(c) CO_3^{2-}
(d) Ag^+
(e) S^{2-}
(f) Cl^-
(g) CrO_4^{2-}
(h) Cu^+
(i) Ni^{2+}
(j) PO_4^{3-}

(k) $Cr_2O_7^{2-}$
(l) Fe^{2+}
(m) HSO_4^-
(n) MnO_4^-
(o) HCO_3^-
(p) OH^-
(q) Mg^{2+}
(r) NH_4^+
(s) Ca^{2+}
(t) K^+

3. Give the formulas for the following ions:

(a) hydrogen sulfide ion
(b) copper (II) ion
(c) chlorate ion
(d) fluoride ion
(e) nitrate ion
(f) sulfite ion

(g) iron (III) ion
(h) acetate ion
(i) oxalate ion
(j) cyanide ion
(k) chlorite ion
(l) chromium(III) ion

4. Give the formulas for the following compounds:

(a) sodium nitrate
(b) tin(II) acetate
(c) manganese(II) sulfate
(d) ammonium acetate
(e) chlorine trifluoride
(f) diarsenic pentasulfide
(g) nitrogen dioxide
(h) osmium(VI) fluoride
(i) iron (II) nitrate
(j) magnesium carbonate

(k) chloric acid
(l) copper (I) oxide
(m) silver phosphate
(n) strontium cyanide
(o) nitric acid
(p) ammonium sulfide
(q) phosphoric acid
(r) radium iodate
(s) zinc chromate
(t) mercury (II) iodide

(u) potassium sulfite
(v) sodium dichromate
(w) carbon disulfide

(x) diarsenic pentasulfide
(y) magnesium chloride
(z) hydrogen fluoride

5. Classify each compound in problem 4 as metal-nonmetal, metal-polyatomic ion, nonmetal-nonmetal, or acid.

6. Classify each of the following compounds and name each using the appropriate nomenclature method:

(a) CCl_4
(b) $FeSO_4$
(c) MgO
(d) Hg_2Br_2
(e) KI
(f) Li_3As
(g) $NaC_2H_3O_2$
(h) $KMnO_4$
(i) $Ni(NO_3)_2$
(j) $KHCO_3$
(k) $CrCl_3$
(l) K_2O

(m) $BaSO_3$
(n) $AgIO_3$
(o) $SnBr_2$
(p) $Zn(OH)_2$
(q) $NaBr$
(r) PuF_6
(s) $AgCl$
(t) NaH_2PO_4
(u) $LiHSO_4$
(v) Cl_2O_7
(w) K_2HPO_4
(x) $K_2Ca(SO_4)_2$

7. Give the names of the missing oxyacids and oxyanions:

(a) $HClO_4$ perchloric acid ClO_4^- _____

(b) HNO_3 _____ NO_3^- nitrate ion

(c) H_2SO_4 sulfuric acid HSO_4^- _____

(d) H_3PO_4 _____ PO_4^{3-} phosphate ion

(e) $HClO$ hypochlorous acid ClO^- _____

(f) $HBrO_3$ _____ BrO_3^- bromate ion

8. Give the chemical name for each of the following compounds that have the trivial names given:
(a) $NaHCO_3$, baking soda
(b) $CaCO_3$, marble
(c) $MgSO_4 \cdot 7H_2O$, epsom salt
(d) N_2O, laughing gas
(e) CaO, lime
(f) $NaOH$, lye
(g) K_2CO_3, potash
(h) $NaNO_3$, saltpeter
(i) $Ca(OH)_2$, slaked lime
(j) $Na_2CO_3 \cdot 10H_2O$, washing soda

9. List the names and formulas of three compounds that have common names that do not describe the chemical composition of the compound.

10. What are hydrates?

11. Name or give the formula for each of the following hydrates:
(a) $AlCl_3 \cdot 6H_2O$

(b) $CaCrO_4 \cdot 2H_2O$
(c) $Ba(ClO)_2 \cdot 2H_2O$
(d) $CdSO_4 \cdot 7H_2O$
(e) $NaBr \cdot 2H_2O$
(f) $Sn(NO_3)_2 \cdot 20H_2O$
(g) $ZnF_2 \cdot 4H_2O$
(h) $NaC_2H_3O_2 \cdot 3H_2O$
(i) ammonium sulfite 1-water
(j) sodium iodide dihydrate
(k) calcium chloride hexahydrate
(l) iron(III) phosphate 8-water
(m) zinc sulfate heptahydrate
(n) nickel(II) acetate 4-water
(o) potassium sulfide pentahydrate
(p) 2-cadmium chloride 5-water

12. Give the correct formulas for the compounds formed by combinations of the positive and negative ions given in the following table (NaCl is given as an example).

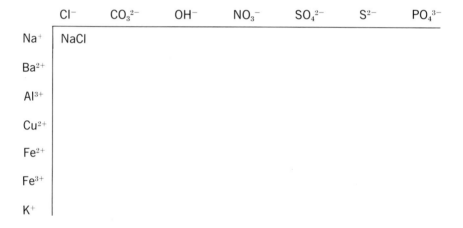

	Cl^-	CO_3^{2-}	OH^-	NO_3^-	SO_4^{2-}	S^{2-}	PO_4^{3-}
Na^+	NaCl						
Ba^{2+}							
Al^{3+}							
Cu^{2+}							
Fe^{2+}							
Fe^{3+}							
K^+							

13. A 2.82-g sample of a hydrate is heated to drive off the water. If the mass after heating is 2.13 g, what is the percent by mass of water in the hydrate?

14. A 24.65-g sample of a hydrate of magnesium sulfate, $MgSO_4$, is heated to drive off the water. If the mass of the $MgSO_4$ remaining after heating is 12.04 g, determine the formula of the hydrate. (Hint: determine the ratio of the number of moles of water to the number of moles of $MgSO_4$.)

15. A 6.95-g sample of a hydrate of copper(II) sulfate, $CuSO_4$, is heated to drive off the water. If the mass of the $CuSO_4$ remaining after heating is 4.45 g, determine the formula of the hydrate. (Hint: determine the ratio of the number of moles of water to the number of moles of $CuSO_4$.)

7

Chapter 7: Gases

Objectives

The student should be able to:

1. Describe a barometer and indicate how it is used.
2. Describe a gas using the kinetic molecular theory or state the postulates of the theory.
3. Describe how a gas exerts a pressure.
4. State the perfect gas law in an algebraic form and define the terms used in the statement.
5. Use the perfect gas law to determine the number of moles in a sample of gas of known temperature, pressure, and volume.
6. Use the basic gas laws to deduce the new condition of a gas corresponding to a given change in conditions.
7. Use the combined gas laws to deduce the new volume of a sample of a gas when both the pressure and temperature are changed.
8. Calculate the molar mass of a gaseous substance from experimental data.

Terms to Know

Boyle's law
Charles' law
Perfect gas law
Torr
Atmosphere
Universal gas Constant, R
Standard temperature and pressure
Molar volume
Dalton's law of partial pressures
Relative humidity
Air shed
Primary air pollutants
Smog
Temperature inversion

7

Gases

7-1 Gases

When you feel the wind blowing in your face or smell the fragrance of perfume from across a room you are experiencing matter in the **gaseous state. Gases** have interested and stimulated the imaginations of scientists for centuries. The fascination of this state is that we can experience it without seeing it. In fact, investigations of gases were fundamental to the development of the atomic theory. As will be discussed in this chapter, the constituent particles (molecules or atoms) of gases are darting about at high speeds, and it is this motion that contributes to the interesting properties of gases.

As we look around us the three states of matter are very apparent. Solid objects around us are firm and occupy definite regions of space. Liquids around us are more mobile. Liquids flow and have forms that depend on the shape of the container; a bottle, or a river bottom. Without a container, liquids evaporate or trickle away. The gaseous form of matter is the most elusive form of matter. We do not see gases, but we experience them in other ways. The wind blowing against us or the firmness of a filled balloon suggest the existence of matter in the form of gases. To trap a gas we must keep it in a closed container. If we open the container the gas quickly escapes into the immediate atmosphere.

In previous chapters we established the idea that matter is made up of atoms or combined atoms in the form of ions or molecules. A solid substance consists of a vast collection of atoms, molecules, or ions. A nail is made up of numerous iron atoms. Ice consists of an aggregation of water molecules. Table salt is made up of visible crystals each of which is an aggregation of sodium ions and chloride ions. Pure liquids are collections of molecules. Liquid water consists of a vast collection of water molecules. Similarly, common gases are collections of molecules.

The description of matter as collections of tiny particles allows us to imagine the submicroscopic makeup of matter but does not provide any explanation of the three possible states of matter. Matter displays various properties and behavior depending on whether it is in the solid state, liquid state, or gaseous state. In fact, it was the observation and description of these properties and behavior that provided evidence of the atomic theory and the particulate nature of matter. However, in addition to providing evidence for the particulate nature of matter, these observations established the view that the particles of matter are in motion. That is, these particles are not frozen in static positions in space but instead are capable of moving about in space. To have a satisfactory description of matter the dynamics of the particles must be included. Such a model or description of gases has been established. Before we describe this model let us consider the behavior and properties of gases.

The gaseous state is a very common state of matter. We are quite familiar with gases, especially the mixture of gases known as air. Substances in the gaseous state are mobile and compressible. A gas will expand to occupy any container into which it is placed. Consider the behavior of a gas sample in a balloon. The balloon serves as a container for the gas. If the balloon is opened the gas will quickly diffuse and dissipate into the surrounding air. If a tiny hole is in the balloon the gas will slowly leak out. More gas can be added to the balloon with an air pump, a compressor, or by just blowing by the mouth. Different gases can mix with one another in any proportions to form gaseous mixtures. When we add more gas to the balloon it will increase in volume until it pops. If we tie off the balloon we will have a sample of a gas having a certain volume. If we warm the balloon the volume increases, and if we cool the balloon the volume decreases. If we squeeze the balloon (increase the pressure) the volume decreases; the gas is compressed. When we stop squeezing (decrease the pressure and release our grip) the volume increases. Many important elements and compounds exist in the gaseous state under normal conditions of room temperature and atmospheric pressure. Furthermore, many substances can be converted to the gaseous state merely by heating them.

If we capture a gas in a container and keep the container closed, we can maintain the sample of the gas indefinitely. To describe such a gas sample we can state the **volume** of the container and the **temperature.** Moreover, the sample is said to exert a specific pressure on the container. When a pressure measuring device is attached to the container, the device will register a specific pressure. We shall find out shortly why gases exert pressures, but first let us consider what a pressure is. The pressure exerted on an object by a gas is the force exerted on a unit area of the object. That is, **pressure** is defined as force per unit area and can be expressed in pressure units as discussed in the next section. To illustrate the idea of pressure place the tip of a pen or pencil on a magazine or notebook and push down gently. Note the slight indentation in the paper resulting from your push (force) exerted over the area of the tip. Now lift the pen or pencil above the magazine or notebook and jab it into the paper with more force. The greater pressure (force per unit area) will result in a larger indentation in the paper. As we shall see, the pressure exerted on objects by a gas results from the force of the constituent particles of the gas constantly colliding with and, thus, "pushing" on the objects.

7-2   Measurement of Gas Pressures

The earth is surrounded by a layer of air contained about the earth by gravitational forces. This air layer is more dense near the lower surfaces of the earth and becomes less dense at distances removed from the earth's surface. The molecules and atoms that are present in the atmosphere exert pressures on all objects exposed to the atmosphere. This pressure is called **air pressure** or **atmospheric pressure,** and it arises from the continuous bombardment of objects by molecules and atoms in the air. Atmospheric pressure is often measured with a device known as a **barometer,** invented by E. Torricelli in the seventeenth century. A barometer can be constructed by filling a glass tube, which is sealed on one end, with liquid mercury. This tube is then inverted and supported in a container of mercury that is open to the atmosphere. See Figure 7-1. The pressure exerted on the surface of the mercury in the container by the atmosphere will support a column of mercury in the tube. The height of the mercury column is directly proportional to the atmospheric pressure. In fact, as the atmospheric pressure changes as a result of temperature and weather changes, the height of the mercury column will fluctuate accordingly. The height of the mercury column supported by any gas is proportional to the pressure of the gas. Consequently, gas pressures are often expressed in terms of the number of **millimeters of mercury (mmHg)** the gas will support. A special unit used to express pressure is the **torr.** A torr (named after Torricelli) is defined as

$$1 \text{ torr} = 1 \text{ mmHg}$$

A gas pressure can be expressed in terms of millimeters of mercury or number of torr. Atmospheric pressures given on weather reports are usually stated in inches of mercury (1 in. = 25.4 mm).

Atmospheric pressures generally decrease with altitude as the air becomes less dense. Atmospheric pressure around sea level is in the

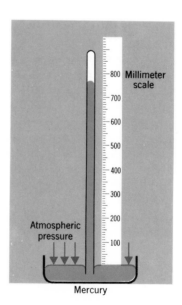

Figure 7-1 A mercury barometer.

Table 7-1   Various Pressure Units

Unit	Abbreviation
Millimeter of mercury	mmHg
Inch of mercury	in. Hg
Torr	torr
Atmosphere	atm
Pounds per square inch	psi
Dynes per square centimeter	dynes/cm^2

Equivalents

1 mmHg = 1 torr
25.4 mmHg = 1 in. Hg
1 atm = 760 torr
1 atm = 14.7 psi
1 atm = 29.9 in. Hg
1 atm = 1.01×10^6 dyne/cm^2
1 atm = 1,033 cm H_2O = 34 ft H_2O

From *Understanding Chemistry: From Atoms to Attitudes* by T. R. Dickson, Wiley, N.Y., 1974.

range of 760 torr. Another unit of pressure called the **atmosphere** is defined as

$$1 \text{ atmosphere (atm)} = 760 \text{ torr}$$

Note that 1 atm is the typical pressure of the atmosphere at sea level. This unit is often used to express high pressures used in industrial and experimental work. For instance, synthetic diamonds can be made by subjecting graphite (a common form of carbon) to high temperatures (2000°C) and pressures around 70,000 atm.

Pressures of contained gas samples are often measured by attaching special gas pressure gauges to the container. Usually these gauges are calibrated so that pressures can be read in units of pounds per square inch (psi) or dynes per square centimeters (dynes/cm^2). These units express pressures in terms of force per unit area that, as you will recall, is how a pressure is defined. The relation between these units and the atmosphere unit is

$$1 \text{ atm} = 14.7 \text{ psi} = 1.0 \times 10^6 \text{ dynes/cm}^2$$

As this shows, typical atmospheric pressure at sea level is 14.7 pounds per square inch. A comparison of the various pressure units and conversion factors that relate them are given in Table 7-1.

Example 7-1   Express the pressure 780 torr in units of atmospheres.
The conversion is accomplished by multiplying the pressure by the conversion factor relating torr to atmospheres.

$$780 \text{ torr} \left(\frac{1 \text{ atm}}{760 \text{ torr}} \right) = 1.03 \text{ atm}$$

7-3   Gas Laws

The systematic study of the behavior of gases has been of interest to scientists for centuries. Many scientists have contributed to our understanding of gases. However, the names of a few scientists are prominent since they stated certain relations concerning the behavior of gases.

In 1662 Robert Boyle, an English scientist, experimented with the compressibility of air. Boyle, using an apparatus similar to that shown in Figure 7-2, observed the relation between the volume and pressure of a sample of gas. He found that if a sample of a gas is kept at a constant temperature then the gas can be compressed by increasing the pressure on the gas. He also noted that when the pressure is released the volume of the gas sample increases. As illustrated in Figure 7-2 Boyle observed that if the pressure on the gas was doubled the gas would compress to one-half the original volume and if the pressure was tripled the volume dropped to one-third of the original volume. Similarly, he found that if the pressure was cut down by one-half the gas would expand to double the volume. To explain the compressibility of air, Boyle suggested that air was made up of particles having weight and a springlike nature. He further suggested that pressure of gases could result from the motion of these particles.

Today after numerous conformations of Boyle's work and additional

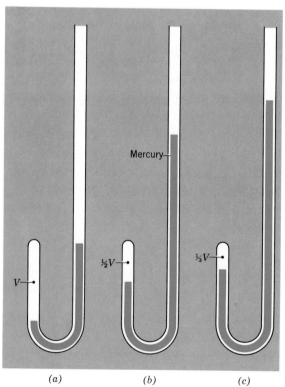

(a)   (b)   (c)

Figure 7-2   Apparatus illustrating Boyle's law. From *Understanding Chemistry: From Atoms to Attitudes* by T. R. Dickson, Wiley, N.Y., 1974.

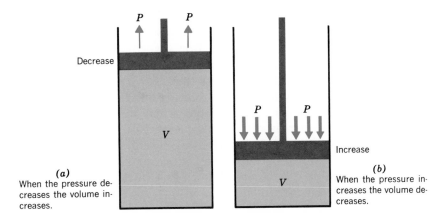

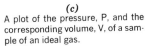

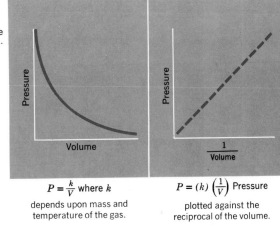

(a)
When the pressure decreases the volume increases.

(b)
When the pressure increases the volume decreases.

(c)
A plot of the pressure, P, and the corresponding volume, V, of a sample of an ideal gas.

$P = \frac{k}{V}$ where k depends upon mass and temperature of the gas.

$P = (k)\left(\frac{1}{V}\right)$ Pressure plotted against the reciprocal of the volume.

Figure 7-3   Boyle's law.

experimentation, a generalization concerning the relation between the volume and the pressure of a gas has become known as **Boyle's law.** A statement of Boyle's law is

The volume of a sample of a gas kept at a constant temperature is inversely proportional to the pressure.

As illustrated in Figure 7-3 this means that if the pressure of such a gas sample increases the volume will decrease proportionately, and if the pressure decreases the volume will increase. If two quantities are inversely proportional, when one increases the other decreases. Consider the filled balloon. When the balloon is squeezed (pressure increased) the volume decreases. When the balloon is released (pressure decreased) the volume increases. A very widely used method to show the relation between two quantities is to construct a graph showing how the two quantities vary with respect to one another. Such a graph illustrating Boyle's law is shown in Figure 7-3.

In 1787, Jacques Charles first investigated the relation between the volume and temperature of a sample of a gas at a constant pressure. He used an apparatus similar to that shown in Figure 7-4. By keeping the pressure constant, the volume is affected only by the temperature change. Charles observed that when the temperature of such a gas sample is increased, the volume increases, and when the temperature decreases the volume decreases. Specifically, he noted a direct linear proportionality between the volume and temperature. Today this proportionality has become known as **Charles' law** and can be stated as follows. (See Figure 7-5.)

> The volume of a sample of a gas at a constant pressure is directly proportional to the Kelvin temperature.

A direct proportion exists between two quantities when as one is increased the other increases, and as one is decreased the other decreases. Charles' law can be remembered if we consider that a filled balloon will increase in volume when warmed and decrease in volume when cooled. A graphical illustration of Charles' law is shown in Figure 7-5.

An additional relation involving gases is that between pressure and temperature of a sample of a gas of constant volume. In such a case, it is observed that when the temperature of the gas sample is increased the pressure increases, and if the temperature is decreased the pressure decreases. This relation can be stated as given on the next page.

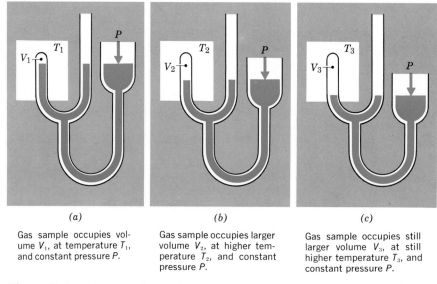

(a)   (b)   (c)

Gas sample occupies volume V_1, at temperature T_1, and constant pressure P.

Gas sample occupies larger volume V_2, at higher temperature T_2, and constant pressure P.

Gas sample occupies still larger volume V_3, at still higher temperature T_3, and constant pressure P.

Figure 7-4   Apparatus illustrating Charles' law. From *Understanding Chemistry: From Atoms to Attitudes* by T. R. Dickson, Wiley, N.Y., 1974.

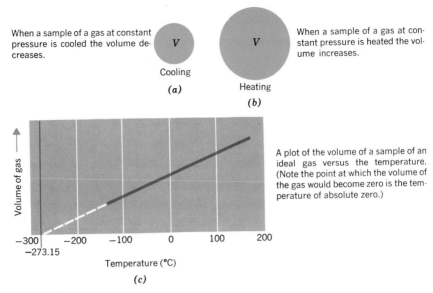

When a sample of a gas at constant pressure is cooled the volume decreases.

V

Cooling

(a)

V

Heating

(b)

When a sample of a gas at constant pressure is heated the volume increases.

A plot of the volume of a sample of an ideal gas versus the temperature. (Note the point at which the volume of the gas would become zero is the temperature of absolute zero.)

Volume of gas

−300
−273.15

−200

−100

0

100

200

Temperature (°C)

(c)

Figure 7-5   Charles' law.

The pressure of a sample of a gas of fixed volume is directly proportional to the Kelvin temperature.

This relation is illustrated in Figure 7-6. As a practical example of this relation consider what happens when a gas-filled aerosol can is heated in a fire. The volume of the can is fixed. Thus, as the tempera-

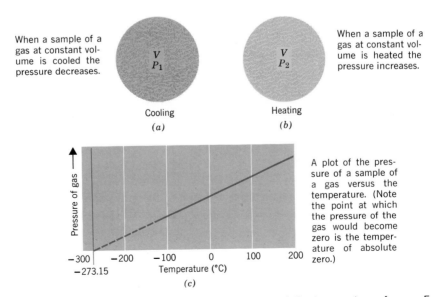

When a sample of a gas at constant volume is cooled the pressure decreases.

V
P_1

Cooling

(a)

V
P_2

Heating

(b)

When a sample of a gas at constant volume is heated the pressure increases.

A plot of the pressure of a sample of a gas versus the temperature. (Note the point at which the pressure of the gas would become zero is the temperature of absolute zero.)

Pressure of gas

−300
−273.15

−200

−100

0

100

200

Temperature (°C)

(c)

Figure 7-6   The relation between the pressure and the temperature of a gas. From *Understanding Chemistry: From Atoms to Attitudes* by T. R. Dickson, Wiley, N.Y., 1974.

ture increases the pressure of the gas increases. When the temperature is high enough, the pressure of the gas is so great that the can ruptures and an explosion results. This is the reason why such cans should never be put into an open fire. Incidentally, numerous people have been injured or killed by exploding aerosol cans. Some consumer groups claim that aerosol can should be equipped with inexpensive pressure relief valves to prevent accidents.

7-4   Kinetic Molecular Theory of Gases

We can observe that gases occupy any volume into which they are placed, are easily compressed, exert pressures, and have measurable temperatures. These observations and the various gas laws can be explained by establishing a general theory of the behavior of gases. The model that is used to explain the dynamic behavior of gases is called the **kinetic molecular theory (KMT).** This theory is based on the idea that gases are composed of particles (atoms or molecules) which are in a state of continuous motion. The kinetic molecular theory can be expressed in terms of the following basic postulates that are illustrated in Figure 7-7.

1. Gases consist of particles (molecules) that are so small and the average distance between them so great that the actual volume occupied by the particles is negligible compared to the empty space between them.

2. There are no attractive or repulsive forces between the particles comprising a gas, and they are considered to behave like very small masses.

3. The particles are in rapid, random, continuous motion and are constantly colliding with one another and with any object in their environment, such as the walls of the container. As a result of this motion, the particles possess **kinetic energy,** KE (KE $= 1/2mv^2$, where m is the mass of a particle and v is the velocity or speed).

4. The collisions result in no net loss in the total kinetic energy of the particles and are said to be perfectly elastic.

These postulates provide a good model of a gas. We can now view a gas as a collection of molecules which are rapidly moving about, constantly colliding with one another and with any object in the vicinity of the particles. Even though you do not feel it, you are constantly being bombarded by the molecules of the air, as are all of the objects within the atmosphere. Gases are quite dynamic, and this will dictate the behavior of gases. A gas will occupy any volume in which it is placed because the particles are in rapid motion and will travel to all parts of the container even if other gas particles are present in the container. As an example of the rapid motion of gas molecules, the average velocity of a hydrogen molecule at room temperature is about 2×10^5 cm/sec or 3600 mi/hr. A sample of a gas can be easily compressed because the sample consists mainly of empty space. The pressure of a gas results

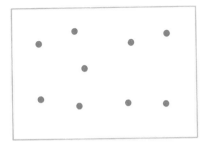

Gases consist of molecules (or atoms) that are so small and so far apart that the volume occupied by the particles is negligible compared to the empty space between them.

There are no appreciable attractive forces between the particles and they behave as tiny masses.

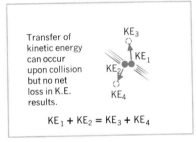

The particles are in rapid, random straight-line motion and are constantly colliding with one another and with the walls of the container. As a result of this motion the particles possess kinetic energy ($KE = 1/2mv^2$ where m is the mass and v is the speed.)

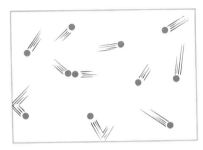

The collisions result in no net loss in kinetic energy of the particles.

Figure 7-7   The postulates of the kinetic molecular theory.

from the collisions of gas particles with the objects with which the gas is in contact. The constant collisions of the particles in a sample of a gas with the walls of a container will result in a certain force per unit area (or pressure) being exerted on the surface of the container. The pressure will be directly related to the rate at which the collisions are occurring and the kinetic energies of the particles. When a thermometer is placed in a sample of gas, the gas particles exchange kinetic energy with the thermometer causing it to register a specific temperature. If the gas sample is heated, increasing the average kinetic energy of the particles, the temperature increases. If the sample is cooled, decreasing the average kinetic energy of the particles, the temperature decreases. In other words, the measurable temperature of a gas is a direct result of the kinetic energy of the particles. In fact, the average kinetic energy of the particles is directly proportional to the temperature.

When a gas is heated, and the speeds of the molecules are increased, the pressure must increase if the volume of the sample is constant. That is, the heating increases the number of collisions of gas molecules with the container and, thus, the pressure increases. If a sample of a gas is heated in a flexible container the increase in the number of collisions causes the volume of the container to increase.

7-5   The Perfect Gas Law

The postulates of the KMT describe an idealized gas that is referred to as a perfect gas. Many real gases can be considered to behave as perfect gases under certain conditions. For our purposes, we can assume that real gases can be described by the postulates of the KMT. The properties of pressure, temperature, volume, and number of moles of a sample of gas are related to the dynamic nature of the gas. Actually, these properties are not independent of one another and can be related. We can use the KMT to develop the relation that exists between the pressure, P, the volume, V, the temperature, T, and the number of moles, n, associated with a sample of a gas. The kinetic energies of the particles of a gas are proportional to the Kelvin temperature ($KE\alpha T$). Now, if a sample of a gas consists of a definite number of moles, n, of gas, since each particle possesses energy, we can say that the energy of the sample is proportional to the product of the number of moles and the Kelvin temperature. (The symbol $\propto$ means "is proportional to".)

$$\text{Energy} \alpha nT$$

If we can determine how the pressure and volume of the sample are related to energy, we can see how the pressure and volume are related to the product nT. The simplest way to find this relationship is to consider the dimensions of pressure and volume. Pressure is force per area, and area has dimensions of length squared (cm², m²). So, the dimensions can be expressed as

$$P = \left(\frac{\text{force}}{\text{length}^2}\right)$$

Volume has dimensions of length cubed (cm³, m³). This can be expressed as

$$V = \text{length}^3$$

Physicists tell us that energy (work) has the dimensions of force times length. Thus, the product of pressure and volume will have the dimensions of energy:

$$PV = \left(\frac{\text{force}}{\text{length}^2}\right)(\text{length}^3) = (\text{force})(\text{length})$$

Consequently, the product PV of a given sample of a gas is proportional to the energy of that sample. Thus, we can state that the product PV is proportional to the product nT, since both products are proportional to the energy

$$PV \alpha nT$$

In order to state an algebraic relation between the two products, all we have to do is to determine the proportionality factor. This factor has been determined experimentally and is called the **universal gas constant;** it is given the symbol **R**. Therefore, the above proportionality can be expressed as

$$PV = nRT \text{ (read "pee-vee equals en-R-tee")}$$

This relation indicates that the product of the pressure and volume of a gas equals the product of the temperature and the number of moles of the gas multiplied by the proportionality constant R. This algebraic relation is very important and is referred to as the **perfect gas law** or the **ideal gas law.** Sometimes the relation is called the equation of state of a perfect or ideal gas. The expression is quite useful because it indicates the relation between the pressure, volume, temperature, and number of moles associated with a sample of a gas. In fact, since R is a known constant, if any three of the properties associated with a gas sample are known, the fourth can easily be determined from the perfect gas equation. The value and units of the gas constant R depend on which units are used to measure the pressure and volume. However, R has the same value for all ideal or perfect gases. The temperature used in the perfect gas law is always the Kelvin temperature. The volume of a sample of a gas is usually expressed in units of liters or, sometimes, milliliters. Its pressure is usually expressed in units of atmospheres or torr.

The units of the gas constant, R, depend on the units used in expressing the other factors involved in the ideal gas law.

Example 7-2   What are the units of R when pressure is measured in atmospheres, volume in liters, and of course, temperature in K?
The n is always expressed as moles.
Solving the ideal gas law, $PV = nRT$, for R gives

$$R = \frac{VP}{Tn}$$

Substituting in the units of V, P, T, and n, the units of R become

$$\text{units of } R = \frac{\ell \text{ atm}}{K \text{ mole}} \text{ (liter atmospheres per Kelvin mole)}$$

The actual value of R can be calculated from the measurements of the pressure, volume, and temperature of a known number of moles of gas.

Example 7-3   A sample of a gas consisting of 1.523 moles of gas is found to have a pressure of 1.500 atm, a volume of 25.0 ℓ, and a temperature of 300 K. Using these data, calculate the value of R in units of (ℓ atm)/(K mole).

$$R = \frac{VP}{Tn} = \frac{(25.0 \ \ell)(1.500 \text{ atm})}{(300 \text{ K})(1.523 \text{ moles})} = \left(\frac{0.0821 \ \ell \text{ atm}}{K \text{ mole}} \right)$$

Example 7-4   How many moles of gas are contained in a 0.500 ℓ sample of a gas that has a pressure of 1.237 atm and a temperature of 27.0°C?
Since the ideal gas law is $PV = nRT$, when we solve for n in terms of other factors, we obtain

$$n = \frac{PV}{RT}$$

The number of moles can be found from this equation by substituting the values of P, V, T, and R. First, however, the temperature should be expressed in units of K (27°C corresponds to 300 K, $273 + 27$). Using a P of 1.237 atm, a V of 0.500 ℓ, a T of 300 K, and an R of 0.0821 ℓ atm/K mole, the solution is

$$n = (0.500 \ \ell) \left(\frac{1.237 \text{ atm}}{300 \text{ K}}\right) \left(\frac{1}{0.0821 \ \ell \text{ atm/K mole}}\right)$$

$$n = (0.500 \ \ell) \left(\frac{1.237 \text{ atm}}{300 \text{ K}}\right) \left(\frac{\text{K mole}}{0.0821 \ \ell \text{ atm}}\right) = 0.0251 \text{ mole}$$

7-6   STP and the Molar Volume

A specific temperature and pressure have been established as standard reference conditions for gases. These standard conditions are

$$0°C \quad \text{or} \quad 273 \text{ K}$$
and
$$760 \text{ torr} \quad \text{or} \quad 1.000 \text{ atm}$$

These values are arbitrarily chosen as convenient typical conditions of gases. They are called the standard temperature and pressure, which is abbreviated STP. What volume does a gas occupy at STP? Actually, this is not a meaningful question because, according to the ideal gas law, the volume of a gas sample at a given temperature and pressure cannot be calculated unless the number of moles of gas is known. A more meaningful question is: What volume does 1 mole of a gas occupy at STP? This can easily be calculated by substituting a P of 1.000 atm, a T of 273 K, an n of 1.000 mole, and an R into the relation $V = nRT/P$. This gives

$$V = \left(\frac{0.0821 \ \ell \text{ atm}}{\text{K mole}}\right) \left(\frac{1 \text{ mole}}{1 \text{ atm}}\right) (273 \text{ K}) = 22.4 \ \ell$$

Thus, 1 mole of an ideal gas will occupy a volume of 22.4 ℓ at STP. This fact can be expressed in terms of the factor

$$\left(\frac{22.4 \ \ell}{1 \text{ mole}}\right)_{\text{STP}}$$

The subscript denotes the conditions at which this relation is useful. This factor is called the molar volume of a gas and is valid at STP. The value 22.4 applies only to gases and not to liquids or solids. The molar volume can be used as a factor to convert the number of moles of a gas to volume at STP or vice versa.

Example 7-5   What is the volume of 2.37 moles of a gas at STP?
Multiplying the number of moles by the molar volume factor gives

$$2.37 \text{ moles} \left(\frac{22.4 \ \ell}{1 \text{ mole}}\right)_{\text{STP}} = 53.1 \ \ell$$

7-7   Gas Law Calculations

A sample of a gas can be described by stating the volume, pressure, and temperature of the sample. The gas laws can be used to determine how any of these factors change when any one of the other factors is changed.

We can use Boyle's law to determine how the volume of a sample of gas is affected by a pressure change if the temperature is constant. Since the product of P and V is a constant for a sample of a gas at a fixed temperature, then, if we change the pressure and, thus, the volume of the gas, the product of the new pressure and volume (P', V') will be equal to the same constant

$$PV = k \qquad P'V' = k$$

Therefore

$$PV = P'V'$$

Solving for the new volume gives

$$V' = V\left(\frac{P}{P'}\right)$$

To obtain the new volume, we can multiply the original volume by a factor involving the two pressures. If we are increasing the pressure of the gas, the original volume will decrease. Therefore, the new volume is found by multiplying the original volume by a ratio of the pressures, which is less than 1. On the other hand, if we are decreasing the pressure of the gas, the original volume will increase. In this case, we multiply the original volume by a ratio of the pressures that is greater than 1.

Example 7-6   What new volume is occupied by 2.50 ℓ of a sample of a gas at 26°C and 780 torr, if the pressure decreases to 760 torr?
Since the pressure decreases (780 torr → 760 torr), the new volume will be greater than 2.50 ℓ. This new volume can be obtained by multiplying the initial volume by a factor involving a ratio of the pressures that is greater than 1 to give a new volume which is larger than the original volume:

$$2.50 \; \ell \left(\frac{780 \text{ torr}}{760 \text{ torr}}\right) = 2.57 \; \ell$$

The new volume that corresponds to a pressure change associated with a fixed amount of gas at a constant temperature can always be obtained by multiplying the initial volume by a ratio of the pressures (using the same units for pressure). Remember, the ratio should be greater than 1 when a decrease in pressure is involved and less than 1 when an increase in pressure is involved.

Example 7-7   What new volume will a 380 ml sample of a gas maintained at 20°C and 0.832 atm occupy if the pressure changes to 760 torr?
Since 760 torr corresponds to 1 atm,

$$760 \text{ torr} \left(\frac{1 \text{ atm}}{760 \text{ torr}}\right) = 1.00 \text{ atm}$$

then the pressure of the sample is increasing (0.832 atm → 1.00 atm) and so the volume must decrease (380 ml → smaller). Thus, the initial volume must be multiplied by a ratio of the pressures that is less than 1 as shown on the next page:

$$380 \text{ ml} \left(\frac{0.832 \text{ atm}}{1.00 \text{ atm}}\right) = 316 \text{ ml}$$

Charles' law can be used to determine how the volume of a sample of a gas is affected by a temperature change if the pressure is constant. It tells us that the ratio of the volume and temperature of a gas is constant for a gas at a fixed pressure ($V/T = k$). When the temperature is changed, the volume changes, and the ratio of the new volume and temperature (V'/T') will equal the same constant:

$$V/T = k \qquad V'/T' = k$$

Therefore

$$V/T = V'/T'$$

Solving for the new volume gives

$$V' = V\left(\frac{T'}{T}\right)$$

To obtain the new volume, we multiply the original volume by a ratio of the temperatures. Since volume is directly proportional to the temperature, this ratio will be greater than 1 for an increase in temperature and less than 1 for a decrease in temperature.

Example 7-8   What new volume is occupied by a 500 ml sample of a gas maintained at 780 torr pressure if it is heated so that the temperature changes from 25°C to 30°C? Kelvin temperatures must be used in Charles' law calculations. Therefore, the Celsius temperatures must first be converted to the Kelvin scale: 25°C is 298 K (25 + 273) and 30°C is 303 K (30 + 273). The temperature of the sample is increasing (298 K → 303 K), so the new volume will be greater than 500 ml. This new volume can be found by multiplying the initial volume by a ratio of the temperatures that is greater than 1:

$$500 \text{ ml} \left(\frac{303 \text{ K}}{298 \text{ K}}\right) = 508 \text{ ml}$$

Example 7-9   What new volume is occupied by a 2.50 ℓ sample of a gas maintained at a constant pressure if the temperature is changed from 500 K to 250 K?
Since the temperature is decreasing (500 K → 250 K), the initial volume should be multiplied by a ratio of the temperatures that is less than 1:

$$2.50 \; \ell \left(\frac{250 \text{ K}}{500 \text{ K}}\right) = 1.25 \; \ell$$

7-8   Combined Gas Laws

How does the volume of a sample of a gas change when both the temperature and the pressure of the sample change? The behavior of a gas corresponding to such changes in conditions can be illustrated by an example. Suppose we had a 1.00 ℓ sample of a gas at 1.00 atm pressure and 273 K (at STP). What volume would the sample occupy if the temperature is changed to 546 K? Since an increase in temperature will increase the volume, the new volume is

$$1.00 \; \ell \left(\frac{546 \text{ K}}{273 \text{ K}} \right) = 2.00 \; \ell$$

What will happen to the volume if the pressure of this 2.00 ℓ sample changes to 0.250 atm? Since the pressure decrease will increase the volume, the new volume becomes

$$2.00 \; \ell \left(\frac{1.00 \text{ atm}}{0.250 \text{ atm}} \right) = 8.00 \; \ell$$

Let us start with the original sample and change the pressure first and then the temperature. The new volume occupied by the 1.00 ℓ sample at STP when the pressure changes to 0.250 atm is

$$1.00 \; \ell \left(\frac{1.00 \text{ atm}}{0.250 \text{ atm}} \right) = 4.00 \; \ell$$

The new volume occupied by this 4.00 ℓ sample when the temperature is changed to 546 K is

$$4.00 \; \ell \left(\frac{546 \text{ K}}{273 \text{ K}} \right) = 8.00 \; \ell$$

For convenience of calculation we can treat the changes separately in any order we desire. That is, experience with gases has shown that, when the temperature and the pressure of a sample of a gas are both changed, we can calculate the final volume by first calculating the change in volume that results from the pressure change and then convert the resulting volume to the volume corresponding to the temperature change or vice versa. This final volume will be the new volume that results from both the pressure and the temperature changes. Let us consider the previous example in which a 1.00 ℓ sample of a gas at STP undergoes a pressure change to 0.250 atm and a temperature change to 546 K. The new volume can be found by multiplying the initial volume by the appropriate ratio involving the pressures; then this product can be multiplied by the appropriate ratio involving the temperatures:

$$1.00 \; \ell \left(\frac{1.00 \text{ atm}}{0.250 \text{ atm}} \right) \left(\frac{546 \text{ K}}{273 \text{ K}} \right) = 80 \; \ell$$

Of course, the temperature factor could be used before the pressure factor and the same result would be obtained. It is permissible to include both a pressure change and a temperature change in such a calculation. The fact that the new volume of a sample of a gas for which the temperature and pressure have changed can be found by multiplying the original volume by both a temperature factor and a pressure factor is called the combined gas law. The pressure and temperature ratios used follow the same reasoning pattern as is used in Boyle's and Charles' laws. As long as the same reasoning is used, the determination of a volume change corresponding to a temperature and a pressure change can correctly be deduced.

Example 7-10   What new volume will a 250 ml sample of a gas maintained at 25°C and 750 torr pressure occupy if the temperature is changed to 20°C and the pressure is changed to 760 torr?

A pressure increase (750 torr → 760 torr) gives a smaller volume, so the pressure factor should be less than 1. A temperature decrease (298 K → 293 K) gives a smaller volume, so the temperature factor should be less than 1:

$$250 \text{ ml} \left(\frac{750 \text{ torr}}{760 \text{ torr}}\right) \left(\frac{293 \text{ K}}{298 \text{ K}}\right) = 243 \text{ ml}$$

Example 7-11   What new volume will a 25.0 ℓ sample of a gas maintained at 0°C and 1.50 atm pressure occupy if the temperature is changed to 100°C and the pressure is changed to 1.00 atm?

A pressure decrease (1.50 atm → 1.00 atm) gives a larger volume, so the pressure factor should be greater than 1. A temperature increase (273 K → 373 K) gives a larger volume so the temperature factor should be greater than 1.

$$25.0 \ \ell \left(\frac{1.50 \text{ atm}}{1.00 \text{ atm}}\right) \left(\frac{373 \text{ K}}{273 \text{ K}}\right) = 51.2 \ \ell$$

7-9   Dalton's Law of Partial Pressures

So far we have been dealing with samples of gases without being concerned with the composition of the sample. We were only interested in the fact that the number of moles (or particles) of gas was constant. Now let us consider the relation that exists between the components of a gas sample consisting of mixture of gases. If a mixture of three gases, for example, were placed in a container of fixed volume, V, each of the gases could be considered to occupy the entire volume. Of course, this is consistent with the postulates of the kinetic molecular theory. That is, if a gas is placed in a container, the molecules of the gas, because of their rapid random motion and small size, can be considered to occupy the container. Thus, each of the three gases comprising the mixture would have the volume, V. Now, if the temperature of the container were maintained at a constant value, T, each of the gases in the mixture would have this temperature. What pressure would each of the components have? The pressure of a given component would be directly related to the number of moles of that component and to the rate at which the particles collide with the walls of the container. Since each of the components has the same volume and temperature, the differences in the pressures exerted by the component would be related to the differences in the number of moles. This is illustrated in Figure 7-8. The pressure that would be exerted by a given component of a mixture of gases if it alone occupied the container is called the partial pressure of that component. Partial pressures can be calculated by applying the ideal gas law to each component. Thus, the partial pressure, P_c, for a component consisting of n_c moles is given by the expression

$$P_c = \left(\frac{n_c RT}{V}\right)$$

As long as the number of moles of each component of a mixture of gases contained in a given volume at a given temperature is known, the partial pressure of each component can be calculated. Since the particles of

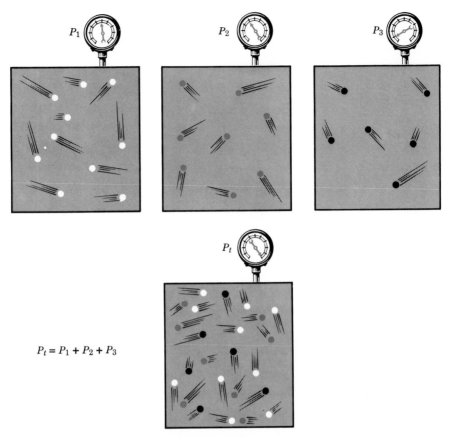

$P_t = P_1 + P_2 + P_3$

Figure 7-8   Dalton's law of partial pressures. Each gas in a separate container exerts a pressure that depends on the amount of gas present. When all of the gases are placed in the same container, each contributes to the total pressure.

each component will behave essentially independently, the total pressure exerted by the mixture will be a result of all of the particles in the mixture. Consequently, the total pressure of the mixture, P_t, will simply be the sum of the partial pressures of the components of the mixture. Expressed algebraically, the total pressure is

$$P_t = P_a + P_b + P_c + \cdots$$

where P_a, P_b, P_c, and so on refer to the partial pressures of the components of a mixture. This relation is known as Dalton's law of partial pressures and indicates that the total pressure of a mixture of gases is the sum of the partial pressures of the components of the mixture. (See Figure 7-8.) Of course, we assume that the components do not chemically react with one another, since this would alter the number of moles that are present. Dalton's law is quite convenient when we want to relate the partial pressures of the components of a mixture to the total pressure.

Example 7-12   What is the partial pressure of oxygen in a mixture consisting of nitrogen and oxygen gas if the mixture contains 0.140 mole of nitrogen gas, occupies a volume of 8.21 ℓ, has a temperature of 27°C, and has a total pressure of 1.000 atm?

According to Dalton's law,

$$P_t = P_{N_2} + P_{O_2}$$

Solving for P_{O_2} gives

$$P_{O_2} = P_t - P_{N_2}$$

Applying the ideal gas law to nitrogen gives

$$P_{O_2} = P_t - \frac{nRT}{V}$$

Since the total pressure is 1.000 atm, the volume is 8.21 ℓ, the temperature is 300 K, and the number of moles of nitrogen is 0.140, the partial pressure of oxygen is

$$P_{O_2} = 1.000 \text{ atm} - 0.140 \text{ mole N}_2 \left(\frac{0.0821 \; \ell \text{ atm}}{\text{K mole N}_2} \right) \left(\frac{300 \text{ K}}{8.21 \; \ell} \right)$$

$$P_{O_2} = 1.000 \text{ atm} - 0.420 \text{ atm} = 0.580 \text{ atm}$$

7-10   Laboratory Preparation of Gases

Many gases can be produced in the laboratory by carrying out a chemical reaction that forms the gas as a product. For example, oxygen gas can be produced in the laboratory by warming a hydrogen peroxide solution in the presence of an iron(III) chloride catalyst. This reaction takes place according to the equation

$$2H_2O_2(aq) \xrightarrow{\text{FeCl}_3} 2H_2O + O_2(g)$$

Typical reactions used in the laboratory to produce samples of some common gases are summarized in Table 7-2. When a gas is produced by a chemical reaction, it is usually evolved from the reaction mixture. In order to collect a sample of a gas produced in this manner, it is necessary to trap the gas in an appropriate container. If the gas is not appreciably soluble in water, a sample of the gas can be collected by displacement of water. The gas produced by a chemical reaction is passed into a container filled with water, and the gas is allowed to replace the water. A typical apparatus used to accomplish the collection of a sample of gas by water displacement is shown in Figure 7-9.

7-11   The Molar Mass of Gases

It is often desirable to determine the molar mass of a gaseous substance. This can be done by applying the gas laws to experimentally determined values of the pressure, volume and temperature of a known mass of a gaseous substance. Furthermore, the molar mass of many liquids can be obtained by boiling a sample of the liquid and determining the volume, pressure and temperature of a known mass of vapor. (See Figure 7-10). From the ideal gas law it is possible to determine the number of moles of a gas given the temperature, pressure and volume. Moreover, if a gas

Table 7-2   Reaction for the Laboratory Preparation of Some Common Gases

Gas	Preparatory Reaction
Oxygen (O_2)	$2H_2O_2 \xrightarrow{\text{Fe}^{3+}} 2H_2O + O_2$ or $2KClO_3 \xrightarrow[\text{heat}]{\text{MnO}_2} 2KCl + 3O_2(g)$ (Danger: This reaction is potentially explosive.)
Hydrogen (H_2)	$Zn(s) + 2H_3O^+(aq) \rightarrow Zn^{2+}(aq) + 2H_2O + H_2(g)$ (Danger: Hydrogen mixed with air can explode.)
Carbon dioxide (CO_2)	$CaCO_3(marble) + 2H_3O^+(aq) \rightarrow Ca^{2+}(aq) + 3H_2O + CO_2(g)$
Acetylene (C_2H_2)	$CaC_2(s) + 2H_2O \rightarrow Ca(OH)_2(s) + C_2H_2(g)$ (Danger: Acetylene mixed with air can explode.)
Ammonia (NH_3)	$NH_3(aq) \xrightarrow{\text{heat}} NH_3(g)$ (This reaction involves the heating of aqua ammonia.)

sample is at STP, the molar volume can be used to find the number of moles. Once the number of moles are known the number of grams per mole is found by dividing the mass of the sample by the number of moles. One approach to the calculation of the molar mass from experimental data involves converting the volume of the sample to STP

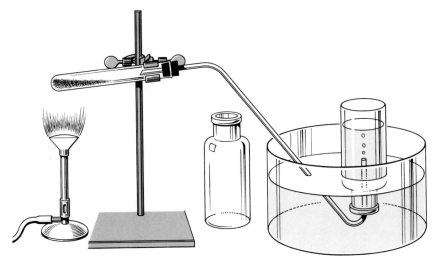

Figure 7-9   Gases prepared in the laboratory can be collected by displacement of water. From *Elements of General and Biological Chemistry* by John Holum, Wiley, N.Y., 1968.

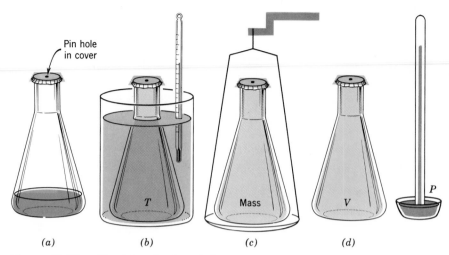

(a)   (b)   (c)   (d)

Figure 7-10   The determination of the number of grams per mole of a substance from experimental measurements.

and using the molar volume to find the number of moles. The mass is then divided by the number of moles to give the molar mass.

Example 7-13   What is the molar mass of a 0.935 g sample of a gas that occupies 0.500 ℓ at 20°C and 776 torr?

The volume can be converted to STP by multiplying by the temperature and pressure ratios corresponding to the changes.

$$0.500 \; \ell \left(\frac{273 \text{ K}}{293 \text{ K}}\right)\left(\frac{776 \text{ torr}}{760 \text{ torr}}\right)\left(\frac{1 \text{ mole}}{22.4}\right) = 2.12 \times 10^{-2} \text{ moles}$$

Next the mass is divided by the number of moles

$$\left(\frac{0.935 \text{ g}}{2.12 \times 10^{-2} \text{ moles}}\right) = \left(\frac{44.0 \text{ g}}{1 \text{ mole}}\right)$$

Another way to calculate the number of grams per mole or molar mass of a gas is to use the ideal gas law, $PV = nRT$. Recall that the n in this relationship is the number of moles. We know that the number of moles of a substance can be determined using the mass, m, and the number of grams per mole factor represented as x

$$n = \frac{m}{x}$$

Substituting this into the ideal gas law for n gives

$$PV = \frac{m \, RT}{x}$$

If we solve for the number of grams per mole, we have a relationship that is quite useful.

$$x = \frac{m \, RT}{PV}$$

Thus, if we know the mass of a sample of a gas and the pressure, temperature, and volume, we can deduce the number of grams per mole of the gas.

Example 7-14   A 0.442 g sample of a gas occupies a volume of 250 ml at 27°C and 0.989 atm pressure. Calculate the number of grams per mole for this gas.
The molar mass can be found by substituting the mass, volume, temperature, and pressure in the relationship

$$x = \frac{m\,RT}{PV}$$

This gives

$$x = \left(\frac{0.442\text{ g}}{.250\;\ell}\right)\left(\frac{0.0821\;\ell\text{ atm}}{\text{K mole}}\right)\left(\frac{300\text{ K}}{0.989\text{ atm}}\right) = 44.0\text{ g/mole}$$

7-12   The Atmosphere

The air of the atmosphere is a mixture of gases that apparently has evolved to the present composition over a period of billions of years. In fact, it is thought that the composition of the atmosphere has not changed for the last 50 million years. However, it now appears that the activities of humans are altering the entire atmosphere to a certain extent. Furthermore, it is obvious that the endeavors of an industrial society alter specific regions of the atmosphere resulting in periodic air pollution episodes.

The vast majority of the estimated 5.5×10^{21} g of air which make up the atmosphere is located within less than 100 km of the surface of the earth. Atmospheric air moves over the earth by a wind system that changes strength and direction daily. Wind activity occurs in the troposphere (up to about 10 km), and a given air parcel can circle the earth in a matter of days. Of course, air parcels move about somewhat randomly resulting in a complete intermixing of the gases of the global atmosphere.

Air is a mixture of 10 to 20 different gases. The major components of dry air in the troposphere in terms of percentage by volume are:

Component	Percentage
Nitrogen, N_2	78.08
Oxygen, O_2	20.95
Argon, Ar	0.934
Carbon dioxide, CO_2	3.14×10^{-2}
Neon, Ne	1.82×10^{-3}
Helium, He	5.24×10^{-4}
Krypton, Kr	1.14×10^{-4}

Note that the air is 99.03% nitrogen and oxygen and 99.96% nitrogen, oxygen, and argon.

Air also contains water vapor, H_2O. However, the concentration of water vapor in the atmosphere is variable and ranges from very small concentrations in desert regions to large concentrations in tropical areas. Typically, the concentration of water vapor in air is 1–3% by volume. Within a local region, the water content varies daily depending on temperature and weather conditions. The water vapor content of the atmosphere within a local region is often expressed in terms of **relative humidity.** Air completely saturated with water vapor at a particular temperature is said to have 100% relative humidity. The percent relative humidity expresses the amount of water vapor in the air as a percentage of the amount of water in the air under completely saturated conditions. For example, if the relative humidity is 60% this means that the air has 60% of the water vapor it could hold under saturated conditions.

7-13   Air Sheds

Weather changes, and winds cause air masses to move from one place to another. However, on land the movement of air masses is dictated to a certain extent by the presence of mountains, hills, valleys, and bodies of water. In other words, the movement of air in the lower altitudes of the troposphere is governed by the topography of the landscape. In fact, as a result of the topography of a region and weather conditions, air masses can be temporarily trapped or isolated within the region. Such a region is called an **air shed** or **air basin.** Temporary immobility of the air in an air shed may last for only a few hours or for several days.

A glance at a topographical map of the United States indicates that major air sheds exist west of the Rocky Mountains, in the Great Plains area of the Midwest and on the East Coast. However, smaller, regional air sheds exist all over the country. A topographical map of the state of California reveals the presence of a huge air shed including the Sacramento and San Joaquin Valleys and smaller air sheds such as the Los Angeles basin bounded by mountains to the north and east and the ocean on the west and south.

The existence of these air sheds has an important effect upon the occurrence of air pollution episodes. Air sheds in various regions of the country are currently being studied in an attempt to discover the boundaries of the sheds, how air flows occur within the sheds, and what weather conditions are responsible for the immobility of the air masses within the sheds.

7-14   Primary Air Pollutants

The activities of an industrial society produce waste gases. Many industrial processes generate by-product gases that are not useful. The automobile produces exhaust gases; most manufacturing processes and the burning of trash produce gases and smoke. When these gaseous products are mixed with the atmosphere, they can become semipermanent com-

Table 7-3   The Origin of the Primary Air Pollutants

Carbon monoxide is formed in the incomplete combustion of fossil fuels.

$$\text{carbon-containing compounds} + O_2 \rightarrow CO + H_2O$$

Sulfur dioxide is formed when sulfur containing impurities in fossil fuels are burned.

$$\text{sulfur-containing compounds} + O_2 \rightarrow SO_2$$

Nitrogen oxides are formed when nitrogen and oxygen combine at higher temperatures of fossil fuel combustion.

$$N_2 + O_2 \rightarrow 2NO$$
$$2NO + O_2 \rightarrow 2NO_2$$

Hydrocarbons result when fossil fuels evaporate or are not completely burned.

$$\text{fossil fuels} \rightarrow HC$$

Hydrocarbons include a variety of hydrogen-carbon compounds represented collectively as HC.

Particulates consist of tiny pieces of solids and liquid droplets that are formed during the combustion process.

---

ponents. Just because the products are released into the air does not mean that they are gone. The fact is, they can produce serious air pollution. It is estimated that, in the United States, over 200 million tons of pollutants are released into the atmosphere each year. Air pollution problems arise when these pollutants accumulate in specific geographical areas. Those gases that are produced by an industrial society and released into the atmosphere are called **primary air pollutants.** There are five major kinds of gaseous primary pollutants. These pollutants are carbon monoxide, sulfur dioxide, nitrogen oxides, hydrocarbons and particulates. The nature and origin of these pollutants is described in Table 7-3. It is worth noting that each of these pollutants is produced in significant amounts by natural biological, volcanic and geological sources. Table 7-4 compares approximate annual production of natural and human made amounts of some of these pollutants.

Table 7-4   Annual Natural and Human-Made Production of Some Gases

Gas	Natural Production (tons per year)	Human-Made Production (tons per year)
Carbon monoxide	3.5×10^9	3×10^8
Nitrogen oxides	1.4×10^9	1.5×10^7
Sulfur oxides (and hydrogen sulfide	1.42×10^8	7.3×10^7

From *Understanding Chemistry: From Atoms to Attitudes* by T. R. Dickson, Wiley, N.Y., 1974.

7-15   ## Air Pollution Sources

Each year around 1 ton of air pollutants are emitted into the atmosphere for each person in the United States. This adds up to over 200
million tons of pollutants annually. These pollutants come mainly from
transportation, electrical power plants, industrial processes, and solid
waste incineration. Table 7-5 lists the sources of pollutants and the percentage each source contributes to the total amount of pollution. Of
course, the effect each source has on air pollution problems depends on
many factors. For instance, in California, transportation emissions provide most problems while power plant emissions are more significant in
parts of the eastern United States. Some experts feel that air pollutants
should be considered with respect to their effects on humans, plants, and
materials rather than a weight-percentage basis. Dr. Lyndon Babcock,
Jr. of the University of Illinois at Chicago has developed a weighted
listing of air pollution sources that considers air pollution standards and
other factors. According to this listing, shown in Table 7-6, particulates
are the pollutants of greatest concern and electrical power plants and
other fuel-burning stationary sources are the most significant pollution
sources. It is interesting to compare Tables 7-5 and 7-6.

7-16   ## Air Pollution Phenomena and Smog

If the 600,000 tons of air pollutants produced in the United States each
day were completely mixed with the entire air mass above the United
States, air pollutants would be present at very low concentrations. Of
course, this does not occur. The accumulation of air pollutants within a

Table 7-5   Sources of Air Pollutants in the United States

Source	Percent by Weight of Grand Annual Total (over 200 million tons)					
	Carbon Monoxide	Sulfur Oxides	Nitrogen Oxides	Hydro-carbons	Partic-ulates	Total
Transportation	29.8	0.4	3.8	7.75	0.56	42.3
Fuel combustion (power plants, space heating, etc.)	0.89	11.4	4.67	0.3	4.2	21.4
Industrial processes	4.5	3.4	0.09	2.1	3.5	13.6
Solid waste disposal	3.6	0.05	0.3	0.75	0.51	5.2
Miscellaneous (forest and agri-cultural fires, etc.)	7.89	0.3	0.79	4.0	4.5	17.5
Total	46.7	15.5	9.6	14.9	13.3	100.0

From *Understanding Chemistry: From Atoms to Attitudes* by T. R. Dickson, Wiley, N.Y.,
1974.

Table 7-6   Sources of Air Pollutants in the United States on a Weighted Basis

| Source | Percentage of Grand Annual Total (over 200 million tons) on Weighted Basis Using Air Quality Standards and Other Factors | | | | | |
	Carbon Monoxide	Sulfur Oxides	Nitrogen Oxides	Hydro-carbons	Partic-ulates	Total
Transportation	3.0	1.0	6.9	4.6	0.9	16.4
Fuel combustion (power plants, industrial space heating, etc.)	0.1	19.8	3.7	1.1	10.2	34.9
Industrial processes	0.3	6.1	0.1	0.1	20.4	27.0
Solid waste disposal	0.2	0.1	0.3	0.2	2.0	2.8
Miscellaneous (forest and agri-cultural fires, etc.)	0.5	0.2	1.3	0.8	16.1	18.9
Total	4.1	27.2	12.3	6.8	49.6	100.0

Adapted from Lyndon Babcock, Jr., and Niren Nagada, University of Illinois at Chicago Circle, Chemical and Engineering News 33(Jan 10, 1972). From *Understanding Chemistry: From Atoms to Attitudes* by T. R. Dickson, Wiley, N.Y., 1974.

restricted geological area can bring about an air pollution episode. As a result of weather fluctuations and winds, air masses can be swept horizontally from an air shed to another region of the atmosphere. When such horizontal movement of air masses is unrestricted, air pollutants are quickly dispersed. However, topological factors of valleys, hills, and mountains can restrict the movement of air masses especially if winds are minimal. Whenever an air mass within an air shed is not replenished by horizontal air movement, it is called a stagnant air mass. Stagnation of air masses can occur when specific unchanging weather patterns exist. A stagnant air mass often remains only for a few hours but such stagnation can last for weeks. Air pollutants can accumulate within a stagnant air mass resulting in an air pollution episode. This type of air pollution is common to many regions of the United States.

7-17   Temperature Inversions

Another means by which air mass movement can occur is by the rising of warm air to higher regions of the atmosphere. When air near the surface of the earth is heated it becomes less dense and rises vertically being replaced by cooler air from the higher regions of the atmosphere. This vertical movement of the surface air can disperse air pollutants into higher regions of the atmosphere. Normally the temperature of the

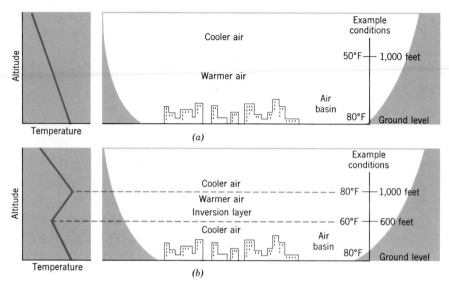

Figure 7-11   Normal and inversion conditions in an air basin.

air decreases with altitude as illustrated in Figure 7-11. The air mass near the earth's surface is warmer. Sometimes a cooler mass of air moves in at a low altitude and underlies the warm air mass. This results in a layer of cooler air under a layer of warm air. As shown in Figure 7-11 the layer of warm air becomes trapped between the cool air mass and the cooler air of higher altitudes. This phenomena is called a **temperature inversion.** When an inversion occurs the temperature of the air decreases with altitude up to the warm air layer. At this level the temperature begins to increase with altitude until the overlying cooler air is reached. Beyond this point the temperature decreases with altitude as usual. A temperature inversion can occur in a sheltered air shed and result in an immobile air mass in which pollutants can accumulate. Often temperature inversions will break up at night but a new inversion may form the next day. Sometimes an inversion may persist for days. Temperature inversions often occur during cloudless sunny days that, as discussed below, compound the air pollution episode.

Once a temperature inversion has formed, primary air pollutants can become trapped and accumulate in localized areas. If the inversion occurs during a warm, cloudless day, the primary air pollutants in the presence of sunlight can form secondary air pollutants. The sunlight induced chemical reactions which produce secondary pollutants are called photochemical reactions. The inversion layer acts as a large reaction vessel in which photochemical reactions and subsequent reactions produce a variety of secondary pollutants known collectively as **smog.** Such **photochemical smog** is characteristic of air pollution episodes in various regions of the country and especially regions of California. Photochemical smog is a complex mixture of chemical compounds produced by chemical reactions between nitrogen oxides, hydrocarbons, particulates, water vapor, and oxygen. The hazy brown appearance of smog is a result of accumulated particulates and nitrogen dioxide. However, most of the dangerous components of smog are colorless gases and are not visible. Nevertheless, their presence is indicated by eye and respiratory irritation in humans and damage to plants.

Questions and Problems

1. State the postulates of the kinetic molecular theory of gases or give a description of a gas in terms of the kinetic molecular theory.

2. Why does a gas exert a pressure on an object or container that is in contact with the gas?

3. Explain the following observations in terms of the kinetic molecular theory of gases:
   (a) If a balloon has a tiny hole in it, the air will slowly leak out.
   (b) When more air is placed in a balloon the volume increases.
   (c) When a sealed balloon is heated the volume increases.
   (d) When you squeeze a sealed balloon the volume decreases.
   (e) When you continue to blow air into a balloon it will break.
   (f) When you release an unsealed balloon it will deflate in a jetlike fashion.
   (g) A parachute usually works.

4. Give an algebraic statement of the perfect gas law and define each term.

5. How is a mercury barometer constructed and how does it function?

6. What is a torr? What is an atmosphere? State the relation between these two pressure units in the form of a factor.

7. If a barometer reads 29.5 in. of mercury, what is the pressure in torr? What is the pressure in atmospheres?

8. What is standard temperature and pressure (STP)?

9. What is the molar volume, and how can it be used to relate the volume of a gas sample to the number of moles of gas in the sample?

10. Give a statement of each of the following gas laws:
    (a) Boyle's law
    (b) Charles' law
    (c) Dalton's law of partial pressures

11. The perfect gas law, $PV = nRT$, can be used to deduce any other gas law. For example, the relation between the pressure and temperature of a gas sample of constant volume is found as follows:

    Solve for P: $P = \dfrac{nRT}{V}$

    $n = $ constant
    $R = $ constant
    $V = $ constant

Thus: $\dfrac{nR}{V} = k = $ constant

Therefore: $P = kT$

The pressure of a gas sample of constant volume is directly proportional to the temperature.

Use the perfect gas law to deduce the following relations for a sample of a gas. (State each relation as an equation and in words.)

(a) relation between volume and pressure at a fixed temperature (Boyle's law)

(b) relation between volume and temperature at a fixed pressure (Charles' law)

(c) relation between volume and number of moles at fixed pressure and temperature

(d) relation between pressure and number of moles at fixed volume and temperature

(e) relation between temperature and number of moles at fixed volume and pressure

12. How many moles of carbon dioxide, CO_2, are contained in a 537 ml sample at 1.75 atm and 312 K temperature?

13. When 10.00 g of water are converted to steam at 100.0°C and 1.000 atm pressure, what volume will the resulting water vapor occupy?

14. If a 2.00 g sample of solid carbon dioxide (dry ice) is placed in a 500 ml sealed container at 25°C, what pressure will the carbon dioxide exert when it sublimes to form a gas?

15. A sample of gas at 240 K and 760 torr occupies a 100 ml volume. What volume will the gas occupy at −75°C if the pressure remains constant?

16. A sample of gas at 15°C and 760 torr is heated to 375 K, and the volume is held constant. What is the new pressure of the gas?

17. In an airliner a gas sample in a flexible container occupies a volume of 2.00 ℓ at 760 torr pressure. When the airliner loses the artificial pressure the volume of the gas changes to 15.2 ℓ. What is the pressure in the airplane if the temperature is constant?

18. An inflated balloon occupies a volume of 3.2 ℓ at 25°C. What volume will the balloon occupy at 50°C if the pressure is constant?

19. A sample of a gas at 20°C and 0.852 atmospheres occupies a volume of 0.800 ℓ. What volume will this gas occupy at STP?

20. A sample of a gas is made up of a mixture of oxygen, carbon dioxide and nitrogen. If the partial pressures of the three gases are 19.8 torr, 7.9 torr, and 297.3 torr, respectively, what is the total pressure of the mixture?

21. A sample of a gas is in a steel container at −75°C and 1.480 atm. What pressure will the sample have when the temperature is changed to 1000°C?

22. A helium-filled balloon occupies a volume of 2.5 ℓ at 300 K and 800 torr pressure. What volume would the balloon occupy at an altitude of 10 km at a temperature of 230 K and a pressure of 100 torr?

23. A sample of gas occupies 3.00 ℓ volume at 1.100 atm pressure and 20°C. What volume will it occupy at −30°C and 1.700 atm?

24. A sample of a gas occupies 4.00 ℓ volume at standard temperature and 750 torr pressure. What volume will it occupy at standard pressure and 25°C?

25. A 500 ml sample of oxygen gas is maintained at 0.945 atm pressure and 20°C. What will the pressure of the sample become when 0.500 g of oxygen gas are added to the container if the temperature and volume are unchanged?

26. A 0.852 g sample of a substance in the vapor phase occupies 153.0 ml at 110°C and 760 torr pressure. Calculate the number of grams per mole of the substance.

27. A 2.00 ℓ sample of gaseous compound has a pressure of 1.500 atm and a temperature of 25°C. If this sample has a mass of 1.987 g, calculate the number of grams per mole of the compound.

28. A 0.391 g sample of a substance in the vapor phase occupies 164.2 ml at 127°C and 760 torr pressure. Calculate the number of grams per mole of the substance.

29. A 500 ml sample of a gaseous compound has a pressure of 1.200 atm and a temperature of 20.0°C. If this sample has a mass of 1.532 g, calculate the number of grams per mole of the compound.

30. If a gas is contained in the left-hand part of the apparatus shown below at 760 torr pressure and 25°C, describe what will happen when the valve is opened and give the new conditions of the gas assuming the temperature is held constant.

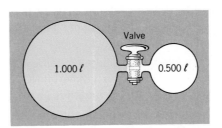

31. The average kinetic energies of gases are proportional to the Kelvin temperatures ($KE\alpha T$ K). The average kinetic energy of a pure gas can be expressed as $KE = 1/2\ mv^2$ where m is the average mass of the molecules and v is the average speed. Thus, when a gas is heated, the average speed increases. If two gases are maintained at the same conditions, the average kinetic energies can be assumed to be equal (KE gas $1 = KE$ gas 2).

Substituting in $1/2\ m_1v_1^2$ for gas 1 and $1/2\ m_2v_2^2$ for gas 2, we can deduce an algebraic relation showing how the ratio of the average speeds are related to the ratio of the average masses of the two gases. (The average masses are given by the molar masses or the number of grams per mole.)

$$\tfrac{1}{2}m_1v_1^2 = \tfrac{1}{2}m_2v_2^2$$

$$m_1v_1^2 = m_2v_2^2$$

$$\frac{v_1^2}{v_2^2} = \frac{m_2}{m_1}$$

Thus

$$\frac{v_1}{v_2} = \sqrt{\frac{m_2}{m_1}}$$

Use this expression to determine the ratio of the speeds of oxygen gas, O_2, and hydrogen gas, H_2. If samples of hydrogen and oxygen gas, under the same conditions, are in containers with very small outlets, which gas will effuse (leak) from the container at a faster rate? Why?

32. Suppose a used aerosol can contains a gas at 760 torr at 25°C. If this can is heated in a fire to 500°C, what will the pressure of the gas be inside the can? (*Danger:* an aerosol can may explode under these conditions. Never put an aerosol can in a fire.)

33. Aerosol cans can cause death by exploding or from use in a confined space. Do you feel that all aerosol can should be fitted with safety release valves and those with potentially dangerous contents should have large warning labels? How many products contained in aerosol cans do you use?

34. What two elements are the main constituents of air?

35. What is relative humidity?

36. What are the five primary air pollutants?

37. What is a temperature inversion?

38. What is photochemical smog, and how is a temperature inversion involved in smog formation?

39. Around 5.5×10^5 tons of pollutants are released into the atmosphere each day in the United States. Calculate the number of tons of air pollutants released per person each year in the United States (assume 200×10^6 people).

40. If a smog-filled atmosphere contains 50 μg/m³ (1 μg $= 10^{-6}$ g) of nitrogen dioxide, NO_2, how many grams of NO_2 are contained in a breath of the polluted air? (Assume a breath is about one-half liter of air.)

41. If a smog-filled atmosphere contains 200 μg/m³ (1 μg $= 10^{-6}$ g) of ozone, O_3, how many grams of O_3 are contained in one cubic mile of the air?

8

Chapter 8: Chemical Stoichiometry

Objectives

The student should be able to:
1. Give the molar interpretation of an equation.
2. Deduce the molar ratios relating the species in a balanced equation.
3. Carry out common stoichiometric calculations.
    (a) mole to mole
    (b) mass to mole
    (c) mole to mass
    (d) mass to mass
    (e) volume to mass
    (f) mass to volume
    (g) volume to volume
4. Calculate the standard enthalphy change for a reaction from given data.

Terms to Know

Molar ratio
Avogadro's hypothesis
Endergonic reaction
Exergonic reaction
Exothermic reaction
Endothermic reaction
Enthalpy change
Fossil fuels
Energy crisis

8

Chemical Stoichiometry

8-1 Chemical Reactions

Countless chemical reactions are continuously taking place in the environment. The growth of plants, the metabolism of animals, and the cycles of nature involve numerous chemical reactions. Chemical technology involves human controlled chemical reactions designed to produce useful chemical products. As we know, chemical reactions can be represented by balanced chemical equations showing the reactants and products. Let us consider some of the most common chemical reactions carried out in our industrial society.

The combustion of fossil fuels is the major source of energy in our society. Combustion or burning refers to the reaction of a hydrocarbon with oxygen to form carbon dioxide and water. The combustion of methane, CH_4, the major component of natural gas is shown by the equation

$$CH_4 + 2O_2 \rightarrow CO_2 + 2H_2O$$

Gasoline is a complex mixture of hydrocarbons. The equation for the combustion of octane, C_8H_{18}, will illustrate a typical reaction that occurs when gasoline burns.

$$2C_8H_{18} + 25O_2 \rightarrow 16CO_2 + 18H_2O$$

The refining of metals from ores is a very common industrial application of chemistry. The major reaction in the refining of iron is represented by the equation

$$Fe_2O_3 + 3CO \rightarrow 2Fe + 3CO_2$$

The overall reaction in the refining of aluminum is given by the equation shown on the next page.

$$2Al_2O_3 + 3C \rightarrow 4Al + 3CO_2$$

Many industrial processes involve a series of chemical reactions that start with raw materials and ultimately produce the desired product. Many chemicals used in industry or used as intermediates in chemical manufacturing are produced or synthesized from common raw material chemicals. Ammonia is synthesized by combining nitrogen obtained from the atmosphere with hydrogen obtained from petroleum products.

$$N_2 + 3H_2 \rightarrow 2NH_3$$

Ammonia is used to prepare explosives and fertilizers. For instance, the fertilizer urea, $(NH_2)_2CO$, is formed by combining ammonia and carbon dioxide.

$$2NH_3 + CO_2 \rightarrow (NH_2)_2CO + H_2O$$

Ethyl alcohol, C_2H_5OH, used in industry can be manufactured by combining ethylene, C_2H_4, from petroleum with water.

$$C_2H_4 + H_2O \rightarrow C_2H_5OH$$

Many medicinal drugs are synthesized using natural products that are obtained from plants or chemicals obtained from petroleum. Such syntheses often involve several chemical reactions. The final reaction in the synthesis of aspirin or acetylsalicylic acid is given by the equation

$$\underset{\substack{\text{salicylic} \\ \text{acid}}}{C_7H_6O_3} + \underset{\substack{\text{acetic} \\ \text{anhydride}}}{C_4H_6O_3} \rightarrow \underset{\text{aspirin}}{C_9H_8O_4} + \underset{\substack{\text{acetic} \\ \text{acid}}}{HC_2H_3O_2}$$

8-2   Molar Interpretations of Equations

The manufacture of chemicals is one of the largest industrial endeavors in the world. Chemical manufacturing industries are the foundation of any industrial society. Nearly everything we purchase today is manufactured by some chemical process or involves the use of chemical products. For economic reasons, such chemical processes and the production of the chemicals must be accomplished in the most convenient manner with as little waste as possible. When a chemical reaction occurs, chemists are often concerned about the amount of product that can be formed from given amounts of reactants. This is important in most research and industrial applications of chemical reactions. As we learned previously, mass is conserved in a chemical reaction. Thus, a specific quantity of reactants will react to form products the mass of which will be equal to the mass of the reactants that has reacted. As long as the equation representing the reaction is known, the mass relationships between individual reactants and products can be deduced. Calculations involving these mass relationships are called stoichiometric (pronounced as stoy-key-oh-met'-rik) calculations. **Stoichiometric** is an interesting word. The -metric part refers to the measuring of masses. The stoichio- is from the Greek and refers to the elements or parts of compounds. Thus, this term refers to the mass relationships in

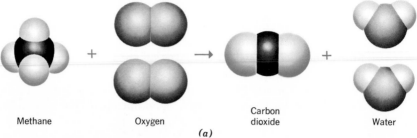

Methane   Oxygen   Carbon dioxide   Water

(a)

This reaction involves one methane molecule reacting with two oxygen molecules to give a carbon dioxide molecule and two water molecules.

To interpret this reaction from a molar point of view it is assumed that the reaction that occurs with individual molecules will be the same if Avogadro's number of molecules are involved. The same reaction from a molar point of view is

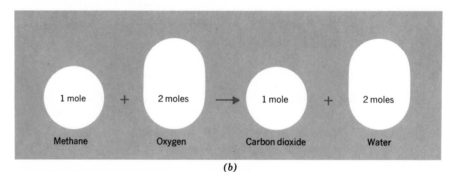

| 1 mole | + | 2 moles | → | 1 mole | + | 2 moles |

Methane   Oxygen   Carbon dioxide   Water

(b)

One mole of methane reacts with two moles of oxygen to give one mole of carbon dioxide and two moles of water.

Figure 8-1   The molar interpretation of an equation.

chemical reactions. When we want to represent a chemical reaction, we write an equation. It is very important to be sure that the equation is an honest representation of the species involved in the reaction. That is, we must be sure that we use, to the best of our knowledge, the proper formulas for all species involved in the reaction. Once a balanced equation is written for a chemical process, the equation can be interpreted from a molar point of view for use in stoichiometric calculations. For example, the equation representing the combustion of methane is

$$CH_4 + 2O_2 \rightarrow CO_2 + 2H_2O$$

How can we interpret such an equation? We could say that the equation indicates that one molecule of methane will react with two molecules of oxygen to produce one molecule of carbon dioxide and two molecules of water. Of course, if this is true for a few molecules, it would be true for a large number, even Avogadro's number. So, a better interpretation of the equation would be on a basis of Avogadro's number of particles or moles of substances. Such an interpretation is called a **molar interpretation.** See Figure 8-1.

From a molar point of view, the above equation indicates that 1 mole of methane reacts with 2 moles of oxygen to produce 1 mole of carbon dioxide and 2 moles of water. It is very important that you understand this molar interpretation. The equation shows which substances are involved in the reaction as reactants and products. The coefficient in front of each formula indicates the number of moles of each substance involved in the reaction. In the absence of a coefficient, of course, a 1 is implied. It is possible to express the relation between any two species involved in the reaction in the form of a **molar ratio.** Such molar ratios would only have meanings related to the equation. The possible molar ratios related to the combustion of methane equation would be

$$\left(\frac{2 \text{ moles } O_2}{1 \text{ mole } CH_4}\right) \quad \left(\frac{1 \text{ mole } CO_2}{1 \text{ mole } CH_4}\right) \quad \left(\frac{2 \text{ moles } H_2O}{1 \text{ mole } CH_4}\right) \quad \left(\frac{1 \text{ mole } CH_4}{2 \text{ moles } O_2}\right)$$

$$\left(\frac{1 \text{ mole } CH_4}{1 \text{ mole } CO_2}\right) \quad \left(\frac{1 \text{ mole } CH_4}{2 \text{ moles } H_2O}\right) \quad \left(\frac{1 \text{ mole } CO_2}{2 \text{ moles } O_2}\right) \quad \left(\frac{2 \text{ moles } H_2O}{2 \text{ moles } O_2}\right)$$

$$\left(\frac{2 \text{ moles } O_2}{1 \text{ mole } CO_2}\right) \quad \left(\frac{2 \text{ moles } O_2}{2 \text{ moles } H_2O}\right) \quad \left(\frac{1 \text{ mole } CO_2}{2 \text{ moles } H_2O}\right) \quad \left(\frac{2 \text{ moles } H_2O}{1 \text{ mole } CO_2}\right)$$

These molar ratios are obtained by using the coefficients given in the balanced equation. They apply only to the specific reaction and can be used as conversion factors to convert from the number of moles of one species involved in the reaction to the number of moles of another species. That is, if we know the number of moles of one species, we can deduce the number of moles of any other species involved in the reaction by using the molar ratio as a factor.

Example 8-1   How many moles of oxygen gas are required to react with 3.27 moles of methane gas if these species react according to the equation

$$CH_4 + 2O_2 \rightarrow CO_2 + 2H_2O$$

From the balanced equation we know that the appropriate molar ratio involving methane and oxygen is

$$\left(\frac{2 \text{ moles } O_2}{1 \text{ mole } CH_4}\right)$$

This factor can be used to convert the number of moles of methane to the corresponding number of moles of oxygen needed to react with the methane:

$$3.27 \text{ moles } CH_4 \left(\frac{2 \text{ moles } O_2}{1 \text{ mole } CH_4}\right) = 6.54 \text{ moles } O_2$$

Example 8-2   How many moles of water will be produced if 0.0203 moles of methane react with oxygen according to the above equation?
The molar ratio involving water and methane is

$$\left(\frac{2 \text{ moles } H_2O}{1 \text{ mole } CH_4}\right)$$

This ratio can be used as a factor to convert the number of moles of methane to the number of moles of water produced by a complete reaction of the methane:

$$0.0203 \text{ mole } CH_4 \left(\frac{2 \text{ moles } H_2O}{1 \text{ mole } CH_4}\right) = 0.0406 \text{ mole } H_2O$$

Example 8-3   How many moles of ammonia are formed when 150 moles of hydrogen react with nitrogen according to the equation

$$N_2 + 3H_2 \rightarrow 2NH_3$$

The molar ratio relating ammonia and hydrogen is

$$\left(\frac{2 \text{ moles NH}_3}{3 \text{ moles H}_2}\right)$$

This ratio can be used to convert the moles of hydrogen to moles of ammonia

$$150 \text{ moles H}_2 \left(\frac{2 \text{ moles NH}_3}{3 \text{ moles H}_2}\right) = 100 \text{ moles NH}_3$$

As we shall see, the molar ratios related to a chemical reaction serve as the basis for many types of stoichiometric calculations. Consequently, it is very important that the correctly balanced chemical equation is known for a chemical reaction so that the proper molar ratios can be deduced. Once the proper molar ratios are known, the interconversion of the number of moles of any two species can easily be made.

8-3　Mass-to-Mole Calculations

If we know the formula of a species involved in a reaction, we can easily deduce the number of grams per mole associated with the species. This number can be used to convert the number of grams of the species to the number of moles of that species or vice versa. See Sections 2-7 and 2-8.

Example 8-4　How many moles of oxygen molecules are contained in a 64.0 g sample of oxygen?

Since oxygen is diatomic, the molar mass is

$$\left(\frac{32.0 \text{ g}}{1 \text{ mole O}_2}\right)$$

The reciprocal of the molar mass can be used as a factor to convert the number of grams of oxygen to the number of moles of oxygen.

$$64.0 \text{ g} \left(\frac{1 \text{ mole O}_2}{32.0 \text{ g}}\right) = 2.00 \text{ moles O}_2$$

How many grams of ammonia, NH_3, are contained in 0.500 moles of ammonia?

The molar mass of ammonia, which contains 1 mole of combined nitrogen (14.01 g/mole) and 3 moles of combined hydrogen (1.008 g/mole), is

$$\left(\frac{17.03 \text{ g}}{1 \text{ mole NH}_3}\right)$$

This factor can be used to convert the number of moles of ammonia to the number of grams of ammonia.

$$0.500 \text{ mole NH}_3 \left(\frac{17.03 \text{ g}}{1 \text{ mole NH}_3}\right) = 8.52 \text{ g}$$

Since the number of grams per mole of a species can be used to convert from grams to moles of the species or vice versa, such conversions can be combined with molar conversions so that the mass of a given reactant or product can be related to the number of moles of any other reactant or product.

Example 8-5　Iron(III) oxide reacts with carbon monoxide to give iron metal and carbon dioxide. How many moles of iron can be produced when 360 g of carbon monoxide are reacted with iron(III) oxide?

The balanced equation for the reaction is

$$Fe_2O_3 + 3CO \rightarrow 2Fe + 3CO_2$$

To determine the number of moles of iron produced, we must work on a mole basis; therefore, the number of grams of carbon monoxide should be converted to the number of moles of carbon monoxide. Then the number of moles of carbon monoxide can be converted to the number of moles of iron with the appropriate molar ratio. The conversions to be made can be represented as

? moles iron = grams carbon monoxide → moles carbon monoxide → moles iron

First, the mass of carbon monoxide is converted to moles:

$$360 \text{ g} \left(\frac{1 \text{ mole CO}}{28.0 \text{ g}} \right)$$

This is then converted to moles of iron:

$$\text{? moles Fe} = 360 \text{ g} \left(\frac{1 \text{ mole CO}}{28.0 \text{ g}} \right) \left(\frac{2 \text{ moles Fe}}{3 \text{ moles CO}} \right) = 8.57 \text{ moles Fe}$$

When dealing with a problem of this type, decide what conversions are needed and then carry out the conversion step by step.

Example 8-6   How many moles of ammonia can be formed if 56.0 g of nitrogen are combined with hydrogen according to the equation

$$N_2 + 3H_2 \rightarrow 2NH_3$$

The conversion sequence to deduce the number of moles of ammonia is to convert the grams of nitrogen to moles of nitrogen and the moles of nitrogen to moles of ammonia.

? moles NH_3 = grams N_2 → moles N_2 → moles NH_3

First, the mass of nitrogen is converted to moles

$$56.0 \text{ g} \left(\frac{1 \text{ mole } N_2}{28.0 \text{ g}} \right)$$

Next, this is converted to moles of ammonia.

$$\text{? moles } NH_3 = 56.0 \text{ g} \left(\frac{1 \text{ mole } N_2}{28.0 \text{ g}} \right) \left(\frac{2 \text{ moles } NH_3}{1 \text{ mole } N_2} \right) = 4.00 \text{ moles } NH_3$$

The reverse problem of calculating the number of grams of a reactant or product corresponding to a given number of moles of another reactant or product can be accomplished in a manner analogous to the above examples.

Example 8-7   Ethane gas, C_2H_6, reacts with oxygen to give carbon dioxide and water. How many grams of ethane are required to react with 14.0 moles of oxygen gas? The balanced equation for the reaction is

$$2C_2H_6 + 7O_2 \rightarrow 4CO_2 + 6H_2O$$

The number of grams of ethane can be found by converting the number of moles of oxygen to the number of moles of ethane and then by applying the number of grams per mole of ethane as a factor to convert from the number of moles to the number of grams of ethane. The conversion sequence is shown on the next page.

$$? \text{ grams } C_2H_6 = \text{moles } O_2 \rightarrow \text{moles } C_2H_6 \rightarrow \text{grams } C_2H_6$$

Since the formula for ethane is C_2H_6, the number of grams per mole of ethane is

$$2(12.01 \text{ g/mole}) + 6(1.008 \text{ g/mole}) = 30.1 \text{ g/mole}$$

The number of grams of ethane involved in the reaction is calculated as follows:

$$? \text{ grams } C_2H_6 = 14.0 \text{ moles } O_2 \left(\frac{2 \text{ moles } C_2H_6}{7 \text{ moles } O_2}\right) \left(\frac{30.1 \text{ g}}{1 \text{ mole } C_2H_6}\right) = 120 \text{ g}$$

If the equation representing a reaction is known, the conversion from the number of grams of one species to the number of moles of another species or vice versa can easily be accomplished by use of the number of grams per mole factor and the molar ratio factor.

Example 8-8   How many grams of nitrogen must be combined with hydrogen to form 4.00 moles of ammonia by the reaction

$$N_2 + 3H_2 \rightarrow 2NH_3$$

The conversion sequence is to convert the moles of ammonia to moles of nitrogen and the moles of nitrogen to grams of nitrogen.

$$? \text{ grams } N_2 = \text{moles } NH_3 \rightarrow \text{moles } N_2 \rightarrow \text{grams } N_2$$

The conversion is accomplished using the molar ratio from the equation and the molar mass of nitrogen (28.0 g/mole N_2):

$$? \text{ grams } N_2 = 4.00 \text{ moles } NH_3 \left(\frac{1 \text{ mole } N_2}{2 \text{ moles } NH_3}\right) \left(\frac{28.0 \text{ g}}{1 \text{ mole } N_2}\right) = 56.0 \text{ g}$$

8-4   Mass-to-Mass Calculations

The conversion from the mass of a given species to the number of moles of that species or from the number of moles to the mass can always be accomplished by use of the number of grams per mole as a factor. The number of moles of reactants and products are related by the molar ratios obtained from the balanced equation. The relation between a given mass of a reactant or product and the corresponding mass of another reactant or product can be deduced by combining the mass to mole and mole to mole conversions. This is a very important type of calculation in chemistry because we usually work with masses of materials. For instance, chemists often ask this question: If this many grams of a reactant are used up in a reaction, how many grams of another reactant are needed, or how many grams of a product are formed? When we want the mass of a substance used or produced in a reaction, we must first find its amount in moles. The moles can then be converted to mass.

Example 8-9   Aluminum oxide is reacted with carbon to form aluminum metal and carbon dioxide.

$$2Al_2O_3 + 3C \rightarrow 4Al + 3CO_2$$

How many grams of aluminum metal are formed from 204 g of aluminum oxide?

To convert the grams of aluminum oxide to grams of aluminum, we first determine the number of moles of aluminum oxide. This can be converted to the number of moles of aluminum, which is then converted to grams of aluminum. The conversion sequence is

$$? \text{ grams Al} = \text{grams Al}_2\text{O}_3 \rightarrow \text{moles Al}_2\text{O}_3 \rightarrow \text{moles Al} \rightarrow \text{grams Al}$$

The conversion is carried out by multiplying the grams of aluminum oxide by the three appropriate factors. Let us consider each step. First, the grams of aluminum oxide is converted to moles using the molar mass [2(27.0 g/mole) + 3(16.0 g/mole) = 102 g/mole].

$$204 \text{ g} \left(\frac{1 \text{ mole Al}_2\text{O}_3}{102 \text{ g}} \right)$$

Second, the moles of aluminum are found by multiplying by the molar ratio from the equation

$$204 \text{ g} \left(\frac{1 \text{ mole Al}_2\text{O}_3}{102 \text{ g}} \right) \left(\frac{4 \text{ moles Al}}{2 \text{ moles Al}_2\text{O}_3} \right)$$

Finally, the mass of aluminum is found by use of the molar mass.

$$? \text{ grams Al} = 204 \text{ g} \left(\frac{1 \text{ mole Al}_2\text{O}_3}{102 \text{ g}} \right) \left(\frac{4 \text{ moles Al}}{2 \text{ moles Al}_2\text{O}_3} \right) \left(\frac{27.0 \text{ g}}{1 \text{ mole Al}} \right) = 108 \text{ g}$$

Example 8-10   Iron(III) oxide reacts with carbon monoxide to produce iron metal and carbon dioxide. How many grams of metallic iron are produced from the reaction of 320 g of iron(III) oxide with carbon monoxide? The balanced equation for the reaction is

$$\text{Fe}_2\text{O}_3 + 3\text{CO} \rightarrow 2\text{Fe} + 3\text{CO}_2$$

The number of grams of iron can be obtained by converting the grams of Fe_2O_3 to the number of moles [2(55.8 g/mole) + 3(16.00 g/mole) = 160 g/mole], which can then be converted to the number of moles of iron and subsequently to the number of grams of iron metal. The conversion sequence should be

$$? \text{ grams Fe} = \text{grams Fe}_2\text{O}_3 \rightarrow \text{moles Fe}_2\text{O}_3 \rightarrow \text{moles Fe} \rightarrow \text{grams Fe}$$

$$320 \text{ g} \left(\frac{1 \text{ mole Fe}_2\text{O}_3}{160 \text{ g}} \right) \left(\frac{2 \text{ moles Fe}}{1 \text{ mole Fe}_2\text{O}_3} \right) \left(\frac{55.8 \text{ g}}{\text{mole Fe}} \right) = 223 \text{ g}$$

8-5   Stoichiometry and the Gas Laws

When dealing with chemical reactions involving one or more reactants or products in the gas phase, the volumes of gases can be related to masses of other species through the combined use of the gas laws and the molar ratios. The gas laws can be used to convert the volumes of gaseous substances maintained at specific conditions to the number of moles of the substances, and then the number of moles can be used in stoichiometric calculations. Gas volumes of gases maintained at STP can be related to the number of moles of a substance by use of the molar volume, (22.4 ℓ/mole)$_{STP}$, while gas volumes at other conditions can be related to the number of moles by use of the ideal gas law. Furthermore, the volumes of gases involved in a reaction can be deduced by determining the number of moles of gas and then converting to the volume at the given conditions.

Example 8-11   How many liters of oxygen gas collected at STP can be obtained by heating 246 g of potassium chlorate so that it decomposes to form potassium chloride and oxygen gas?

The balanced equation for the reaction should be

$$2KClO_3 \rightarrow 2KCl + 3O_2$$

The liters of oxygen gas produced at STP can be found by converting the mass of $KClO_3$ to the number of moles of $KClO_3$ that can be converted to the number of moles of oxygen by use of the proper molar ratio. Finally, the number of moles of oxygen gas can be converted to the volume of oxygen at STP by use of the molar volume. The conversion steps should be

$$? \text{ liters } O_2 = \text{grams } KClO_3 \rightarrow \text{moles } KClO_3 \rightarrow \text{moles } O_2 \rightarrow \text{liters } O_2$$

Consider the steps involved in this problem. First, we deduce the number of moles of potassium chlorate involved by use of the molar mass $[39.1 \text{ g/mole} + 35.5 \text{ g/mole} + 3(16.0 \text{ g/mole}) = 123 \text{ g/mole}]$

$$246 \text{ g} \left(\frac{1 \text{ mole } KClO_3}{123 \text{ g}} \right)$$

Second, the moles of potassium chlorate can be converted to the moles of oxygen using the factor from the balanced equation:

$$246 \text{ g} \left(\frac{1 \text{ mole } KClO_3}{123 \text{ g}} \right) \left(\frac{3 \text{ moles } O_2}{2 \text{ moles } KClO_3} \right)$$

Finally, the moles of oxygen can be converted to the volume at STP:

$$\text{liters } O_2 = 246 \text{ g} \left(\frac{1 \text{ mole } KClO_3}{123 \text{ g}} \right) \left(\frac{3 \text{ moles } O_2}{2 \text{ moles } KClO_3} \right) \left(\frac{22.4 \text{ } \ell}{1 \text{ mole } O_2} \right)_{STP} = 67.2 \text{ } \ell \text{ of } O_2 \text{ at STP}$$

Example 8-12   How many liters of hydrogen gas maintained at 298 K and 1.000 atm pressure are needed to react with a 128 g sample of oxygen gas to produce water?

The balanced equation is

$$2H_2 + O_2 \rightarrow 2H_2O$$

The number of liters of hydrogen required can be deduced by converting the mass of the oxygen to the number of moles of oxygen and then using the molar ratio to convert to the number of moles of hydrogen. The number of moles of hydrogen can then be converted to the volume of hydrogen at STP using the molar volume factor. The volume of hydrogen at STP can be converted to the volume of hydrogen at 298 K by multiplying by the proper ratio of the temperatures. This ratio should be greater than 1, since the temperature change would be from 273 K (standard temperature) to 298 K. No pressure correction is needed because the pressure is 1 atm. The conversion sequence should be

$$? \text{ liters } H_2 = \text{mass } O_2 \rightarrow \text{moles } O_2 \rightarrow \text{moles } H_2 \rightarrow \text{volume } H_2 \text{ at STP} \rightarrow$$

$$\text{volume } H_2 \text{ at 298 K and 1 atm}$$

Consider the steps involved in this problem. First, find the number of moles of oxygen.

$$128 \text{ g} \left(\frac{1 \text{ mole } O_2}{32.0 \text{ g}} \right)$$

Second, convert this to moles of hydrogen using the factor from the equation:

$$128 \text{ g} \left(\frac{1 \text{ mole } O_2}{32.0 \text{ g}} \right) \left(\frac{2 \text{ moles } H_2}{1 \text{ mole } O_2} \right)$$

Third, convert the moles of hydrogen to volume at STP:

$$128 \text{ g} \left(\frac{1 \text{ mole O}_2}{32.0 \text{ g}}\right) \left(\frac{2 \text{ moles H}_2}{1 \text{ mole O}_2}\right) \left(\frac{22.4 \ \ell}{1 \text{ mole H}_2}\right)_{\text{STP}}$$

Finally, convert this volume to the volume at 298 K:

$$? \text{ liters H}_2 = 128 \text{ g} \left(\frac{1 \text{ mole O}_2}{32.0 \text{ g}}\right) \left(\frac{2 \text{ mole H}_2}{1 \text{ mole O}_2}\right) \left(\frac{22.4 \ \ell}{1 \text{ mole H}_2}\right)_{\text{STP}} \left(\frac{298 \text{ K}}{273 \text{ K}}\right) = 196 \ \ell$$

Essentially, the same approach that is used in the above examples can be used to determine the mass of a species that is related to a certain volume of another species in the reaction. The calculations involved in this type of problem can be accomplished by converting the volume of a gas to the number of moles that can then be converted to the number of moles of the other species. Finally, the number of moles of this species can be converted to the number of grams. Of course, the first conversion of the volume to the number of moles can be accomplished by use of the gas laws.

Example 8-13   How many grams of copper metal can be produced by reacting 250 ml of hydrogen gas, measured at 15°C and 1.02 atm pressure, with sufficient copper(II) oxide, if the copper(II) oxide and the hydrogen react to form copper metal and water?
The balanced equation for the reaction is

$$CuO + H_2 \rightarrow Cu + H_2O$$

The number of grams of copper produced can be found by converting the volume of hydrogen at the given conditions to the volume at STP by use of the proper temperature and pressure ratios. Then, the volume of hydrogen at STP can be converted to the number of moles of hydrogen by use of the molar volume factor; and the number of moles of hydrogen can be converted to the number of moles of copper, and then to grams of copper. The conversion sequence should be

? grams Cu = initial volume $H_2 \rightarrow$ volume H_2 at STP $\rightarrow$ moles $H_2 \rightarrow$ moles Cu $\rightarrow$

mass Cu

To obtain the moles of hydrogen we must first convert the volume of hydrogen to STP by use of the proper temperature and pressure factors: (288 K $\rightarrow$ 273 K, 1.02 atm $\rightarrow$ 1.00 atm)

$$0.250 \ \ell \left(\frac{273 \text{ K}}{288 \text{ K}}\right) \left(\frac{1.02 \text{ atm}}{1.00 \text{ atm}}\right)$$

Now the moles of hydrogen can be found by use of the molar volume:

$$0.250 \ \ell \left(\frac{273 \text{ K}}{288 \text{ K}}\right) \left(\frac{1.02 \text{ atm}}{1.00 \text{ atm}}\right) \left(\frac{1 \text{ mole H}_2}{22.4 \ \ell}\right)_{\text{STP}}$$

Next, the moles of copper can be found using the factor from the balanced equation:

$$0.250 \ \ell \left(\frac{273 \text{ K}}{288 \text{ K}}\right) \left(\frac{1.02 \text{ atm}}{1.00 \text{ atm}}\right) \left(\frac{1 \text{ mole H}_2}{22.4 \ \ell}\right)_{\text{STP}} \left(\frac{1 \text{ mole Cu}}{1 \text{ mole H}_2}\right)$$

Finally, this can be converted to the mass of copper using the molar mass:

$$? \text{ grams Cu} = 0.250 \ \ell \left(\frac{273 \text{ K}}{288 \text{ K}}\right) \left(\frac{1.02 \text{ atm}}{1.00 \text{ atm}}\right) \left(\frac{1 \text{ mole H}_2}{22.4 \ \ell}\right) \left(\frac{1 \text{ mole Cu}}{1 \text{ mole H}_2}\right) \left(\frac{63.5 \text{ g}}{\text{mole Cu}}\right) = 0.685 \text{ g}$$

8-6   Volume-to-Volume Calculations

For reactions that involve two or more gaseous reactants or products, it is sometimes necessary to determine how the volume of one species is related to the volume of another. This kind of calculation can often be solved quite readily if we recall how the volumes of gases, kept at constant temperature and pressure, are related to the number of moles. The ideal gas law, $PV = nRT$, indicates that the volume of a gas is directly related to the number of moles if the pressure and temperature are constant:

$$V = \frac{RT}{P}\, n \quad \text{or} \quad V = kn \text{ (at constant } T \text{ and } P)$$

Notice that this indicates that, if we had two gases maintained at the same temperature and pressure and both had the same volume, then the number of moles of each gas would be equal:

$$V_1 = k_1 n_1 \qquad\qquad V_2 = k_2 n_2$$

$$\text{and}$$

$$k_1 = k_2 \qquad\qquad V_1 = V_2$$

Therefore

$$V_1 = k_1 n_1 = V_2 = k_2 n_2$$

and $$n_1 = n_2$$

This relationship was first theorized by A. Avogadro in 1811 and is often stated in a form called **Avogadro's hypothesis.**

Equal volumes of gases at the same temperature and pressure contain the same number of molecules.

Avogadro's hypothesis is illustrated in Figure 8-2. What good does this relationship do us? Well, since the number of moles of a gas (constant T and P) is directly related to the volume, we can say that what is true for the number of moles of a gas in a reaction will be true for the number of liters of the gas. This means that it is possible to interpret a reaction involving gases at a constant temperature and pressure on a volume basis rather than a molar basis. Thus, the volumes of gases involved in a reaction can be related through **volume ratios** that are analogous to molar ratios. **This will be true as long as the gases are maintained at the same temperature and pressure.** For example, in the reaction represented by the equation

$$H_2(g) + Cl_2(g) \rightarrow 2HCl(g)$$

one mole of hydrogen reacts with one mole of chlorine to give two moles of hydrogen chloride. If the temperature and pressure of the gases are the same, the relative volumes of the gases involved will be the same as the molar amounts that are involved.

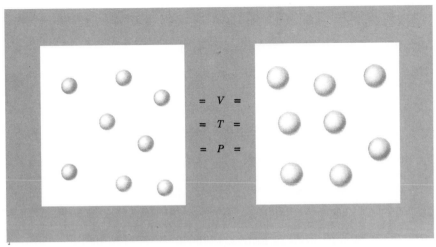

Figure 8-2   Avogadro's hypothesis—equal volumes of gases at the same pressure and temperature contain the same number of molecules.

We can express the volume ratios in the same manner that we expressed molar ratios. The molar ratios for the above reaction are

$$\left(\frac{1 \text{ mole H}_2}{1 \text{ mole Cl}_2}\right) \qquad \left(\frac{1 \text{ mole H}_2}{2 \text{ moles HCl}}\right) \qquad \left(\frac{1 \text{ mole Cl}_2}{1 \text{ mole H}_2}\right)$$

$$\left(\frac{2 \text{ moles HCl}}{1 \text{ mole H}_2}\right) \qquad \left(\frac{1 \text{ mole Cl}_2}{2 \text{ moles HCl}}\right) \qquad \left(\frac{2 \text{ moles HCl}}{1 \text{ mole Cl}_2}\right)$$

and the corresponding volume ratios would be

$$\left(\frac{1 \ \ell \text{ H}_2}{1 \ \ell \text{ Cl}_2}\right) \qquad \left(\frac{1 \ \ell \text{ H}_2}{2 \ \ell \text{ HCl}}\right) \qquad \left(\frac{1 \ \ell \text{ Cl}_2}{1 \ \ell \text{ H}_2}\right)$$

$$\left(\frac{2 \ \ell \text{ HCl}}{1 \ \ell \text{ H}_2}\right) \qquad \left(\frac{1 \ \ell \text{ Cl}_2}{2 \ \ell \text{ HCl}}\right) \qquad \left(\frac{2 \ \ell \text{ HCl}}{1 \ \ell \text{ Cl}_2}\right)$$

These volume ratios can be deduced from the balanced equation representing a reaction involving two or more species in the gaseous phase. These ratios will be valid as long as the gases are maintained at the same conditions of temperature and pressure. Of course, species not in the gaseous phase cannot be related to other species by such volume ratios. These ratios only apply to gases. The volume ratios can be used as factors to convert the volume of one gaseous species involved in a reaction to the volume of another gaseous species.

Example 8-14   How many liters of carbon dioxide gas are produced at STP when 30.0 ℓ of oxygen gas at STP react with sufficient methane to give carbon dioxide and water?
The balanced equation is

$$CH_4 + 2O_2 \rightarrow CO_2 + 2H_2O$$

The volume of CO_2 can be found by multiplying the volume of O_2 by the proper volume ratio. From the balanced equation we can see that the ratio to be used is

$$\left(\frac{1 \ \ell \text{ CO}_2}{2 \ \ell \text{ O}_2}\right)$$

Using this factor, the problem would be solved as follows:

$$30.0 \; \ell \; O_2 \left(\frac{1 \; \ell \; CO_2}{2 \; \ell \; O_2}\right) = 15.0 \; \ell \; CO_2$$

8-7   Energy

Energy is simply defined as the capacity for doing work. When work is done, energy is expended. Energy can be stored and transformed from one form to another. For instance, the kinetic energy or energy of motion of an automobile comes from the mechanical energy output by the engine, which uses the chemical energy of burning gasoline. Before the sources and uses of energy are discussed, it is important to consider how to express amounts of energy. A truck can do more work than a car; a match gives off less heat energy than a fire. Just as amounts of mass can be measured in grams, kilograms, pounds, or tons, various terms have been established for the measurement of energy. We often associate energy with heat. An automobile engine becomes hot; a burning match is hot; an electric lightbulb becomes hot as it lights; a pan of water becomes hot on the stove. Heat is not the only form of energy, but since it is a familiar and easily observed form, it serves as a means to define a unit for the measurement of energy. When a sample of water is heated, it absorbs the heat energy which changes the temperature of the water. The amount of heat absorbed is directly related to the amount of water and the increase in the temperature of the water. It is possible to measure the amount of heat absorbed by measuring the increase in temperature of a given amount of water.

A unit of heat or energy commonly used is the **calorie (cal),** which is defined as the amount of heat required to raise the temperature of 1 gram of water by 1°C. It would take about 75,000 calories to heat 1 quart of water from room temperature to the boiling point. The calorie is used to express amounts of energy in various forms. A kilocalorie (kcal) is 1000 calories.

Another unit of energy used in the United States is the **British thermal unit (BTU).** The British thermal unit is defined as the amount of heat needed to raise the temperature of 1 pound of water by 1°F (1 pound of water is about 0.12 gallons). One British thermal unit is equivalent to 252 calories. Thus, about 300 British thermal units are needed to heat 1 quart of water from room temperature to the boiling point.

Physicists often use another unit of energy called the joule. The **joule** is the defined unit of energy in the metric system of measurement. The magnitude of the joule can be understood if you consider that you would expend approximately 2 joules of energy if you lifted a 1 pound object from the floor to waist height. One calorie is equivalent to 4.184 joules. Consequently, it requires about 320,000 joules to heat 1 quart of water to the boiling point.

8-8   Stoichiometry and Energy Changes

Chemical reactions involve the breaking and forming of chemical bonds. When these processes occur, energy exchanges are involved. Consequently, when a chemical reaction occurs a corresponding energy change occurs. Some chemical reactions release energy to the surroundings and are called **exergonic reactions.** Other reactions require a source of energy to occur and are called **endergonic reactions.** The energy changes that accompany reactions often take the form of heat energy. A reaction in which heat is released is an **exothermic reaction.** A reaction that absorbs heat from the surroundings is an **endothermic reaction.**

Many chemical reactions are carried out for the sole purpose of releasing energy. The sources and uses of energy in our society are discussed in Section 8-10. The sources of the energy changes in chemical reactions are the chemical species involved in the reaction. Each species has a chemical potential energy and, when a chemical reaction occurs, there is an overall change in the chemical potential energy. If the energy content of the products is greater than that of the reactants, an external source of energy is needed so that the reaction can occur. Such a reaction that requires energy is an endergonic reaction. If the energy content of the products is less than the energy content of the reactants, the excess energy is released. This type of reaction is an exergonic reaction. In a chemical reaction that is opened to the atmosphere, the energy involved which takes the form of heat energy is called the **enthalpy change** of the reaction and is given the symbol ΔH.* Actually, the enthalpy change of a reaction depends on the conditions at which the reaction is carried out. Generally, the heat involved in a chemical reaction occurring at 25°C and 1.000 atm pressure is called the heat of the reaction or the standard enthalpy change of the reaction, symbolized as $\Delta H°$. For example, 94.1 kcal of heat are released when 1 mole of carbon in the form of graphite and 1 mole of oxygen gas at 25°C and 1.000 atm pressure react to produce 1 mole of carbon dioxide at the same conditions. Thus, the standard enthalpy change of a reaction is expressed in terms of the number of kilocalories involved per mole of product. Furthermore, by convention, the enthalpy change of an exothermic process is given a negative sign, and the enthalpy change of an endothermic process is given a positive sign. We can express the standard enthalpy change involved in the reaction between carbon and oxygen to produce carbon dioxide as

$$C(\text{graphite}) + O_2(g) \rightarrow CO_2(g) \quad \Delta H° = -\left(\frac{94.1 \text{ kcal}}{1 \text{ mole } CO_2}\right)$$

The $\Delta H°$ expresses the number of kilocalories involved with the production of a mole of carbon dioxide in the reaction, and the negative sign indicates that the reaction is exothermic. The enthalpy change of this reaction can be used as a factor to determine the heat involved when a certain number of moles of CO_2 are produced or to determine the number of moles of CO_2 if the heat evolved is known.

Example 8-15   How many kilocalories of heat are evolved when 2.56 moles of carbon dioxide are produced in the reaction of graphite with oxygen at 1.000 atm and 25°C? The enthalpy change for the reaction can be used as a factor to convert the number of

---

* The symbol Δ (delta) is used to indicate change. Thus, a change is enthalpy, H, represented as ΔH (read as delta H.)

moles of CO_2 to the number of kilocalories evolved:

$$2.56 \text{ moles } CO_2 \left(\frac{94.1 \text{ kcal}}{1 \text{ mole } CO_2} \right) = 241 \text{ kcal}$$

Example 8-16   When 4.79 g of hydrogen gas react with oxygen gas to produce water at 25°C and 1.000 atm pressure, 137.0 kcal of heat are evolved. What is the standard enthalpy change for this reaction?
The balanced equation is

$$2H_2 + O_2 \rightarrow 2H_2O$$

We want to determine the number of kilocalories involved per mole of water produced. We know that 137.0 kcal of heat are produced when 4.79 g of hydrogen react with oxygen. This information can be expressed as

$$\left(\frac{137.0 \text{ kcal}}{4.79 \text{ g}} \right)$$

The grams of H_2 can be converted to moles of H_2 that, in turn, can be converted to moles of H_2O by use of the molar factor obtained from the equation. The conversion sequence should be

$$\frac{\text{kilocalories}}{\text{grams } H_2} \rightarrow \frac{\text{kilocalories}}{\text{moles } H_2} \rightarrow \frac{\text{kilocalories}}{\text{moles } H_2O}$$

$$\left(\frac{137.0 \text{ kcal}}{4.79 \text{ g}} \right) \left(\frac{2.02 \text{ g}}{1 \text{ mole } H_2} \right) \left(\frac{2 \text{ moles } H_2}{2 \text{ moles } H_2O} \right) = \left(\frac{57.8 \text{ kcal}}{1 \text{ mole } H_2O} \right)$$

Since the heat is evolved during the reaction, the enthalpy change is negative and the reaction can be represented as

$$2H_2(g) + O_2(g) \rightarrow 2H_2O(g) \quad \Delta H° - \left(\frac{57.8 \text{ kcal}}{1 \text{ mole } H_2O} \right)$$

Example 8-17   How many grams of methane gas must be reacted with oxygen to produce 100 kcal of heat at 25°C and 1 atmosphere? The reaction and standard enthalpy change are

$$CH_4 + 2O_2 \rightarrow 2H_2O + CO_2 \quad \Delta H° = - \left(\frac{192 \text{ kcal}}{1 \text{ mole } CH_4} \right)$$

The enthalpy change can be used as a factor to convert the kilocalories to moles of methane that can then be converted to grams of methane by use of the molar mass. The conversion sequence is

$$? \text{ g methane} = \text{kcal} \rightarrow \text{moles methane} \rightarrow \text{g methane}$$

$$? \text{ g } CH_4 = 100 \text{ kcal} \left(\frac{1 \text{ mole } CH_4}{192 \text{ kcal}} \right) \left(\frac{16.0 \text{ g}}{1 \text{ mole } CH_4} \right) = 8.33 \text{ g}$$

8-9   Fossil Fuels

Today our major source of energy for our industrial society comes from the combustion or burning of gasoline, oils, coal, and natural gas. These substances, originating from prehistoric plants and animals that be-

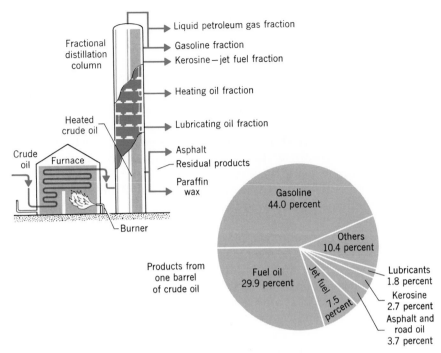

Figure 8-3   Petroleum refining tower. From *Understanding Chemistry: From Atoms to Attitudes* by T. R. Dickson, Wiley, N.Y., 1974.

came entrapped in the lithosphere by geological processes, are called **fossil fuels.** Fossil fuels are composed of hundreds of different carbon-containing compounds. Compounds of carbon are called organic compounds. There are so many organic compounds that an entire branch of chemistry called organic chemistry is devoted to their study. See Chapter 13 for a discussion of organic chemistry. The majority of the organic compounds found in fossil fuels contain only carbon and hydrogen. Such hydrogen and carbon compounds are called hydrocarbons.

Petroleum (petra = rock and oleum = oil) or oil is a complex liquid made up of numerous organic compounds. It is not known exactly how petroleum deposits came about in the lithosphere. It is thought that petroleum was formed millions of years ago when plants and animals living in shallow waters died and became buried in muddy sediments. Certain microorganisms caused these components of the biosphere to undergo partial decay. The sediment layers built up into sedimentary rocks. As a result of the forces of microorganism decay, pressure, and heat from geological phenomena, the plant and animal molecules were converted to petroleum and natural gas deposits. Crude oil, as it is extracted from underground deposits, is a viscous liquid. Crude oil is made up chiefly of hydrocarbon (94 to 99%) with some organic compounds containing sulfur, nitrogen or oxygen. Typical crude oil contains over 500 different compounds. Crude oil composition varies depending upon the region of the lithosphere from which it is obtained.

The refining of petroleum is the process in which the crude oil is separated into useful components by distilling fractions of various boiling point ranges. The **refining process** is illustrated in Figure 8-3. The fraction of crude oil ranging in boiling point from 0° to 200°C is the gasoline fraction. The other fractions are kerosene and jet fuel (b.p. 175 − 274°C), fuel oil and diesel oil (b.p. 250 − 400°C) and lubricating oils. The solids separated from the liquids are paraffin wax and asphalt. Lubricating oils are purified and mixed with various additives before use.

Natural gas consists of gaseous hydrocarbons produced from fossil fuels that have accumulated in pockets in the lithosphere. Natural gas is withdrawn from gas wells, processed, and piped long distances for use as fuel. Natural gas varies in composition but is made up chiefly of methane gas, CH_4, with some ethane, C_2H_6, propane, C_3H_8, and butane, C_4H_{10}. Before natural gas is used most of the hydrocarbons other than methane are removed for use in the manufacture of other organic chemicals.

Coal is solidified plant material that has been deposited in rock layers, has undergone partial decay, and has been subjected to geological heat and pressure. Most coal is thought to have been derived from peat bogs. In fact, the various forms of coal, lignite, bituminous (soft) coal, and anthracite (hard) coal are ancient peat in various stages of decay and compaction. Coal contains carbon and a variety of hydrocarbons. Coal can be burned in air and used as a fuel. However, coal does contain some compounds containing sulfur, oxygen and nitrogen. When soft coal is heated in the absence of air, hydrocarbon substances are given off leaving a residue of impure carbon called coke. Vast amounts of coke are used in the manufacture of steel. The hydrocarbon substances given off during the coking of coal are cooled down and separated into a fraction that condenses to a liquid called coal tar and a fraction that remains gaseous called coal gas. Coal gas consists of hydrogen, H_2, and methane, CH_4, along with a variety of other hydrocarbons and gases. Coal gas is used as a fuel. Coal tars contain a wide variety of organic compounds used in the manufacture of other organic compounds.

8-10   Energy Use and Energy Sources

Each year the United States uses about 1.8×10^{19} cal of energy. As an indication of how much energy this is, consider each of the over 200 million Americans burning one hundred and twenty 100-watt light bulbs for 24 hours each day every day of the year. The energy consumption per person by a country is directly related to the level of industrialization and affluence. The United States, Canada, and Western European countries have high per capita energy consumption and underdeveloped countries, such as India and China, have very low energy consumption. In fact, the United States with about 6% of the people in the world, consumes about one-third of all of the energy used throughout the world each year. The use of energy in the United States has been increasing each year as can be seen from Figure 8-4, which shows the increase in energy use over a period of 100 years along with the per capita increase. The per capita energy use of the world has been

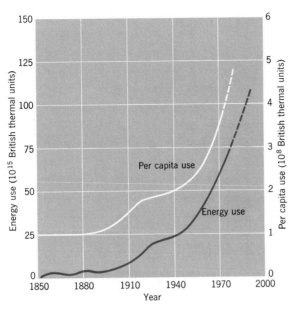

Figure 8-4 The increase in energy use and the per capita increase in the United States. From *Understanding Chemistry: From Atoms to Attitudes* by T. R. Dickson, Wiley, N.Y., 1974.

increasing at a rate somewhat greater than that of the United States.

Figure 8-5 shows the sources and uses of energy in the United States. As can be seen, fossil fuels provide over 95% of the energy. Coal provides 20% of our energy. Somewhat more than half of the coal is used to produce electricity and the remainder is used by industry. Natural gas accounts for about 35% of our energy. Only about one-sixth of this gas is used to produce electricity, and the rest is used directly as a heating fuel. Petroleum and natural gas liquids (butane and propane) contribute about 40% of our energy. Of this only about one-tenth is used to produce electricity and the rest is used directly as fuel. A small portion of our energy comes from hydroelectric sources (around 3.8%) and nuclear fuels (around 0.3%). Both of these sources are used entirely in the production of electricity.

Overall, about 26% of our energy sources are used to produce electricity, and 74% is consumed directly mainly as fuel, but a certain amount finds nonenergy use such as lubricating oils, coke, and petrochemical products. Consumption of electricity is growing at a faster rate than any other use of energy. Consequently, it is expected that an increasing fraction of our energy sources will be used to produce electricity. However, it is thought that nuclear energy sources will provide much of the energy for electricity in the future. See Chapter 15 for a discussion of nuclear energy.

Referring to Figure 8-5 let us now consider the uses of energy in the United States. Industry uses over 40% of the energy in producing consumer goods. The largest fraction of this energy is used in blast furnaces and metal refining to produce metals such as iron and aluminum. Other large industrial energy consumers include oil refineries, mines, chemical plants, glass factories, food processing plants, and paper plants.

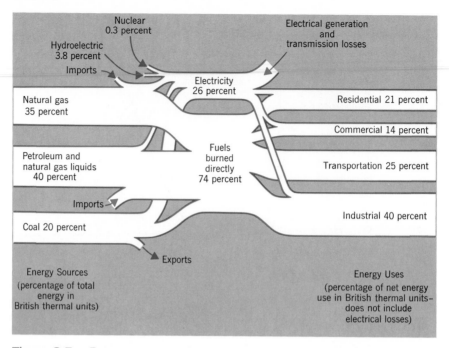

Figure 8-5   Energy sources and energy uses in the United States. From *Understanding Chemistry: From Atoms to Attitudes* by T. R. Dickson, N.Y., 1974.

Commercial establishments (stores, offices, hotels, etc.) account for about 14% of energy use. Half of the energy used commercially is for space heating (heaters) and air conditioning. Transportation uses 25% of our energy. Most of this comes from the burning of gasoline and diesel oil in cars, buses, and trucks. Trains, ships, and airplanes are additional energy-consuming transportation modes.

Domestic uses consume around 20% of our energy. Around half of this energy goes for space heating using natural gas and oil heaters. Other domestic uses include electricity for cooking and heating of water. The rapid growth of residential energy use that has occurred over the last few decades is due to increased electricity demands.

8-11   The Energy Crisis

National concern has developed over what has been termed the **"energy crisis."** The problem revolves around the diminishing supply of domestic energy sources and the increasing demand for energy. The problem of supply is partly due to the fact that fossil fuel sources in the United States are being depleted, and new sources are becoming more difficult to find and more expensive to develop. A potential crisis is predicted based on the rising demand for energy that has been increasing at a rate of over 4% each year. This rate of increase is more than twice the rate of the long-term U.S. population growth.

Three possible approaches to the energy dilemma are apparent. 1. Energy demand can be curtailed by increased energy conservation practices and by enforced limitation of energy use. 2. Energy sources needed to meet energy demands can be imported. It is estimated that by 1985 about 30% of our energy sources will be imported. 3. Domestic energy sources can be developed to meet energy needs. Exploitation of domestic resources may include the large-scale use of public lands for strip mining of coal, the mining of uranium ores, and the extraction of oil and natural gas. Furthermore, more offshore oil leases will be needed, and wildlife sanctuaries may be encroached upon. Alaskan oil and gas reserves will have to be rapidly developed, and numerous fossil fuel and nuclear power plants will be needed.

The national approach to energy use and future energy sources will have profound political, social, economic, and environmental manifestations. In fact, decisions concerning energy use and sources will fundamentally affect the environmental dilemmas of air pollution, water pollution, and strip mining.

8-12   Future Energy Sources

Let us now discuss the possibilities of some future energy sources. (See Figure 8-6.) Fossil fuels will remain the major energy sources throughout this century. By 1985 some 57% of our petroleum and 28% of our natural gas may have to be imported. It is expected that in 1985 around 8% of imported natural gas will be shipped as **liquid natural gas (LNG)** in special tankers. Domestic coal production may double between now and 1985. More efficient use of fossil fuels is being developed including new combustion methods and antipollution techniques. One promising approach that is now being researched is the use of **magnetohydrodynamic (MHD) generators** that convert heat from combustion reactions directly into electricity. MHD appears to be a low pollution alternative to the generation of electricity and has the added advantage of using coal as a fuel.

In the late 1970s it is expected that diminishing natural gas supplies will be supplemented by **synthetic natural gas (SNG)** prepared by chemical processes from petroleum or coal. Figure 8-7 gives a schematic of the production of SNG from petroleum or coal. As can be seen, petroleum or coal gasification involves mixing these carbon-containing substances with steam and converting to a mixture of carbon monoxide (CO) and hydrogen (H_2). These two gases are then combined in the presence of a special catalyst to form methane gas (SNG) and water. One attractive feature of SNG production is that sulfur-containing compounds responsible for some air pollution problems can be removed. This is termed desulfurization. Petroleum gasification is most easily accomplished and will be the first source of SNG. However, coal gasification is a promising long-term source of SNG.

Other long-term sources of oil are petroleum products produced from oil in tar sands and oil shale. Large oil-shale (shale rock formations impregnated with petroleum) deposits are located in the western United States. These deposits represent a large supply of oil, but the petroleum products are difficult to remove from the shale. Furthermore, large amounts of shale would have to be processed and the spent shale disposed of after use.

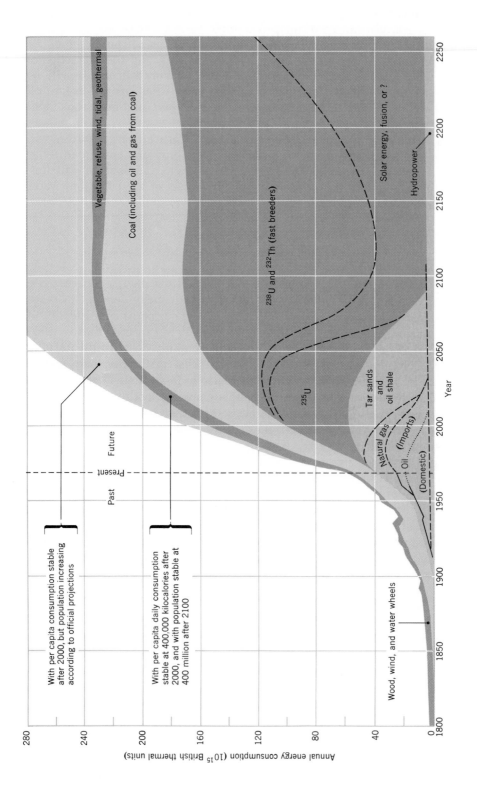

Figure 8-6   Possible future energy sources in the United States. From Earl Cook, *Chemical and Engineering News* **28**, (Jan. 10, 1972).

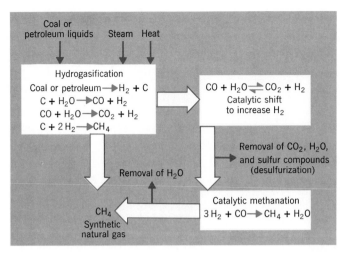

Figure 8-7   Schematic of synthetic natural gas production. From *Understanding Chemistry: From Atoms to Attitudes* by T. R. Dickson, Wiley, N.Y., 1974.

Hydroelectric power sources are limited and, since few new dam sites can be developed, this will be a source of decreasing importance. It is expected that hydroelectric power will decrease from the current 15% of electricity generated in the United States to around 8% in 1985. **Geothermal** sources are now beginning to be exploited. Certain regions of the earth (especially those of recent volcanic activity) have geological formations in which ground water comes into contact with hot rocks and produces hot water or steam that rises to the surface. In certain cases the hot water or steam can be contained and used to turn a turbine to generate electricity. Geothermal sources occur in many regions of the western United States and can be used to supplement the energy needs of certain western states. However, geothermal energy is not expected to contribute much to the total energy needs of the nation.

Other potential energy sources include tides and winds. **Tidal energy,** which uses the tidal forces to generate power, and **wind energy** obtained from windmills could be used in certain regions of the country. They are not expected to contribute significantly to the national energy needs.

It is expected that nuclear energy will supply around half of our electricity by 1985 if development of nuclear power plants continues without delay. The continued use of nuclear energy, however, will depend on the development of commercial breeder reactors as discussed in Chapter 15.

Two long-range possible future energy sources are nuclear fusion and solar energy conversion. Nuclear fusion is promising but is just now in the basic research stage and neither the technological nor commercial feasibility has been demonstrated. See Chapter 15 for a discussion of fusion, fission and nuclear energy.

8-13   Solar Energy

Solar energy represents a large energy source. In fact, the **solar energy** incident on about 0.1% of the land area of the United States would meet

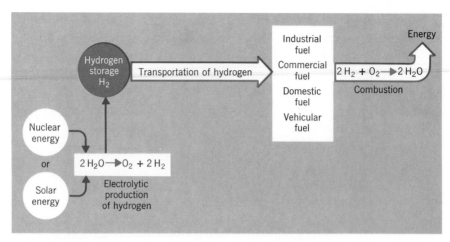

Figure 8-8   The use of hydrogen gas as an energy source. From *Understanding Chemistry: From Atoms to Attitudes* by T. R. Dickson, Wiley, N.Y., 1974.

all of our current energy needs. Some experts feel that solar energy will have to be our future energy source. It is a continuous, environmentally sound source. However, sunlight is diffuse, is interrupted by night, and depends on weather conditions. Use of solar energy will require large land areas for collection and some means of energy storage.

The methods of capturing solar energy and converting it to other energy forms are currently being researched. One method uses **photovoltaic cells** to convert sunlight to electricity. Solar-conversion cells seem to work well to supply energy for spacecraft, but the technology of solar conversion is not developed to the extent that large amounts of energy could be produced by this means. Nevertheless, photovoltaic conversion remains a long-range possibility that would involve energy plants in which several square kilometers of land area are covered with solar cells. Other methods of solar energy conversion are being investigated that involve the use of mirrors or lenses to concentrate the sunlight. The concentrated sunlight would then be used to heat circulating fluids that could then be used to generate electricity.

Even though solar conversion as a large-scale energy source appears to be unlikely in the near future, some scientists claim that solar energy for domestic use will be commercially available by 1980. This will involve the use of **solar energy collection systems** on the roofs of homes. The captured solar energy can then be used for heating and cooling of the homes. Of course, this source of energy could only be used in certain regions of the country and would have to be backed up by normal fossil fuel supplies.

One of the problems of obtaining energy from solar sources or from large-scale nuclear sources is storing energy. That is, since solar sources are intermittent and nuclear sources generate electricity, some means of storing energy for future use would be needed. Several energy storage methods are possible but one involving the use of hydrogen gas is being considered as a very likely possibility. As shown in Figure 8-8 the idea is to use electrical energy to decompose water into hydrogen and oxygen. The hydrogen then would be stored or transported through gas lines, such as existing natural gas lines, for use. **Hydrogen** is an excellent and environmentally sound fuel. Combustion of hydrogen produces only water as a product. Hydrogen gas could be used as a source of industrial energy as well as a fuel for domestic use. It is even possible to use hydrogen as a fuel for automobiles. Hydrogen is highly explosive, but with proper handling it should be no more dangerous than natural gas or gasoline.

Questions and Problems

(*Note:* A ton used in any of the following problems refers to a metric ton. One metric ton is 10^3 kg or 10^6 g. One metric ton equals 1.1 English tons or 2200 pounds.)

1. Butane gas burns in air to form carbon dioxide and water as shown by the equation

$$2C_4H_{10}(g) + 13O_2(g) \rightarrow 8CO_2(g) + 10H_2O(g)$$

Give the molar interpretation of this equation and give the molar ratios that relate the following pairs of reactants and products:
(a) C_4H_{10} and O_2
(b) C_4H_{10} and CO_2
(c) C_4H_{10} and H_2O
(d) O_2 and CO_2

2. Balance the following equation in which acetylene gas and oxygen gas combine to form carbon dioxide and water.

$$C_2H_2(g) + O_2(g) \rightarrow CO_2(g) + H_2O(g)$$

(a) How many moles of oxygen gas are required to react with 4.93 moles of acetylene?
(b) How many grams of carbon dioxide are produced when 2.30 moles of acetylene react with oxygen?
(c) How many grams of acetylene are needed to react with 4.52 moles of oxygen?
(d) How many grams of water are produced when 5.62 g of acetylene react with oxygen?
(e) How many liters of oxygen gas measured at STP are required to combine with 6.93 g of acetylene?
(f) How many liters of oxygen gas measured at 25°C and 1 atm pressure will react with 5.00 ℓ of acetylene gas measured at 25°C and 1 atm pressure.

3. One of the first steps in the refining of lead is the reaction of lead(II) sulfide with oxygen to give sulfur dioxide and lead(II) oxide. Balance the equation for this reaction using the following unbalanced equation:

$$PbS(s) + O_2(g) \rightarrow PbO(s) + SO_2(g)$$

(a) How many moles of oxygen gas are needed to react with 6.25 moles of PbS?

(b) How many grams of PbO are produced from 75.0 moles of PbS?

(c) How many grams of PbO are produced from 50.0 kg of PbS?

(d) How many liters of oxygen gas measured at 30°C and 750 torr pressure are needed to react with 1.00×10^6 g (1 ton) of PbS?

(e) How many grams of PbO are produced when 20.0 ℓ of oxygen gas measured at 250°C and 760 torr pressure react with PbS?

(f) How many liters of SO_2 measured at STP are produced when 500 ℓ of oxygen gas measured at STP react with PbS?

4. Water can be electrolytically decomposed to form hydrogen and oxygen gas. Write the balanced equation and deduce how many hydrogen molecules can be prepared by decomposing 10.0 g of water.

5. The burning of natural gas can be represented by the equation

$$CH_4(g) + 2O_2(g) \rightarrow CO_2(g) + 2H_2O(g)$$

(a) About 2.4×10^{13} cubic feet of CH_4 measured at 20°C and 760 torr are burned in the United States each year. Assuming that the gases are at the same conditions, how many cubic feet of oxygen are needed to combine with the methane?

(b) If 1 ft³ = 28.3 ℓ and the density of methane gas at 20°C and 760 torr is 0.67 g/ℓ, determine the number of grams of CH_4 that is burned in the United States each year [see part (a) for volume of CH_4 used].

(c) How many grams of carbon dioxide are formed each year when the number of grams of methane from part (b) are burned?

6. Gasoline is a complex mixture of hydrocarbons. The burning of gasoline involves numerous reactions. The burning of octane to form carbon dioxide and water can represent the burning of gasoline for sake of discussion.

$$2C_8H_{18}(\ell) + 25O_2(g) \rightarrow 16CO_2(g) + 18H_2O$$

(a) Using a density of 0.703 g/ml for octane, and the fact that 1 gal equals 3.79 ℓ, determine the mass in grams of 1 gal of gasoline (assuming the gasoline to be octane).

(b) How many grams of oxygen are needed to burn 1.00 gal of gasoline (octane) to carbon dioxide and water. See part (a) for the mass of a gallon of gasoline.

(c) How many liters of oxygen at 20°C and 152 torr pressure are used in the burning of 10.0 gal of gasoline. See part (a) for the mass of a gallon of gasoline.

(d) How many gallons of gasoline have to be burned to produce 1 ton of carbon dioxide? [*Hint:* find the grams of octane needed and convert to gallons using the grams per gallon factor from part (a).]

(e) If 1.00×10^{11} gal of gasoline are burned in the United States each year determine the number of liters of carbon dioxide produced at 20°C and 760 torr pressure.

7. Around 30×10^6 tons of sulfuric acid, H_2SO_4, are manufactured in the United States each year. Calculate the number of grams of sulfur used in the manufacture of sulfuric acid. Note that there is 1 mole of sulfur per mole of H_2SO_4.

8. Around 13×10^6 tons of ammonia are produced in the United States each year according to the equation

$$3H_2(g) + N_2(g) \xrightarrow[425°C]{500 \text{ atm}} 2NH_3(g)$$

   (a) How many liters of hydrogen at 425°C and 500 atmospheres are used in the manufacture of ammonia?
   (b) How many liters of nitrogen at 425°C and 500 atmospheres are used in the manufacture of ammonia? The results of part (a) can be used to solve this problem.

9. Nitric acid, HNO_3, is manufactured by the following series of reactions:

$$4NH_3(g) + 5O_2(g) \rightarrow 4NO(g) + 6H_2O$$
$$2NO(g) + O_2(g) \rightarrow 2NO_2(g)$$
$$3NO_2(g) + H_2O \rightarrow 2HNO_3(\ell) + NO(g)$$

   How many grams of nitric acid can be formed from 17 tons of ammonia? (*Hint:* use of the molar ratios from the equations and follow the conversion sequence: $gNH_3 \rightarrow$ moles $NH_3 \rightarrow$ moles $NO \rightarrow$ moles $NO_2 \rightarrow$ moles $HNO_3 \rightarrow gHNO_3$.)

10. The overall reaction in the refining of aluminum is

$$2Al_2O_3 + 3C \rightarrow 4Al + 3CO_2$$

   (a) If 5.0×10^6 tons of aluminum are produced in the United States each year, how many tons of Al_2O_3 are used?
   (b) If bauxite ore averages 50% Al_2O_3, how many tons of ore are used in the United States each year? See part (a).
   (c) How many tons of C are consumed in the production of 1.0 ton of aluminum?

11. When a carbohydrate like glucose, $C_6H_{12}O_6$, is used in body metabolism the overall reaction is

$$C_6H_{12}O_6 + 6O_2 \rightarrow 6CO_2 + 6H_2O$$

   How many grams of oxygen, O_2, are needed for each gram of glucose that reacts?

12. The "phosphate" in detergents is usually sodium tripolyphosphate, $Na_5P_3O_{10}$. If a detergent contains 30% $Na_5P_3O_{10}$, what is the percentage phosphorus in the detergent? Note that there are 3 moles P per mole $Na_5P_3O_{10}$ and 30% means 30 g per 100 g of detergent.

13. A sample of unknown material is analyzed for bromide ion by dissolving the sample and carrying out the following reaction:

$$Ag^+(aq) + Br^-(aq) \rightarrow AgBr(s)$$

   If a 2.37 g sample of the unknown material produces 1.627 g of AgBr what is the percentage of bromide ion in the original sample? (*Hint:* find the grams of Br⁻ from the grams of AgBr and divide by the sample mass to get the percentage.)

14. Give a statement of Avogadro's hypothesis.

15. A balloon containing nitrogen and a balloon containing carbon dioxide have identical volumes, pressures and temperatures. Which balloon contains the greater amount of gas? Explain your answer.

16. Two containers each of which has a volume of 2.0 ℓ are connected by a valve and maintained at 300 K. Hydrogen gas is in one container at a pressure of 760 torr, and oxygen is in the other container at a pressure of 760 torr. The gases are mixed and allowed to react according to the equation

$$2H_2(g) + O_2(g) \rightarrow 2H_2O(\ell)$$

After the reaction occurs and the water forms a liquid describe the contents of the containers in terms of the number of grams of water present and the gas that remains in the container. Remember that $PV = nRT$ can be used to find moles.

17. Define the following terms.
    (a) exergonic reaction
    (b) endergonic reaction
    (c) exothermic reaction
    (d) endothermic reaction
    (e) enthalpy change of a reaction

18. A 25.0 g sample of fluorine gas at 25°C and 1 atm pressure reacts with hydrogen gas at the same conditions to produce hydrogen fluoride gas and 84.4 kcal of heat. Determine the standard enthalpy change for the reaction. Write the balanced equation for the reaction and give the enthalpy change with the proper sign. If 86.3 g of fluorine react with hydrogen at 25°C and 1 atm pressure, how many kilocalories of heat are produced?

19. The combustion reactions and the standard enthalpy changes for methane, propane and butane are

$$CH_4(g) + 2O_2(g) \rightarrow CO_2(g) + 2H_2O(g) \qquad \Delta H = -\left(\frac{192 \text{ kcal}}{1 \text{ mole methane}}\right)$$

$$C_3H_8(g) + 5O_2(g) \rightarrow 3CO_2(g) + 4H_2O(g) \qquad \Delta H = -\left(\frac{489 \text{ kcal}}{1 \text{ mole propane}}\right)$$

$$2C_4H_{10}(g) + 13O_2(g) \rightarrow 8CO_2(g) + 10H_2O(g) \qquad \Delta H = -\left(\frac{636 \text{ kcal}}{1 \text{ mole butane}}\right)$$

Determine which is the best fuel in terms of the number of kilocalories produced per gram of fuel. Do this by converting the enthalpy changes from kcal per mole to kcal per gram of compound.

20. Hydrogen is a potential fuel of the future. The combustion reaction of hydrogen and the enthalpy change are

$$2H_2(g) + O_2(g) \rightarrow 2H_2O(g) \qquad \Delta H = -\left(\frac{28.9 \text{ kcal}}{1 \text{ mole } H_2}\right)$$

(a) How many kilocalories of heat are released in the combustion of 5.0 tons of hydrogen?

(b) How many kilocalories of heat are released per gram of hydrogen? How does hydrogen compare to methane in terms of kilocalories per gram? See problem 19 for methane.

21. The combustion reaction for methane is

$$CH_4(g) + 2O_2(g) \rightarrow CO_2(g) + 2H_2O(g) \qquad \Delta H = - \left(\frac{192 \text{ kcal}}{1 \text{ mole } CH_4} \right)$$

(a) If a methane heater produces 30,000 BTU of heat per hour (30×10^3 BTU/hr) and 1 BTU $= 0.252$ kcal, how many grams of methane are burned per hour?

(b) How many liters of methane at 22°C and 770 torr pressure are used in 1 hr by a 30,000 BTU/hr heater? See part (a).

22. Food contains proteins, fats, and carbohydrates. When proteins, fats, and carbohydrates are used by the body for energy a certain number of calories is obtained from each. The food value of proteins, fats, and carbohydrates is an expression of the number of kilocalories obtained per gram. The average food values are

protein	9.0 kcal/g
fat	4.0 kcal/g
carbohydrate	9.0 kcal/g

A dietic Calorie is equivalent to 1 kcal. The Calorie content of a food depends on the percentages of proteins, fats, and carbohydrates in the food. The Calorie content of a food sample is found by determining the number of grams of protein, fats, and carbohydrates in the sample and multiplying each by the food value. The total number of kilocalories gives the Calorie content. Determine the Calorie content in 100 g of the following foods.

	Protein	Fats	Carbohydrates
Cheddar cheese	28	37	4.0
Round steak	21	17	—
Bacon	10	65	—
Eggs	13	11	—
Peanuts	26	39	22
Apples	0.4	0.5	13

23. Describe the three common types of fossil fuels.

24. Describe some possible energy sources of the future.

9

Chapter 9: Liquids and Solids

Objectives

The student should be able to:

1 Describe the liquid state.
2 Describe how liquids can evaporate.
3 Describe the solidification or freezing of a liquid.
4 Give a description of the six types of solids (ionic, polar molecular, nonpolar molecular, atomic, metallic, and polymeric).
5 Calculate the energy change involved in the phase change of a certain amount of a substance.

Terms to Know

Hydrogen bond
Liquefaction
Critical temperature and pressure
Viscosity
Surface tension
Evaporation
Dynamic Equilibrium
Vapor pressure
Boiling point
Normal Boiling Point
Melting point
Heat of fusion
Heat of crystallization
Heat of vaporization
Heat of condensation
Specific Heat

9

Liquids and Solids

9-1   Collections of Molecules

Before discussing the properties and natures of solids and liquids, it is necessary to consider the forces involved in collections of molecules. In gases, the molecules do not have a great influence on one another. However, in liquids and solids the molecules do interact. Consequently, in order to explain the solid and liquid state, it is necessary to understand the forces of interaction between chemical particles. These are called intermolecular forces of interaction. We know that the constituent atoms of molecules are held in aggregation by covalent bonds. Intermolecular forces are weaker forces than covalent bonds, but such forces of interaction are important when we consider the behavior of collections of molecules. The four common types of **molecular interactions** referred to as **Van der Waals forces** are described in Table 9-1. Let us consider some examples of these forces.

When we try to compress a sample of a liquid we find that liquids resist compression. This resistance to compression results from the **repulsive forces** that arise when liquid molecules are pushed too closely together. We take advantage of this repulsive force in hydraulic systems such as automobile brakes. Pushing on the brake pedal applies a compressive force to the brake fluid. Since the fluid does not readily compress, the force is transmitted through the fluid to the wheel brakes.

A second type of interaction arises between polar molecules. Since polar molecules have separate centers of positive and negative charge, polar molecules can be attracted to one another by electrostatic forces. This is called **dipole-dipole** interaction. Thus, polar molecules can intermix readily in the liquid state. In fact, since water molecules are quite polar, water readily dissolves polar substances. For instance, polar substances like ammonia and simple alcohols are quite soluble in water.

A third type of intermolecular force occurs between polar and non-

Table 9-1   Types of Molecular Interaction

Type of Force	Nature of Force	
Repulsive	The contact of the orbitals of molecules give rise to strong repulsive force.	
Dipole-dipole attraction	Oppositely charged centers of polar molecules cause attraction between molecules.	
Dipole-induced dipole attraction	Charge center of a dipole molecule attracts electrons of nonpolar molecule.	
Instantaneous-induced dipole attraction	Attraction between nuclei of molecules for electrons of other molecules.	
Hydrogen Bonds	Strong dipole-dipole attraction between hydrogens of one molecule and highly electronegative atoms (F, O, N) of other molecules.	

polar molecules. The positive charge centers of the polar molecules may attract the mobile electrons of the nonpolar molecules. This results in an induced dipole in the nonpolar molecule. That is, the positive end of the polar molecule can attract the electrons of a nonpolar molecule. Such **dipole-induced dipole interactions** arise when a nonpolar substance such as oxygen dissolves in a polar substance such as water. However, these interactions are weak, which correlates with the fact that nonpolar substances are only slightly soluble in water. In fact, many nonpolar substances are essentially insoluble in water.

The fourth type of molecular interaction results from the instantaneous attraction of the nuclei of one molecule for the electrons of another and can occur with any type of molecule. This interaction is illustrated on the next page and is called an **instantaneously induced dipole interaction.** These interactions can be formed and broken very quickly, and they occur repeatedly in a collection of molecules.

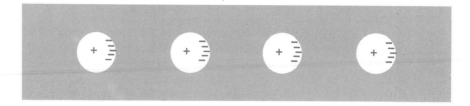

A collection of molecules can then be regarded as consisting of aggregates of molecules that interact by the possible intermolecular forces that are summarized in Table 9-1. However, such a collection is quite dynamic, and the extent and duration of the aggregation depends on the temperature and other factors. If the temperature is high enough so that sufficient energy of motion is available, the forces result in instantaneous aggregations that are constantly being broken up to form new ones. As we shall find out later, these forms of interaction are involved in the formation of the solid and liquid states.

9-2   The Hydrogen Bond

Pure water is a colorless, odorless, tasteless liquid that boils at 100°C and freezes at 0°C. The covalent bonds in water involve the sharing of electrons between the hydrogen and oxygen. The oxygen attracts the shared electrons more strongly than the hydrogen. This results in the electrons being pulled toward the oxygen. Since electrons are negatively charged, this produces a slight negative charge around the oxygen and a positive charge around the hydrogens due to the positive protons in the hydrogen nuclei. This separation of charges makes water molecules polar. That is, the molecules have a positive end and a negative end, much like a magnet has a north pole and a south pole. Molecules of many other substances are polar, but water molecules are highly polar. The polarity of water molecules accounts for some of the unique properties of water as discussed in Section 10-4.

The polar nature of water molecules results in strong intermolecular attraction. The hydrogen ends of the water molecules, with partial positive charges, are attracted to the oxygen region of neighboring molecules, a region of partial negative charge. Perhaps you have noticed an analogous attraction between two bar magnets. When the two like poles of the magnets are brought together, the magnets repel, but when opposite poles are brought together, the magnets attract to form a loose attachment. Water molecules interact in a somewhat similar manner except that the attractive and repulsive forces result from partial electrical charges and not magnetic forces. The attraction between water molecules is strong enough to result in a loose aggregation or clumping of the molecules. In fact, this force of attraction between the hydrogen of one molecule and the oxygen of another is strong enough to be considered to be a type of chemical bond called the **hydrogen bond.** The hydrogen bond is also common to collections of other kinds of molecules in which hydrogen is bonded to such elements as oxygen, nitrogen, and fluorine. Hydrogen bonding is important in certain biological molecules.

Liquid water can be viewed as a loose aggregate of water molecules that are hydrogen bonded to one another:

Perspective of part of above network

These hydrogen bonds are continuously being formed and broken in a collection of water molecules.

9-3   The Liquid State

According to the kinetic molecular theory, no attractive forces exist between the gaseous particles of an ideal gas. Most real gases exhibit ideal behavior only under certain conditions. One of the reasons real gases do not always behave as ideal gases is that certain attractive forces between particles in the gas phase do exist. These attractive forces are the Van der Waals forces.

When a gas is cooled, the average kinetic energy of the particles decreases. At high pressures the gaseous particles move closer together, and the attractive forces become important. Under the proper conditions of temperature and pressure, a gas can be liquefied; converted from the gaseous to the liquid state. **Liquefaction** occurs when the attractive forces between molecules overcome the forces of kinetic motion, which causes the molecules to aggregate into the liquid state. The liquefaction process is illustrated in Figure 9-1. The **liquid state** consists of molecules or, in certain cases, atoms or ions, that are randomly packed in a relatively close manner. However, the liquid state is dynamic, and the molecules are moving about randomly with short distances between collisions with other molecules and the container. These collisions are perfectly elastic with no net loss in kinetic energy. The average kinetic energy of the molecules is directly related to the Kelvin temperature of the liquid. The attractive forces are more important at lower temperatures than at higher ones. A good thing to remember about the liquid state is that it consists of an aggregation of closely packed particles that are in constant random motion. Figure 9-2 gives an illustration of the liquid state.

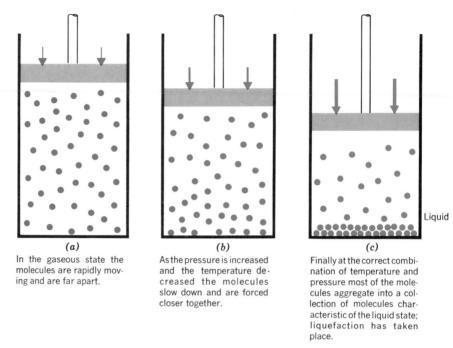

(a)

In the gaseous state the molecules are rapidly moving and are far apart.

(b)

As the pressure is increased and the temperature decreased the molecules slow down and are forced closer together.

(c)

Finally at the correct combination of temperature and pressure most of the molecules aggregate into a collection of molecules characteristic of the liquid state; liquefaction has taken place.

Figure 9-1   The liquefaction of a gas.

If the conditions of a gas sample are adjusted to the proper combination of pressure and temperature, the gas can be liquefied. However, for each real gas there is a temperature above which the kinetic energy or energy of motion of the particles is so great that liquefaction cannot occur. This temperature is called the **critical temperature** of the gas. The pressure required to liquify a gas at the critical temperature is called the **critical pressure** of the gas. If a gas is maintained at a temperature below the critical temperature, it may be liquefied at a correspondingly lower pressure. A gas that is maintained at a temperature below the critical temperature is sometimes called a vapor. To liquefy a gas, it must be cooled below the critical temperature.

9-4   Properties of Liquids

The simple view of the liquid state as a collection of molecules that are constantly moving about and mutually interacting can be used to describe and rationalize some of the properties of liquids. As a result of the freedom of movement and the attractive forces that exist between molecules, all liquids can flow. For example, a sample of a liquid could be made to flow (poured) from one container to another. The attractive forces allow the aggregation of molecules to be transferred as a body. However, these same attractive forces cause a liquid to exhibit a certain internal resistance to flow, which is called **viscosity.** Some liquids, such as oils, can be quite viscous while others have very low viscosities and flow readily. The viscosity of a liquid depends on the nature of its par-

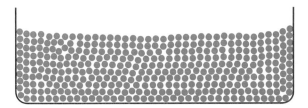

Figure 9-2   The liquid state can be viewed as an aggregation of closely packed particles that move about randomly with short distances between collisions. The particles intermix and move about the aggregation readily.

ticles. At lower temperatures, the viscosity of a liquid is usually greater since the attractive forces rather than the kinetic forces are dominant. The attractive forces have a great influence on the behavior of the liquid state. Within a liquid a given molecule is attracted equally by the molecules around it. The molecules at the surface boundary of the liquid are attracted only by the molecules below them and beside them. As a result of this, unbalanced attractive forces exist. The unbalanced attraction results in a net inward pull tending to draw the surface molecules into the body of the liquid. Consequently, the liquid surface is under a certain strain or tension, which is called the **surface tension** of the liquid. See Section 10-4 for a discussion of the surface tension of water. When a force is exerted on the surface of a liquid, a very small change in volume occurs. Liquids are essentially incompressible because the amount of free space between molecules in the liquid state is quite small.

9-5 Evaporation

In any collection of molecules in the gaseous, liquid, or solid state, all of the molecules do not have the same kinetic energy. In fact, because of

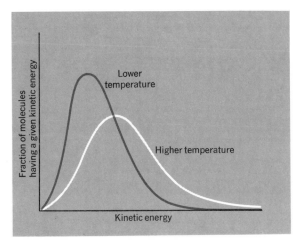

Figure 9-3   A representation of a Maxwell-Boltzmann distribution.

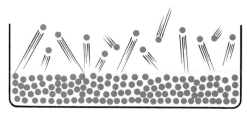

Figure 9-4   Evaporation. Some of the molecules near the surface of a liquid have enough energy to break away from the collection of molecules and enter the vapor state. When these molecules escape from the liquid state, the process is called evaporation.

the kinetic motion and collisions of the particles, exchanges in kinetic energy between molecules can occur. Most particles have similar kinetic energies although some may be very high or very low. At a given temperature, the kinetic energies of the particles in a sample of a substance will fit a certain pattern of distribution. This pattern can be predicted and is often represented in the form of a **Maxwell-Boltzmann distribution** plot as illustrated in Figure 9-3. This distribution is a plot of the fraction of molecules having a particular value of kinetic energy versus the kinetic energy. The plot illustrates the distribution for a collection of molecules at two different temperatures. The greatest fraction of molecules have kinetic energies equal to or near the average. However, as can be seen from the plot, a certain fraction has high kinetic energies and some have very low kinetic energies. Notice also that at a higher temperature, the average kinetic energy of the collection is greater. Of course, we would expect this because the average kinetic energy of such a collection is directly proportional to the temperature.

The kinetic energies of the molecules in a liquid tend to oppose the attractive forces. Some molecules that possess sufficient kinetic energy and are on or near the surface of the liquid can break away from the attractive forces of the other molecules and escape into the vapor state. This is called **evaporation** (see Figure 9-4). Since the higher energy molecules are those that can evaporate, the average kinetic energy of the molecules in the liquid decreases, and thus the temperature of the liquid phase is lowered by the evaporation process. This is why you often feel cooler when water evaporates from your skin. If the environment of a liquid is correctly maintained, it is possible for the liquid to evaporate completely. The rate of evaporation depends on the amount of surface area of the liquid and the distribution of kinetic energies of the molecules. So, if we heat a liquid or spread it out so that it has more surface area, it should evaporate faster.

9-6   Vapor Pressure

Consider a sample of a liquid in a closed container as shown in Figure 9-5. In such a closed container, the molecules that escape into the vapor state by evaporation cannot escape from the container. Under these conditions, a certain amount of evaporation of the liquid occurs, but, eventually, the amount of liquid in the container becomes constant. This can be explained by the fact that as more and more molecules enter the space above the liquid by evaporation, the more often a collision of a vapor molecule with the liquid can occur. Such collisions can result in the vapor molecules reentering the liquid state. This change from the vapor state to the liquid state is called condensation. The evaporation process and the condensation process begin to act in opposition to one another, and, finally, a point is reached at which the rate of condensation equals the rate of evaporation. At this point, the two competing processes produce no net change in the amount of liquid present even though evaporation and condensation are continuously taking place. Such a situation in which two processes are acting in equal and opposite manners is called a state of **dynamic equilibrium.** The term **dynamic** refers to the fact that the processes are taking place, and the term **equilibrium** refers to the fact that the rates of the processes are equal. Dynamic equilibrium is a very important concept in chemistry. Keep in mind that when such a state of equilibrium exists, certain dynamic processes are continuously taking place, but, since the changes oppose one another, no apparent change occurs in the system involving the equilibrium. Other examples of dynamic equilibrium in chemistry are discussed in Chapter 11. Once a state of dynamic equilibrium has been attained, the relative amounts of the liquid and vapor remain constant at a given temperature. In any case, the number of molecules in the vapor phase will be such that the vapor will exert a definite constant pressure at a fixed temperature. The pressure of the vapor in dynamic equilibrium with a liquid at a given temperature is called the **vapor pressure** of the liquid at that temperature. Vapor pressures depend on the nature of the liquid, are usually expressed in units of torr, and can

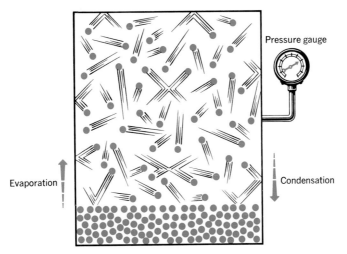

Figure 9-5   Vapor pressure of a liquid. Evaporation of the liquid molecules and condensation of vapor molecules occur continuously. When the rates of evaporation and condensation are equal, the vapor exerts a specific pressure called the vapor pressure of the liquid.

be experimentally determined. Liquids can be described according to their volatility or readiness to evaporate. A liquid that is said to be volatile will have a high vapor pressure. For example, heavy oils are nonvolatile, liquid water is moderately volatile (vapor pressure at 25°C is 23.7 torr), and liquid carbon disulfide (CS_2) is highly volatile (vapor pressure at 25°C is 400 torr).

9-7   Boiling

The vapor pressure of a liquid varies with the temperature. This variation can be represented by plotting the vapor pressure of a liquid versus the temperature. Such a plot is called a **vapor pressure curve,** and is illustrated in Figure 9-6. Note that, as we would expect, the vapor pressure of a liquid increases as the temperature of the liquid increases. Let us consider what happens when we heat a sample of a liquid in an appropriate container that is open to the atmosphere. As the liquid is heated, the temperature will increase. The increase in temperature corresponds to an increase in the kinetic energy of the particles in the liquid. Of course, as the temperature increases, the vapor pressure of the liquid increases. As the heating is continued, eventually a point is reached at which further heating does not increase the temperature of the liquid, and bubbles of vapor are rapidly formed throughout the entire liquid volume. These bubbles rise to the surface and burst as the vapor escapes. When this point is reached, the liquid is said to be **boiling.** Boiling will not take place until the vapor pressure of the liquid is equal to the prevailing atmospheric pressure to which the liquid is exposed. When the vapor pressure of the liquid equals the atmospheric pressure, there is essentially no inhibition of the evaporation process anywhere within the liquid, and very rapid evaporation (boiling) takes place. As boiling takes place, the temperature of the

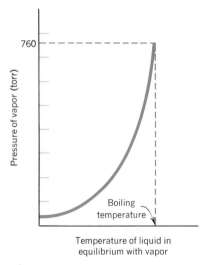

Figure 9-6   A typical vapor pressure curve.

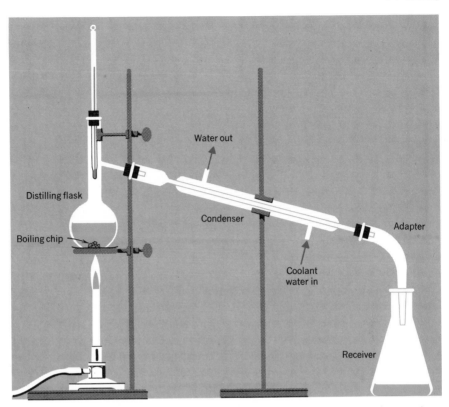

Water out

Distilling flask

Condenser

Boiling chip

Adapter

Coolant
water in

Receiver

Figure 9-7 A liquid is boiled in the flask. The vapors pass into the condenser where they are cooled and condensed back to the liquid form. This liquid flows into the receiver flask. Cold water is continuously passed through the outer jacket of the condenser. The condensation of the vapor occurs in the inner tube.

liquid remains constant. This temperature at which the vapor pressure of the liquid is equal to the atmospheric pressure is called the **boiling point** of the liquid. Each liquid will have a characteristic boiling point, which depends on the volatility of the liquid. The **normal boiling point** of a liquid is the temperature at which the vapor pressure of the liquid is 760 torr or 1 atm. Consequently, normal boiling points are often used in the identification of liquids. The location of the boiling point on a typical vapor pressure curve is illustrated in Figure 9-6. The addition of more heat to a boiling liquid will not increase the temperature of the liquid but will only increase the rate of the evaporation or boiling. As long as the liquid is constantly heated, the boiling process will continue until all of the liquid has been boiled off. Once a liquid has been evaporated to the vapor state, the vapor can escape into the atmosphere. However, if the vapor is collected in some way it can be cooled and caused to condense back to the liquid state. Special laboratory devices are designed to allow for the boiling of a liquid and the subsequent condensation of the vapors. The process of boiling a liquid and condensing the vapors is called distillation. This process can be used to separate liquids of differing volatilities and to purify liquids contaminated with nonvolatile impurities. A typical apparatus used for distillation is illustrated in Figure 9-7.

9-8   The Solid State

As we have seen in the previous section, the liquid state can be considered to consist of a definite grouping of particles in which the particles have a certain freedom of motion within the group. The liquid state represents a situation in which a balance exists between the kinetic forces arising from the kinetic motion of the particles and the attractive forces that exist between particles. When the kinetic force becomes dominant, the importance of the attractive forces diminishes and the liquid state changes to the gaseous state. This change of state can be accomplished by heating. Now, let us consider what happens when a liquid is cooled rather than heated. As the liquid is cooled, the kinetic motion of the particles decreases and the attractive forces become more and more dominant. As cooling is continued, the attractive forces become so dominant and the kinetic motion diminishes to such an extent that the particles begin to occupy relatively fixed positions in space. In fact, as the liquid cools, a point is reached when, in order for the particles to best satisfy the dominant attractive forces, the particles become arranged in a definite three-dimensional pattern in which they occupy definite spatial positions. When this occurs, the liquid is said to have changed to the solid state. The phase change from the liquid to the solid state is called **solidification, crystallization,** or **freezing.** See Figure 9-8 for an illustration of the freezing process. The crystalline

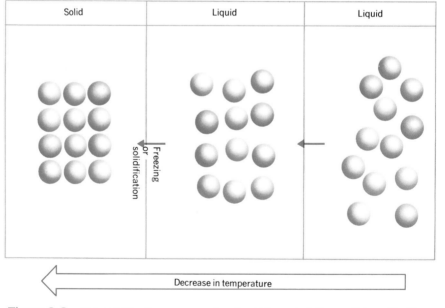

Figure 9-8   The solidification process. As a liquid is cooled, the particles will ultimately arrange into a definite pattern in which the particles have fixed positions in space. When this occurs, the liquid has solidified.

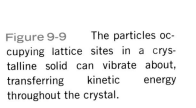

Figure 9-9   The particles occupying lattice sites in a crystalline solid can vibrate about, transferring kinetic energy throughout the crystal.

solid state is characterized by a definite three-dimensional distribution of particles occupying relatively fixed positions in space. Such a three-dimensional arrangement is called a **crystal lattice** or simply a **crystal.** The positions that the particles occupy in the crystal lattice are called **crystal lattice sites.** The particles occupying the lattice sites are held in a definite arrangement by the attractive forces. However, these particles still possess kinetic energy so they can be considered to be vibrating about their positions in the crystal lattice. See Figure 9-9. This vibratory motion allows for the transfer of kinetic energy throughout the solid. As a result of this, there exists a certain distribution of the kinetic energies of the particles in the solid state that is analogous to the distribution of kinetic energies characteristic of particles in the liquid or gaseous states. At a given temperature, a certain number of particles near the surface of a solid have high enough kinetic energies so that they may break away from the solid state and enter the vapor state. Consequently, a solid will have a specific vapor pressure at a given temperature. However, the vapor pressures of solids are usually much lower than the vapor pressures of liquids. As we would expect, the average kinetic energy of the particles in a sample of a solid substance is directly proportional to the Kelvin temperature. If a solid is heated, it is possible to add enough energy so that the particles will break away from the fixed positions in space and form the liquid state. For a given substance, this transition from the solid to the liquid state will occur at a definite temperature. This temperature is called the **melting point** of the solid. The melting point of a pure solid is often used to identify the substance.

The nature of the particles that occupy the crystal lattice sites depends on the nature of the solid. That is, the particles may be ions, atoms, or molecules. In any case, the particles will arrange in a manner dictated by the attractive forces that exist between the particles. These forces may be very strong, such as the electrostatic force of attraction between oppositely charged ions or the covalent bonding force involving the mutual sharing of electrons between combined atoms. On the other hand, these forces may involve intermolecular forces, such as hydrogen bonding or Van der Waals forces. Solids can be visualized as a definite spatial arrangement of chemical particles which give a specific shape and substance to the solid state. Some solids are soft and pliable, some are brittle and others are rigid and resist deformation.

9-9   Types of Solids

For sake of discussion, we can classify solids according to the nature of the particles that occupy the lattice sites of the crystal. Six types of solids distinguished in this manner are described below. See Table 9-2.

1. Ionic solids are made up of positive and negative ions arranged in a geometrical pattern which is dictated by the shapes, sizes and charges of the ions. Ionic bonds hold the particles in the crystal.

2. Polar molecular solids consist of polar molecules arranged in a crystal lattice according to the mutual attraction between the oppositely charged ends of the molecules.

3. Nonpolar molecular substances form crystals in which the molecules are held in position by instantaneously induced dipole attraction.

Table 9-2   Types of Solids

Type	Particles Occupying Lattice Sites	Forces Between Particles	Properties of Solids	Examples
Ionic solids	Positive and negative ions	Attraction between ions of opposite charge (ionic bonds)	High melting points Nonvolatile Hard, but shatter easily	$NaCl(s)$ $CaF_2(s)$ $(NH_4)_2SO_4(s)$
Polar molecular solids	Polar molecules	Dipole-dipole attraction between molecules	Moderate melting points Moderate volatility and hardness	$HCl(s)$ $H_2S(s)$ $SO_2(s)$
Nonpolar molecular solids	Nonpolar molecules	Instantaneous induced-dipole attraction	Low melting points Low volatility Soft	$I_2(s)$ $CH_4(s)$ $O_2(s)$
Atomic solids	Atoms	Covalent bonds between atoms	Very high melting points Very hard and non-volatile	$C(s)$ diamond $Ge(s)$ $SiC(s)$
Metallic solids	Metal atoms	Attraction between outer energy level electrons and positive atomic centers	Variable melting Low volatility Variable hardness	$Na(s)$ $Mg(s)$ $Al(s)$
Polymeric solids	Macromolecules (many atom molecules called polymers)	Van der Waals attraction and hydrogen bonds	Variable melting Low volatility Variable hardness	polystyrene polyethylene starch protein

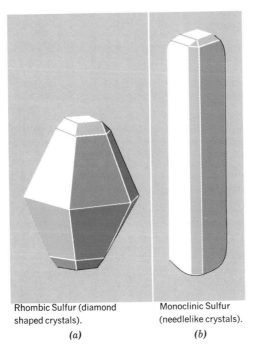

Rhombic Sulfur (diamond shaped crystals).

(a)

Monoclinic Sulfur (needlelike crystals).

(b)

Figure 9-10   Two crystalline forms of sulfur.

4. Atomic solids consist of network of covalently bonded atoms. These solids are not common but some are noteworthy. Diamond is pure carbon in which the atoms are arranged in a three-dimensional pattern of covalent bonds such that the entire crystal can be considered as a giant molecule.

5. Metals form solids in which the metal atoms occupy lattice sites. The nature of bonding in **metallic** solids is not completely understood, but most likely involve the outer level electrons of metal atoms interacting with the nuclei of other metal atoms.

6. Polymers are made up of macromolecules consisting of hundreds or thousands of covalently bonded atoms. In **polymeric** solids the macromolecules are distributed in a pattern dictated by intermolecular attraction. Polymeric solids include plastics, rubbers, some carbohydrates, proteins and some other biological substances.

The six types of solids discussed above are convenient generalizations of the nature of the solid state. A summary of these solid types is contained in Table 9-2. Some substances form solids that take on forms that do not fit one of these six classifications but rather are intermediate between two or more forms. Nevertheless, many substances in the solid state are quite similar to one of these six solid types. Some substances may crystallize in two or more forms depending on the environmental conditions. These forms may merely involve different geometrical configurations of particles or completely different crystal forms. The ability of a substance to form more than one crystal form is called **polymorphism** and the different forms are called **polymorphic** forms of the substance. Sulfur, S_8, usually exists in one of two polymorphic crystalline forms. The difference between the two forms is that the S_8 molecules have different spatial arrangements in the crystal lattice. See Figure 9-10.

9-10   Changes of State

When a solid is heated, the constituent particles of the solid gain kinetic energy. When the kinetic energy of the particles is great enough, the solid can melt and become a liquid. Of course, this change from the solid state to the liquid state occurs at the melting point of the solid. All of the solid does not immediately melt when the solid is heated to the melting point. Actually, when the solid is heated to the melting point a certain amount of heat is required to convert all of the solid to the liquid state. That is, when the melting point is reached, further heating does not change the temperature but is used up in the process of melting. Such a phase change requires a certain amount of energy to allow the particles to break away from the crystal lattice sites and enter the liquid state in which the positions of the particles are less definite. When a change occurs in which the positions of the particles become less definite, we say that the change involves transfer from an ordered state to a more random state. The amount of energy required to convert a specific amount of a solid to the liquid state at a given temperature is called the **heat of fusion** of the solid. Most solids have unique heats of fusion that depend on the nature of the crystal state and the attractive forces that exist between particles. Heats of fusion are often expressed in units of calories per gram of the substance or in kilocalories per mole and are represented by the symbol ΔH_{fusion} (read as delta H fusion). We are familiar with the melting of ice so let us consider water as an illustrative example of the solid to liquid phase change. When ice is heated to 0°C and begins to melt, it requires 79.7 cal of heat to melt each gram of water. This can be expressed as

$$\left(\frac{79.7 \text{ cal}}{1 \text{ g}} \right)$$

Example 9-1   If the heat of fusion of ice at 0°C is 79.7 cal/g, what is the heat of fusion of ice in units of kilocalories per mole of water? (The heat of fusion expressed in the units of kilocalories per mole is called the molar heat of fusion.)

Since we know the number of calories required to melt a gram of ice at 0°C and we know the number of grams per mole of water, the molar heat of fusion can easily be found by multiplying the number of calories per gram by the number of grams per mole. Of course, the number of calories has to be converted to kilocalories as a final step.

$$\left(\frac{79.7 \text{ cal}}{1 \text{ g}} \right) \left(\frac{18.02 \text{ g}}{1 \text{ mole H}_2\text{O}} \right) \left(\frac{1 \text{ kcal}}{10^3 \text{ cal}} \right) = \left(\frac{1.44 \text{ kcal}}{1 \text{ mole H}_2\text{O}} \right)$$

Using the same sign convention established for the enthalpies of reactions (see Section 8-8), this phase change can be represented as

$$\text{H}_2\text{O(s)} \rightarrow \text{H}_2\text{O}(\ell) \quad \Delta H_{\text{fusion}} = + \left(\frac{1.44 \text{ kcal}}{1 \text{ mole H}_2\text{O}} \right)$$

The positive sign of the heat of fusion indicates that energy is required for the phase change from the solid state to the liquid state.

When a liquefied solid is cooled, it will usually freeze at the same temperature at which the solid melted. This temperature is called the freezing point of the substance. The freezing or solidification process is essentially the reverse of the melting process. As the particles become arranged in the crystal lattice of the solid, the excess kinetic energy of the particles is given off. The amount of heat evolved when a certain amount of a liquid is solidified at a specific temperature is called the **heat of crystallization** of the substance, ΔH_{cryst}. Since the crystallization or solidification process is the reverse of the fusion or melting process, the heat of fusion and the heat of crystallization have the same value. The difference is that melting requires energy (ΔH_{fusion} is positive) and crystallization gives off energy (ΔH_{cryst} is negative). If we know the value of the heat of fusion of a solid at a given temperature, we will know the value of the heat of crystallization at that temperature. For example, the solidification of liquid water at 0°C can be represented as

$$H_2O(l) \rightarrow H_2O(s) \quad \Delta H_{\text{cryst}} = -\left(\frac{1.44 \text{ kcal}}{1 \text{ mole } H_2O}\right)$$

Notice that the sign of the heat of crystallization indicates that 1.44 kcal of heat are evolved for every mole of water that freezes at 0°C.

Whenever a phase change occurs in which the particles of the substance are more randomly distributed, energy is required. On the other hand, when the phase change involves a change to a more ordered arrangement from a more random distribution, energy is evolved. When a liquid is evaporated or boiled, the particles change from the more ordered liquid state to the more random vapor state. Since energy is required for this vaporization process, a certain amount of heat is required to change a specific amount of liquid to the vapor state at a given temperature. This amount of heat is called the **heat of vaporization** of the liquid, ΔH_{vap}. Each liquid has a unique heat of vaporization. For example, the heat of vaporization of liquid water at 100°C is 540 cal/g or 9.72 kcal/mole H_2O. The vaporization of water at 100°C can be represented as

$$H_2O(l) \rightarrow H_2O(g) \quad \Delta H_{\text{vap}} = +\left(\frac{9.72 \text{ kcal}}{1 \text{ mole } H_2O}\right)$$

When a vapor condenses to a liquid, the particles change from a more random state to a more ordered state. Thus we would expect that the condensation process releases energy. In fact, the condensation process can be considered to be the reverse of the vaporization process. The amount of energy evolved when a certain amount of a vapor condenses at a given temperature is called the **heat of condensation** of the vapor, ΔH_{cond}. Each liquid will have a unique heat of condensation. Of course, since condensation is the reverse of vaporization, the heat of condensation of a liquid will be of the same value as the heat of vaporization of the liquid at a specific temperature. The difference would be that vaporization requires energy and condensation releases energy. The condensation of steam at 100°C can be represented as

$$H_2O(g) \rightarrow H_2O(l) \quad \Delta H_{cond} = -\left(\frac{9.72 \text{ kcal}}{1 \text{ mole } H_2O}\right)$$

The negative sign on the heat of condensation indicates that 9.72 kcal of heat are evolved when a mole of water vapor condenses at 100°C.

The ΔH values associated with a given phase change can be used to determine the total energy change that occurs when a certain amount of a substance undergoes a phase change.

Example 9-2   How many calories of heat are needed to vaporize 273 g of water at 100°C?

Since we know how many calories are needed to vaporize a gram of water at 100°C, we can multiply the mass of water by this factor to find the total number of calories needed.

$$273 \text{ g} \left(\frac{540 \text{ cal}}{1 \text{ g}}\right) = 1.47 \times 10^5 \text{ cal}$$

9-11   Specific Heat

If you have ever heated water on a stove or held a piece of metal in a flame you know that it is possible to change the temperature of a substance by heating it. When a substance is exposed to a source of heat energy the particles of the substance gain kinetic energy, and the temperature increases. Moreover, if a hot object is placed in a cooler environment it will lose heat energy, and the temperature will decrease.

It is possible to experimentally determine the amount of heat required to change the temperature of a specific amount of a substance. The **specific heat** of a substance is the number of calories needed to change the temperature of one gram by one degree Celsius. The specific heats of substances are quite variable. For instance the specific heat of water is 1 calorie per gram-degree Celsius and that of aluminum is 0.212 calories per gram-degree Celsius. Recall from Section 1-13 that the calorie is defined as the amount of heat required to change the temperature of one gram of water by one degree Celsius. This of course establishes the specific heat of water as

$$\left(\frac{1.00 \text{ cal}}{1 \text{ g } 1°C}\right)$$

Specific heats of substances can be used to determine the number of calories required to change the temperature of a given mass of the substance by a specified number of degrees.

Example 9-3   How many calories are required to change the temperature of 25.0 g of water from 25.0°C to 28.0°C?

The number of calories is found by multiplying the grams of water and the temperature change $(28.0°C - 25.0°C = 3.0°C)$ by the specific heat of water.

$$(25.0 \text{ g})(3.0°C) \left(\frac{1 \text{ cal}}{1 \text{ g } 1°C}\right) = 75 \text{ cal}$$

It will require 75 cal of heat to warm the 25.0 g of water by 3.0°C.

The specific heat can be used in conjunction with the heats of phase changes to determine the number of calories involved in the heating or cooling of a substance and an accompanying phase change.

Example 9-4   How many calories are required to melt 50.0 g of ice at 0°C, heat the resulting water from 0°C to 100°C, and vaporize the water to steam at 100°C? The heat of fusion of water (79.7 cal/g) is used to find the calories needed for melting. Next the calories needed to heat the water by 100°C (0°C to 100°C) are found using the specific heat of water. Then the calories needed to vaporize are found using the heat of vaporization of water (540 cal/g). The total number of calories needed is the sum of these three values.

$$\text{? calories} = 50.0 \text{ g} \left(\frac{79.7 \text{ cal}}{1 \text{ g}}\right) + (50.0 \text{ g})(100°C) \left(\frac{1 \text{ cal}}{1 \text{ g } 1°C}\right) + 50.0 \text{ g} \left(\frac{540 \text{ cal}}{1 \text{ g}}\right)$$

$$\underbrace{\qquad}_{\text{melting step}} \qquad \underbrace{\qquad}_{\text{water-warming step}} \qquad \underbrace{\qquad}_{\text{vaporizing step}}$$

$$\text{? calories} = 3.98 \times 10^3 \text{ cal} + 5.00 \times 10^3 \text{ cal} + 27.0 \times 10^3 \text{ cal} = 36.0 \times 10^3 \text{ cal}$$

Questions and Problems

1. Describe the interaction that can take place between polar molecules.

2. Describe the interaction that can take place between polar and nonpolar molecules.

3. Describe the instantaneously induced dipole interaction that can occur between molecules.

4. What is a hydrogen bond? Give an example.

5. Describe the liquid state in terms of a collection of molecules.

6. Describe the liquefaction of a gas in terms of the particles comprising the gaseous and liquid states.

7. Two properties of the liquid state are viscosity and surface tension. Explain these two properties in terms of the interactions between particles within a liquid.

8. How does a liquid evaporate?

9. Why does the skin become cooler when water or some other liquid evaporates from the skin?

10. Give definitions for the following:
    (a) vapor pressure of a liquid
    (b) boiling point of a liquid

11. What is the normal boiling point of a liquid?

12. Why is the normal boiling point of water exactly 100°C and the freezing point of water exactly 0°C?

13. A student measures the boiling point of pure water in the laboratory and

finds it to be 100.2°C. If her thermometer was working correctly, why does she obtain this value for the boiling point of water?

14. Describe what happens in terms of molecular motion when a liquid solidifies or freezes to a solid.

15. Give a definition for melting point.

16. Give a brief description of each of the following types of solids by indicating the particles occupying the lattice sites and the forces holding them in position. Also, give an example of each type.
   (a) ionic solid
   (b) polar molecular solid
   (c) nonpolar molecular solid
   (d) atomic solid
   (e) metallic solid
   (f) polymeric solid

17. Indicate the type of solid you would expect to be formed by each of the following (See Table 9-2).
   (a) potassium chloride, KCl
   (b) krypton, Kr
   (c) bromine, Br_2
   (d) polyvinyl chloride plastic
   (e) gold, Au
   (f) hydrogen fluoride, HF
   (g) carbon tetrachloride, CCl_4
   (h) sodium phosphate, Na_3PO_4
   (i) magnesium, Mg

18. Give definitions for the following:
   (a) heat of fusion
   (b) heat of crystallization
   (c) heat of vaporization
   (d) heat of condensation

19. What is specific heat?

20. How many calories of heat are required to heat 2.0 quarts of water from 20°C to 100°C and to convert the water at 100°C to vapor? Assume a specific heat for water of (1.0 cal/g °C) and 1 quart = 0.946 ℓ.

21. An ice cube tray contains 500 g of water. How many calories of heat are lost when the water is cooled from 25°C to 0°C and then frozen to ice at 0°C? Assume a specific heat for water of (1.0 cal/g °C).

22. Some persons place large tubs of water in greenhouses during frosty weather. How does this help to prevent frost damage to plants?

23. Why is a steam burn often more serious than a burn by another gas that is of the same temperature of the steam?

24. One calorie is the amount of heat required to change one gram of water by one degree Celsius (1 cal/1 g 1°C). A British Thermal Unit or BTU is the amount of heat required to change one pound of water by one degree Fahrenheit (1 BTU/1 lb 1°F). Using the relationship between a degree Celsius and a degree Fahrenheit and the factor (453.6 g/1 lb) determine the number of calories per 1 BTU.

25. If 150.5 cal of heat are released when 5.0 g of benzene freeze at the normal freezing point, what is the heat of crystallization of benzene, C_6H_6, in units of calories per mole? Give an equation representing the freezing of benzene and the energy change.

26. What is the heat of fusion of benzene at the freezing point of benzene? (See problem 25).

10

Chapter 10: Water and Solutions

Objectives

The student should be able to:

1 Describe the water cycle.
2 Calculate the molarity of a solution.
3 Use molarity as a mole-to-volume or volume-to-mole factor.
4 Calculate the number of grams of a solute needed to prepare a specific volume of a solution of known molarity.
5 Calculate the volume of solution needed to prepare a lower molarity solution by dilution.
6 Calculate the volume to which a sample of solution should be diluted to prepare a solution of lower molarity.
7 Calculate the concentration of a solution in terms of percentage by mass, parts per million, or molality.
8 Calculate the volume of a solution of known molarity needed to react with a specified amount of a substance.
9 Calculate the amount of a substance that will react with a specified volume of a solution of known molarity.

Terms to Know

Water table
Aquifer
Solution
Solvent
Solute
Molarity
Standard solution
Colligative properties
Osmosis
Salinity
Chlorinity

10

Water and Solutions

10-1 Water

Water, the ubiquitous compound that covers nearly three-quarters of the surface of the earth, is thought to have existed in the celestial material from which the earth might have been formed billions of years ago. Water, through erosion and glacial action, has played a significant part in forming the physical features of the earth. Photographs from Apollo spacecraft have emphasized the beauty and turbulence of the everchanging cloud cover shrouding the globe. Water is the only common substance that exists in the environment in all of the three states of matter: solid, liquid, and gas.

The waters of the primeval oceans are thought to be the birthplace of primitive life forms. All forms of life are dependent on water and, to varying degrees, are composed of water. Ancient civilizations developed in regions of abundant water. The Egyptian civilization of the Nile and Mesopotamian civilization of the Tigris and Euphrates river valleys are familiar examples. Thousands of years ago similar civilizations developed around Indus and Ganges rivers of India and the Yellow and Yangtze rivers of China. Of course, today these two areas support about 40% of the entire population of the world.

Water has served as a source of energy and a medium of energy transfer that has allowed the development of industrial societies. The waterwheel and steam engine were fundamental to the Industrial Revolution. Today, hydroelectric power, steam engines, and steam turbines are basic to our industrial society.

10-2 Water in the Ecosphere

The waters of the hydrosphere are continuously being transferred from the oceans to the land areas and back to the oceans in cyclic process called the water cycle. Since relatively little new water is generated by volcanic activity, the same water has been passing through the water cycle for ages. The same water molecule found in a drop of your saliva may have spent millions of years in the primeval oceans and thousands of years in a polar ice cap. It may have been carried through the atmosphere as vapor and deposited in the Nile River during the reign of a great pharaoh. The same molecule may have drifted in the Mediterranean Sea and through evaporation and condensation entered a Roman aqueduct where it was consumed by Caesar, becoming part of his body water. It may have left his decaying body and spent centuries in the local water cycles of medieval Europe. In 1492, this molecule may have traveled with Columbus to be urinated in the New World. It may have become part of the underground water supply of the American colonies and been carried West by the pioneers. The molecule may have circled the globe countless times or been caught in local water cycles finally entering your body yesterday.

Before discussing the details of the water cycle let us consider the distribution of the earth's water. It is estimated that the total volume of the water of the **hydrosphere** is 1.5 billion cubic kilometers. As shown in Table 10-1, over 97% of this water is in the oceans that cover nearly three-quarters of the surface of the earth. Slightly more than 2% is in the form of ice and snow in the polar icecaps and glaciers. The remainder is made up of the surface and subsurface waters of the lithosphere and the water vapor and clouds of the atmosphere. Only about 0.001% of the earth's water is in the atmosphere.

We are familiar with ocean water and surface waters in lakes, rivers, and streams, but most of the water in the lithosphere is **subsurface ground water.** When water falls on soil it wets the soil. When the soil is saturated, the excess water may run off as streams and some percolates down through the soil particles and porous subsurface rocks until it reaches some impermeable rock. At this point it accumulates forming a saturated subsurface zone. The top of this zone is called the **water table** as shown in Figure 10-1. Ground water can flow horizontally when it accumulates in porous geological formations called **aquifers.** Aquifers may be a few meters to hundreds of meters thick and may underlie an area of a few hundred square meters or many square kilo-

Table 10-1   The Distribution of Water in the Hydrosphere

	Percentage of 1.5 Billion Cubic Kilometers
Oceans	97.2
Polar icecaps and glaciers	2.15
Subsurface water (soil moisture and ground water)	0.63
Surface water (fresh-water lakes, rivers, and streams, and saline lakes and landlocked seas)	0.019
Atmospheric water	0.001

From *Understanding Chemistry: From Atoms to Attitudes* by T. R. Dickson, Wiley, N.Y., 1974.

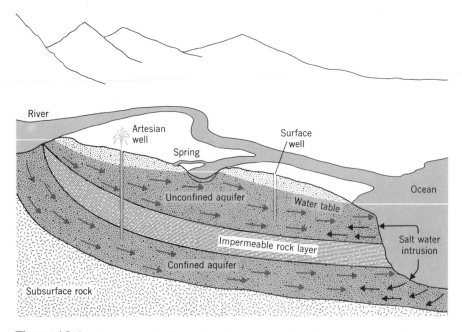

Figure 10-1   Ground water, the water table, and aquifers. From *Understanding Chemistry: From Atoms to Attitudes* by T. R. Dickson, Wiley, N.Y., 1974.

meters. In fact, within the aquifer ground water can flow for hundreds of miles to fill spring-fed lakes or to empty into the ocean. However, ground water movement is quite slow and is often measured in meters per year as compared to river flows of meters per second or minute. Some relatively confined aquifers contain water which was deposited 10,000 to 30,000 years ago.

When the land surface dips below the water table, the water may become surface water as a spring or accumulate as a lake as illustrated in Figure 10-1. An **unconfined aquifer** may be tapped by use of a surface well; a shaft sunk into the ground to allow access to the subsurface water table. A **confined aquifer** is one in which the vertical movement of the water is prevented by an impermeable rock layer (Figure 10-1). Often water in a confined aquifer is under pressure, and when a well is sunk into the aquifer the pressure will drive the water upward as an **artesian well.**

10-3   The Water Cycle

The **water cycle** is illustrated in Figure 10-2. The oceans of the world are exposed to large amounts of solar radiation. About one-half of the solar radiation absorbed by the sea results in the evaporation of the ocean water. Solar radiation is the source of energy for the water cycle and the weather fluctuations and turbulence that is associated with the cycle. Water vapor formed by evaporation from the oceans, lakes, rivers, or soil by solar radiation or transpired from the leaves of plants may travel long distances in the atmosphere. **Transpiration** is the phenomenon by which plants release excess water to the atmosphere through tiny leaf pores called stomata that also act as exchange pores for the oxygen and carbon dioxide involved in photosynthesis.

Weather fluctuations are accompanied by increasing and decreasing concentrations of atmospheric water vapor. Some of the water vapor condenses into clouds of water droplets or ice crystals. Interestingly, the formation of clouds occurs when condensation occurs on tiny particulates of dust, smoke, and sea salt in the air. **Precipitation** depends on these particulates that serve as nuclei upon which the cloud particles grow. When the particles are large enough, they fall to the earth as rain, hail, or snow.

Much of the precipitation occurs over the ocean and does not reach the land surface. However a portion of the evaporated waters of the oceans is deposited as fresh water on the land. The water that reaches the lithosphere may almost immediately reevaporate or accumulate in lakes, streams, or rivers as surface water. However, much of the precipitation percolates through the soil and becomes ground water. A portion of the precipitation is captured in the snow packs of mountains and in the polar ice-caps and glaciers. Some of the water passes through the life cycle of animals and plants serving as a source of hydrogen in photosynthesis and as a component of living cells.

Ultimately, much of the fresh water of the lithosphere flows back into the oceans to complete the water cycle (see Fig. 10-2). Of course, before the water reaches the oceans it may be used numerous times for agricultural irrigation, industrial needs, domestic purposes, and to carry away our wastes and sewage.

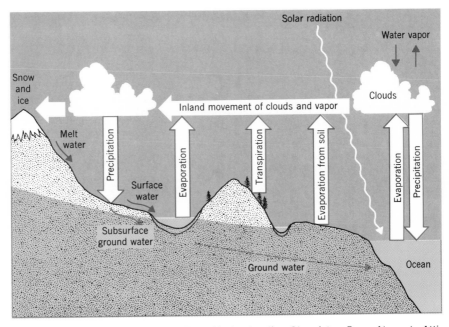

Figure 10-2   The water cycle. From *Understanding Chemistry: From Atoms to Attitudes* by T. R. Dickson, Wiley, N.Y., 1974.

10-4   The Properties of Water

Recall that polar molecules of water interact by hydrogen bonding. Water can be viewed as consisting of ever changing aggregations of water molecules that clump together by hydrogen bonds. With this view of water in mind let us consider some of the properties of water.

Hydrogen bonding accounts for the fact that ice floats on water. Most substances expand when heated and contract when cooled. Water follows this behavior except that at 4°C it no longer contracts but begins to expand. As it is cooled further and freezes into ice it suddenly expands by 10% to form a solid that is less dense than the water from which it was formed. Let us consider how hydrogen bonding explains this behavior. Hot water contains numerous clumps of hydrogen bonded molecules. These clumps break and reform in fractions of seconds. As the water is cooled, the kinetic energies of the molecules decrease and larger hydrogen bonded aggregations are formed. The molecules come closer together and the liquid contracts in volume. At 4°C the kinetic energies of the water molecules are low enough so that hydrogen bonding becomes the dominant force that dictates the arrangements of the molecules in the liquid. The molecules begin to arrange themselves in a pattern in which the maximum amount of hydrogen bonding can occur. As freezing occurs, the water molecules form an open cagelike network of hydrogen bonded water molecules, characteristic of ice. This structure allows for the maximum amount of hydrogen bonding in which each water molecule is bonded to others. The structure of ice is such that the water molecules are further apart than they are in liquid water, and, thus, the ice is less dense than the water. As ice melts to a

Figure 10-3

The boiling and freezing behavior of water. The temperature range between the freezing and boiling points of water deviates from the expected behavior of water compared to chemically similar compounds. The deviation is attributed to hydrogen bonding effects which allow the water molecules to clump together and behave as more massive molecules Ammonia, NH_3, and hydrogen fluoride, HF, also display this anomolous behavior when compared to chemically similar compounds. Hydrogen bonding accounts for the deviant properties of these molecules. From *Understanding Chemistry: From Atoms to Attitudes* by T. R. Dickson, Wiley, N.Y., 1974.

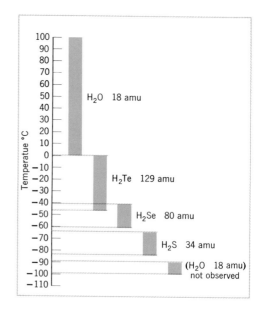

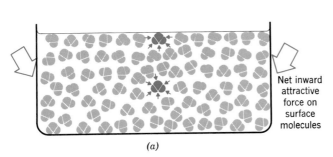 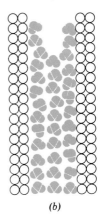

Net inward attractive force on surface molecules

(a)

(b)

Surface tension—A molecule in the body of the liquid is attracted equally by the molecules around it. The molecules at the surface of the liquid are attracted only by the molecules below them and beside them. As a result, unbalanced attractive forces exist on the surface molecules. The unbalanced attraction results in a net inward pull tending to draw the surface molecules into the body of the liquid. Consequently, the liquid surface is under a certain strain or tension called the surface tension.

Capillary action— Water wets glass as a result of an attraction of water molecules to the ions in the glass. In a narrow diameter tube (capillary tube) the water molecules wet the tube, and surface molecules attract other water molecules a certain distance up the tube. This ascension of water up a narrow tube is called capillary action. In a very narrow tube water may ascend many centimeters.

Figure 10-4   Some properties of water. From *Understanding Chemistry: From Atoms to Attitudes* by T. R. Dickson, Wiley, N.Y., 1974.

liquid, some of the hydrogen bonds break, the open, cagelike structure is disrupted, and the water molecules move closer to one another.

Incidentally, this uncommon behavior of water is of great importance to us. If ice were more dense than water, it would sink to the bottom of lakes and rivers. Ice would build up from the bottom and ultimately freeze over the lakes, rivers, and oceans of the world.

Hydrogen bonding causes the boiling point and freezing point of water to be much higher than expected. The boiling and freezing points of most chemically similar compounds follow a pattern of increasing as the molecular weights of the compounds increase. As shown in Figure 10-3, if water fit this pattern when compared to chemically related compounds, it would boil at about −90°C and freeze at −101°C. The clumping together of water molecules causes the shift in the boiling and freezing points.

Some other properties of water such as its ability to dissolve many substances, its surface tension, and capillary action can be explained on the basis of the polar nature of water molecules. The dissolving process is discussed in Section 11-1 and explanations of the other two properties are illustrated in Figure 10-4.

10-5   Solutions

A **solution** is a mixture of two or more components that are so intimately mixed that the particles making up the components are intermingled on an atomic, molecular, or ionic basis. The types of particles
involved depend on the natures of the components. As a result of the
great degree of intermingling of the particles of the components of a solution and the forces of interaction that exist between them, the components of a solution cannot be separated by filtration. In fact, solution
components can only be separated by more involved separation methods.
For instance, if we want to get the salt out of sea water, we have to
heat the solution to evaporate the water.

Solutions are very important in chemistry and industry and are used
in everyday life. Many food products and medicines are solutions, as are
many household chemicals such as cleaning fluids and rubbing alcohol.
The gasoline we use in our car is a solution, and the water we drink
from the tap is a solution. Special solutions are often prepared for specific uses while many occur naturally.

A solution composed of two substances is called a **binary solution.** A
ternary solution has three components, and it is possible to have a solution consisting of four or more substances. However, not all substances
mix with one another to form solutions. Usually, solutions are prepared
by dissolving one substance in another. When a solution is prepared
using two substances of different phases, the substance that is of the
same phase as the resulting solution is called the **solvent,** and the substance that has been dissolved in the solvent is called the **solute.** For
example, if we prepare a liquid solution by dissolving sugar in water,
the water is the solvent, and the sugar is the solute. If the two substances are of the same phase, the solvent is usually considered to be the
component present in the greater amount, but in such solutions the dis-
tinction is not important. In a solution made up of water and alcohol,
the water is considered to be the solvent if it is present in the greater

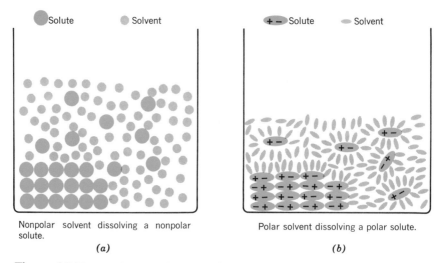

Nonpolar solvent dissolving a nonpolar solute.

(a)

Polar solvent dissolving a polar solute.

(b)

Figure 10-5   Idealized solution formation.

amount. It is possible to have liquid-phase solutions, solid-phase solutions, and gaseous-phase solutions. The most commonly encountered type is the liquid-phase solution. Since water is one of the most useful solvents, the most common solutions are aqueous solutions. All solutions involving water as the solvent are called **aqueous solutions,** while liquid solutions not involving water are called nonaqueous solutions.

It is difficult to specifically predict which substances dissolve in one another. However, because of the forces of interaction between particles, a polar solvent such as water will readily dissolve a polar solute and a nonpolar solvent such as carbon tetrachloride will readily dissolve a nonpolar solute. See Figure 10-5.

10-6   Methods of Describing Solution Concentrations

An aqueous solution can be described in terms of the kinds of solutes present. This is a qualitative description and states nothing about how much solute is present in the solution. Of course, the amount of solute in a solution depends on the amount of solution considered. The ocean as a large salt solution contains an amount of dissolved salt totaling about 4×10^{22} g while a cup of sea water contains about 8 g of dissolved salt. Obviously, just stating the amount of solute in a solution is not a useful description.

The quantitative makeup of a solution is best described by stating the concentration of the solute. **Concentration** is an expression of the amount of solute contained in a unit amount of solution. For instance, the concentration of dissolved salt in sea water can be expressed as 30 g of sodium chloride per liter of sea water. The concentration of a solution is independent of the amount of solution. The concentration of a large volume of a given solution is the same as the concentration of a small volume. The concentration of dissolved salt in a cup of sea water is the same as the concentration of salt in a bucket of sea water.

It is important to distinguish between amount and concentration. The amount of dissolved gold in the oceans of the world is great enough to provide each person in the world with around 2 kg of gold. However, gold is present in sea water at such low concentration that it is not economically feasible to extract it.

Solution concentrations provide a basis of comparison. Sea water contains 30 g of dissolved salt per liter, and typical drinking water contains less than 0.4 g of dissolved salt per liter. The sea water is more concentrated in salt than the drinking water. The common methods of expressing concentration are discussed in the following sections.

10-7   Molarity

A common and useful expression of the concentration of a solute is known as molarity. **Molarity** expresses the number of moles of solute per unit liter of solution. Molarity, which is represented by the symbol, M, can be defined as given on the next page.

$$\text{molarity (M)} = \left(\frac{\text{number of moles of solute}}{\text{liter of solution}}\right)$$

The molarity of a solute can be determined if we know the amount of solute dissolved in enough solvent to produce a specific volume of solution. The mass of solute can be converted to the number of moles of solute, and then the number of moles of solute can be divided by the volume of solution in liters express the molarity. Keep in mind that molarity is independent of the amount of solution and merely serves as a convenient expression of the amount of solute per unit volume of solution. For instance, we may have a 10-ml sample of one solution and a 5-ℓ sample of another, but both solutions could have the same number of moles per liter concentration.

Example 10-1   A solution is prepared by dissolving 117 g of sodium chloride in enough water to make 500 ml of solution. What is the molarity of the sodium chloride in the solution?

The molarity of NaCl can be determined by converting the number of grams of NaCl to the number of moles and then dividing by the volume of solution in liters.

First, determine the number of moles using the molar mass (22.99 + 35.45 = 58.4)

$$117 \text{ g} \left(\frac{1 \text{ mole NaCl}}{58.4 \text{ g}}\right) = 2.00 \text{ moles NaCl}$$

Next, divide by the volume in liters.

$$\left(\frac{2.00 \text{ moles NaCl}}{0.500 \ \ell}\right) = \left(\frac{4.00 \text{ moles NaCl}}{1 \ \ell}\right)$$

Such a solution is said to be 4.00 molar NaCl or 4.00 M NaCl. The M indicates molar, molarity, or number of moles per liter.

The concentrations of solutions used in the laboratory are often expressed in terms of molarity. Many substances are conveniently stored as solutions, and, when a sample is needed, a portion of the solution can be measured out. In fact, as long as the molarity of a solution is known, it is possible to obtain a specific number of moles of the substance in solution by measuring out a specific volume of the solution. The molarity of the solution can be used as a factor to interconvert moles of solute and volume of solution. This is analogous to the use of density as a conversion factor. Recall that density is defined as the mass per unit volume of a substance.

$$D = \frac{m}{v} \qquad \begin{array}{l} m = \text{mass} \\ v = \text{volume} \end{array}$$

The mass of a given volume of a substance is found by multiplying by the density.

$$m = vD$$

The volume of a given mass of a substance is found by multiplying by the inverse of the density.

$$v = m \left(\frac{1}{D}\right)$$

Molarity is defined as the number of moles of solute per unit volume of solution.

$$M = \frac{n}{V} \qquad \begin{array}{l} n = \text{number of moles} \\ V = \text{volume in liters} \end{array}$$

The number of moles in a given volume of solution is found by multiplying the volume in liters by the molarity.

$$n = VM$$

The volume of solution needed to contain a specific number of moles of solute is found by multiplying the moles by the inverse of the molarity.

$$V = n \left(\frac{1}{M}\right)$$

Example 10-2   How many moles of dissolved NaOH are contained in 25.0 ml of a 6.00 M NaOH solution?

The number of moles can be found by multiplying the volume in liters by the molarity (6.00 moles NaOH/1 ℓ). The conversion sequence should be

? moles NaOH = solution volume → liters solution → moles NaOH

$$? \text{ moles NaOH} = 25.0 \text{ ml} \left(\frac{1 \ \ell}{10^3 \text{ ml}}\right) \left(\frac{6.00 \text{ moles NaOH}}{1 \ \ell}\right) = 0.150 \text{ moles NaOH}$$

Example 10-3   How many milliliters of a 2.00 M NaCl solution are needed to contain 0.250 moles of NaCl?

The number of moles can be multiplied by the inverse of the density (1 ℓ/2.00 moles NaCl) to give the volume in liters that can be converted to milliliters. The conversion sequence should be

? ml = moles NaCl → liters solution → milliliters solution

$$? \text{ ml} = 0.250 \text{ moles NaCl} \left(\frac{1 \ \ell}{2.00 \text{ moles NaCl}}\right) \left(\frac{10^3 \text{ ml}}{1 \ \ell}\right) = 125 \text{ ml}$$

10-8   Standard Solutions

Storing substances in the solution phase so that desired amounts can be easily measured out by measuring volumes of solution is a useful technique. Solutions of known molarity are called **standard solutions.** They are often prepared by weighing out a specific amount of solute and dissolving the sample in enough solvent to produce a specific volume of solution. A special container, called a **volumetric flask,** which is calibrated to contain a specific volume of liquid, is often used to prepare standard solutions. The use of the volumetric flask is illustrated in Figure 10-6. In order to prepare a standard solution, it is necessary to calculate the mass of solute needed to prepare a certain volume of solution of a specific molarity. The amount of solute needed can be deduced by determining the number of moles of solute needed. This is done by converting the volume of solution to the number of moles of solute by

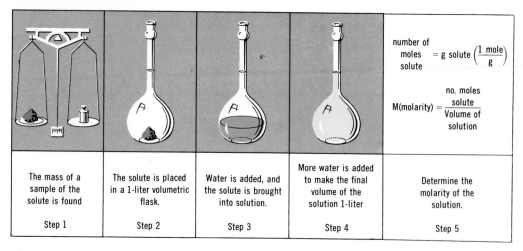

			More water is added	number of moles solute $= $ g solute $\left(\dfrac{1\ \text{mole}}{\text{g}}\right)$
The mass of a sample of the solute is found	The solute is placed in a 1-liter volumetric flask.	Water is added, and the solute is brought into solution.	to make the final volume of the solution 1-liter	$M(\text{molarity}) = \dfrac{\text{no. moles solute}}{\text{Volume of solution}}$
				Determine the molarity of the solution.
Step 1	Step 2	Step 3	Step 4	Step 5

Figure 10-6 The preparation of $1\,\ell$ of a standard solution.

use of the molarity. Then, the number of moles of solute can be converted to the number of grams of solute.

Example 10-4   How many grams of sodium chloride are needed to prepare 250 ml of a 0.500 M solution of NaCl?

The number of moles of NaCl that is needed can be found by converting the volume of solution to the moles of NaCl using the molarity as a factor. Then, the number of moles of NaCl can be converted to the number of grams of NaCl using the number of grams per mole as a factor. The conversion sequence should be

$$? \text{ grams NaCl} = \text{volume solution} \rightarrow \text{moles NaCl} \rightarrow \text{grams NaCl}$$

First, the moles of NaCl are found by multiplying the volume in liters (0.250 ℓ) by the molarity (0.500 moles NaCl/1 ℓ).

$$0.250 \ \ell \left(\frac{0.500 \text{ moles NaCl}}{1 \ \ell}\right)$$

Next, the grams of sodium chloride are found by multiplying by the molar mass.

$$? \text{ grams NaCl} = 0.250 \ \ell \left(\frac{0.500 \text{ moles NaCl}}{1 \ \ell}\right)\left(\frac{58.4 \text{ g}}{1 \text{ mole NaCl}}\right) = 7.30 \text{ g}$$

The standard solution would be prepared by weighing this amount of sodium chloride, placing it in a 250-ml volumetric flask and dissolving it in enough water to form 250 ml of solution.

10-9   Dilution

Sometimes it is necessary to prepare a solution of a certain molarity by **dilution** of a specific amount of a more concentrated solution. When preparing solutions by this process, the questions that arise are either what volume of the more concentrated solution should be used to prepare a specific volume of the diluted solution or to what volume should a specific volume of the more concentrated solution be diluted to produce the weaker solution. These questions can be answered by considering the relationship between molarity and volume. From the definition of molarity as the number of moles of solute per liter of solution,

$$M = \left(\frac{moles}{volume} \right)$$

we can see that the molarity of a solution and the volume are inversely related. This is illustrated in Figure 10-7. That is, to decrease the molarity of a solution the volume can be increased (i.e., the solution can be diluted to decrease the molarity). Thus we can deduce the volume to which a specific volume of a more concentrated solution should be diluted to give a solution of lower molarity by multiplying the original volume by a factor involving the molarities that is greater than 1. Dilution problems can be solved in a manner analogous to the gas law problems involving pressure and volume.

Example 10-5   To what volumes should 10.0 ml of a 15.00 M aqua ammonia be diluted to produce a 2.00 M solution?

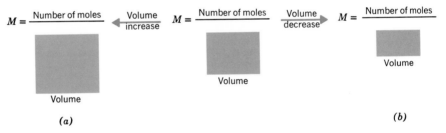

(a)

When the volume of a solution is increased (the solution is diluted), the molarity decreases. Since the number of moles is constant, dividing by a larger volume gives a smaller molarity.

(b)

When the volume of a solution is decreased (the solvent is evaporated), the molarity increases. Since the number of moles is constant, dividing by a smaller volume gives a larger molarity.

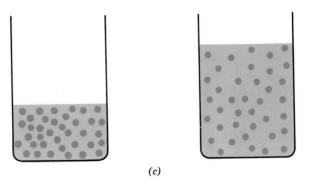

(c)

Figure 10-7 The molarity of a solution is inversely proportional to the volume.

The volume to which the 10.0 ml sample should be diluted is found by multiplying the 10.0 ml by a ratio of the molarities that is greater than 1:

$$10.0 \text{ ml} \left(\frac{15.00 \text{ M}}{2.00 \text{ M}}\right) = 75.0 \text{ ml}$$

Thus, to prepare the 2.00 M solution from the 15.00 M solution, it is necessary to add enough water to the 10.0 ml sample of the 15.00 M solution so that a *total* volume of 75.0 ml is obtained.

Sometimes we want to prepare a specific amount of the more dilute solution. In this case the volume of the more concentrated solution needed can be determined by multiplying the desired volume of the dilute solution by a ratio of the molarities that is less than 1.

Example 10-6   How many milliliters of 6.00 M hydrochloric acid should be used to prepare 100.0 ml of a 0.500 M hydrochloric acid solution by dilution?
The required amount of the 6.00 M solution is found by multiplying the 100.0 ml by a ratio of the molarities that is less than 1:

$$100.0 \text{ ml} \left(\frac{0.500 \text{ M}}{6.00 \text{ M}}\right) = 8.33 \text{ ml}$$

Thus, to prepare the 0.500 M solution, it is necessary to dilute 8.33 ml of 6.00 M hydrochloric acid with enough water to produce 100.0 ml of solution.

10-10   Other Solution Concentration Terms

Several other methods of expressing concentrations of solutions are described below.

Parts Per Million. There are about 1.4×10^{21} g of dissolved aluminum in the oceans of the world. This is indeed a large amount of aluminum, but its concentration is so low we find it convenient to refer to the concentration in terms of parts per million. Solutes found in very low concentrations in solutions are sometimes expressed in terms of the number of milligrams of solute per kilogram of solution, or the number of milligrams of solute per liter of solution. The densities of dilute aqueous solutions are approximately 1 g/ml, so a kilogram of solution and a liter are approximately the same. A milligram is one-millionth part of a kilogram, thus, this expression of concentration corresponds to parts per million concentration. The parts per million (ppm) concentration of a species is found by determining the number of milligrams of the species in a known number of liters or kilograms of the solution. For instance, if a 0.50 ℓ sample of a water solution is found to contain 2.2 mg of fluoride ion. The parts per million of fluoride is

$$\text{ppm F}^- = \left(\frac{2.2 \text{ mg F}^-}{0.50 \text{ } \ell}\right) = \left(\frac{4.4 \text{ mg F}^-}{1 \text{ } \ell}\right) = 4.4 \text{ ppm F}^-$$

Percentage by Mass (or Weight) Composition. The amount of each component in a solution can be expressed in terms of a percentage of a given mass of the solution. This percentage can be obtained by dividing the mass of a given component present in a given amount of solution by the mass of this amount of solution and then multiplying by 100. Such a percentage can be represented as

$$\text{percent component} = \left(\frac{\text{mass component}}{\text{mass solution}}\right) 100$$

Percentage compositions are sometimes used in industry and to express concentration of commercial solutions.

Example 10-7 A solution consists of 20.0 g of sugar dissolved in 53.5 g of water. What is the percentage by mass solute in the solution?

The mass of sugar is 20.0 g, and the mass of the solution containing this amount of solute is found by adding the mass of the solute to the mass of the solvent:

$$\text{mass solution} = 53.5 \text{ g} + 20.0 \text{ g} = 73.5 \text{ g}$$

We do this because the total mass of solution will be the sum of the mass of the solute used and the mass of the solvent used. So, the percentage by mass solute is found by dividing the mass of the solute by the mass of the solution and then multiplying by 100:

$$\text{percent solute} = \left(\frac{20.0 \text{ g}}{73.5 \text{ g}}\right) 10^2 = 27.2\%$$

Molality. Another way in which the concentration of a solute can be expressed is in terms of the number of moles of solute per kilogram of solvent. This concentration term indicates the number of moles of solute per unit mass of solvent. The unit mass that is used is the kilogram. This expression of concentration is called molality and is represented by the symbol m. Molality is defined as

$$\text{molality (m)} = \left(\frac{\text{number of moles of solute}}{\text{kilograms solvent}}\right)$$

The molality of a solute can be deduced if we know the mass of the solute dissolved in a given mass of solvent. The number of moles of the solute can be calculated from the mass using the number of grams per mole factor, and then the number of moles of solute can be divided by the mass of solvent. Keep in mind that molality expresses the amount of solute per unit amount of solvent and is independent of the actual amount of solution with which we are dealing. That is, two solutions may have the same molality with respect to a given solute, while there may only be a tiny amount of one solution and a very large amount of the other.

Example 10-8 A solution contains 68.4 g of sucrose, $C_{12}H_{22}O_{11}$ (342 g/mole), dissolved in 150 g of water. What is the molality of the sucrose?

The grams of sucrose can be converted to the number of moles of sucrose and the grams of water can be converted to kilograms of water. Then, the number of moles of sucrose per 1 kg of water can be found by dividing the number of moles of sucrose by the number of kilograms of water.

First, the moles of sucrose are found by multiplying the grams by the inverse of the molar mass as shown on the next page.

$$\text{moles sucrose} = 68.4 \text{ g} \left(\frac{1 \text{ mole } C_{12}H_{22}O_{11}}{342 \text{ g}}\right) = 2.00 \times 10^{-1} \text{ moles } C_{12}H_{22}O_{11}$$

Next, the moles are divided by the mass of water in kilograms (150 g = 0.150 kg)

$$\text{sucrose molality} = \left(\frac{2.00 \times 10^{-1} \text{ moles } C_{12}H_{22}O_{11}}{0.150 \text{ kg}}\right) = 1.33 \text{ m } C_{12}H_{22}O_{11}$$

Concentrations in terms of mass percent or molality express the amount of solute per mass of solution or solvent while molarity expresses the amount of solute per volume of solution. The concentration of a solute in terms of mass percent or molality can be converted to molarity if the density of the solution is known. The density is needed since it relates the mass of the solution to the volume of solution. Thus, if we know the amount of solute in a specific mass of solution, the density can be used to convert this to the amount of solute per liter of solution.

Example 10-9   Concentrated aqua ammonia contains 29.2% by mass NH_3 in water and has a density at 20°C of 0.896 g/ml. What is the molarity of NH_3 in concentrated ammonia?

From the mass percent we know that there are 29.2 g of NH_3 per 100 g of solution. Now, by use of the density, the number of grams of ammonia per 100 g of solution can be converted to the number of grams of ammonia per milliliter of solution, and then to the number of grams of ammonia per liter of solution. Finally, the number of grams of ammonia can be converted to the number of moles of ammonia by use of the inverse of the molar mass. The conversion sequence should be

$$? \text{ M } NH_3 = \left(\frac{g\ NH_3}{g\ solu}\right) \rightarrow \left(\frac{g\ NH_3}{ml\ solu}\right) \rightarrow \left(\frac{g\ NH_3}{\ell\ solu}\right) \rightarrow \left(\frac{moles\ NH_3}{\ell\ solu}\right)$$

Thus, the solution to the problem is

$$? \text{ M } NH_3 = \left(\frac{29.2 \text{ g } NH_3}{100.0 \text{ g}}\right)\left(\frac{0.896 \text{ g}}{1 \text{ ml}}\right)\left(\frac{10^3 \text{ ml}}{1 \ell}\right)\left(\frac{1 \text{ mole } NH_3}{17.03 \text{ g } NH_3}\right) = \left(\frac{15.4 \text{ moles } NH_3}{1 \ell}\right)$$

10-11   Solution Stoichiometry

As we know the molarity of a solution can be used to determine the number of moles of solute from the volume or the volume needed to contain a specific number of moles of solute. Chemicals can be stored as standard solutions of known molarity. These are called reagent solutions. Very often chemical reactions are carried out in solutions using these reagents. Sometimes stoichiometric calculations are applied to these solution reactions. For instance we may want to determine the volume of some reagent solution that is needed to react with a specific number of grams of a substance or the number of grams of a substance that will react with a specific volume of reagent solution. Calculations for such problems involve usual stoichiometric conversions coupled with the molarity of the reagent solution to convert volume to moles or moles to volume.

Example 10-10   How many milliliters of 6.00 M NaOH solution are nee
with 0.300 moles of acetic acid, $HC_2H_3O_2$ according to the equation

$$OH^-(aq) + HC_2H_3O_2(aq) \rightarrow H_2O + C_2H_3O_2^-(aq)$$

The number of moles of hydroxide ion needed is found by converting the moles of acetic
acid by use of the molar ratio from the equation. Then the reciprocal of the molarity is used
to find the volume of NaOH solution in liters that can be converted to milliliters. The conversion sequence is

$$\text{ml } OH^- = \text{moles } HC_2H_3O_2 \rightarrow \text{moles } OH^- \rightarrow \ell\ OH^- \rightarrow \text{ml } OH^-$$

First find the moles of hydroxide with the molar ratio.

$$0.300 \text{ moles } HC_2H_3O_2 \left(\frac{1 \text{ mole } OH^-}{1 \text{ mole } HC_2H_3O_2}\right)$$

Next find the volume of solution by use of the molarity and convert to milliliters.

$$\text{ml } OH^- = 0.300 \text{ moles } HC_2H_3O_2 \left(\frac{1 \text{ mole } OH^-}{1 \text{ mole } HC_2H_3O_2}\right)\left(\frac{1\ \ell}{6.00 \text{ moles } OH^-}\right)\left(\frac{10^3 \text{ ml}}{1\ \ell}\right) = 50.0 \text{ ml}$$

Example 10-11   How many milliliters of 2.00 M hydrochloric acid are needed to react
with 3.27 g of zinc metal according to the equation

$$2H_3O^+(aq) + Zn(s) \rightarrow Zn^{2+}(aq) + H_2(g) + 2H_2O$$

The number of moles of hydronium ion needed is found by converting the grams of zinc to
moles and then using the molar ratio from the equation. Next the reciprocal of the molarity
is used to find the volume of hydrochloric acid in liters that can be converted to milliliters.
The conversion sequence is

$$\text{ml } H_3O^+ = \text{g } Zn \rightarrow \text{moles } Zn \rightarrow \text{moles } H_3O^+ \rightarrow \ell\ H_3O^+ \rightarrow \text{ml } H_3O^+$$

First find the moles of zinc by use of the molar mass.

$$3.27 \text{ g} \left(\frac{1 \text{ mole } Zn}{65.4 \text{ g}}\right)$$

Next find the moles of hydronium ion with the molar ratio.

$$3.27 \text{ g} \left(\frac{1 \text{ mole } Zn}{65.4 \text{ g}}\right)\left(\frac{2 \text{ moles } H_3O^+}{1 \text{ mole } Zn}\right)$$

Finally find the volume of solution by the use of the molarity and convert to milliliters.

$$\text{ml } H_3O^+ = 3.27 \text{ g} \left(\frac{1 \text{ mole } Zn}{65.4 \text{ g}}\right)\left(\frac{2 \text{ moles } H_3O^+}{1 \text{ moles } Zn}\right)\left(\frac{1\ \ell}{2.00 \text{ moles } H_3O^+}\right)\left(\frac{10^3 \text{ ml}}{1\ \ell}\right) = 50.0 \text{ ml}$$

Example 10-12   How many grams of aluminum metal will react with 20.0 ml of 6.00 M
hydrochloric acid according to the equation

$$6H_3O^+(aq) + 2Al(s) \rightarrow 2Al^{3+}(aq) + 3H_2(g) + 6H_2O$$

The moles of hydronium ion needed can be found from the volume of solution in liters and
the molarity. The moles of hydronium ion can then be converted to moles of aluminum
using the molar ratio. Finally the grams of aluminum can be found using the molar mass
of aluminum. The conversion sequence is

$$\text{g } Al = \text{ml } H_3O^+ \rightarrow \ell\ H_3O^+ \rightarrow \text{moles } H_3O^+ \rightarrow \text{moles } Al \rightarrow \text{g } Al$$

First find the volume in liters and then the moles of hydronium ion by use of the molarity

$$20.0 \text{ ml} \left(\frac{1 \ \ell}{10^3 \text{ ml}}\right) \left(\frac{6.00 \text{ moles } H_3O^+}{1 \ \ell}\right)$$

Next convert to moles of aluminum by the molar ratio

$$20.0 \text{ ml} \left(\frac{1 \ \ell}{10^3 \text{ ml}}\right) \left(\frac{6.00 \text{ moles } H_3O^+}{1 \ \ell}\right) \left(\frac{2 \text{ moles Al}}{6 \text{ moles } H_3O^+}\right)$$

Finally convert to mass of aluminum using the molar mass

$$\text{g Al} = 20.0 \text{ ml} \left(\frac{1 \ \ell}{10^3 \text{ ml}}\right) \left(\frac{6.00 \text{ moles } H_3O^+}{1 \ \ell}\right) \left(\frac{2 \text{ moles Al}}{6 \text{ moles } H_3O^+}\right) \left(\frac{27.0 \text{ g}}{1 \text{ mole Al}}\right) = 1.08 \text{ g}$$

10-12   Colligative Properties of Solutions

Certain properties of solutions, called **colligative properties,** are related more to the presence of solute particles dispersed among the solvent particles than to the identities of the particles. The term colligative refers to the collective effect of the solute particles. These properties are

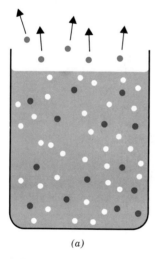

○ Solute
● Solvent

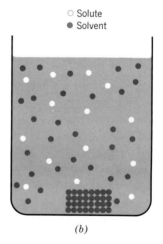

(a)

(b)

Boiling point elevation—The presence of solute molecules interferes with the ability of the solvent molecules to evaporate. Consequently, a higher temperature is needed to make the solution boil. The increase in the boiling point of a solvent caused by the addition of a nonvolatile solute is called boiling point elevation.

Freezing point depression—The presence of solute particles interferes with the ability of the solvent molecules to freeze to ice crystals. Consequently, a lower temperature is needed to reach the freezing point. The decrease in the freezing point of a solvent caused by the addition of a solute is called freezing point depression. Freezing point depression is used in the making of homemade ice cream. Salt is added to an ice–water mixture to lower the freezing temperature enough to freeze the ice cream mixture.

Figure 10-8   Boiling-point elevation and freezing-point depression. From *Understanding Chemistry: From Atoms to Attitudes* by T. R. Dickson, Wiley, N.Y., 1974.

generally the same no matter what solute is involved, and they depend on the concentration and not the nature of the solute particles.

One important colligative property is the effect of a nonvolatile solute on the freezing point and boiling point of the solvent. When we dissolve a nonvolatile solute in a solvent, the resulting solution will boil at a higher temperature than the pure solvent and will freeze at a lower temperature than the pure solvent. These phenomena are called **boiling point elevation** and **freezing point depression** (See Figure 10-8). We take advantage of the freezing point depression of water by a solute when we put antifreeze in the radiator of our car. The antifreeze is a solute that causes the freezing point of the water in the radiator to be lowered so that it does not freeze in cold weather.

Another colligative property of aqueous solutions that is very important in chemical and biological processes is osmosis. **Osmosis** is a phenomenon involving solutions separated by a membrane. The membrane acts as a barrier between the solutions and has the property of allowing certain types of molecules to pass through while preventing the passage of other species in solution. This membrane is called a semipermeable membrane since it is only permeable to selected species. Semipermeable membranes that allow the passage of the solvent but not the solute are called osmotic membranes. When a solution is separated by an osmotic membrane from a sample of pure solvent (or a like solution of lower concentration), the solvent molecules will spontaneously penetrate the membrane from both directions but not at equal rates. There is a net migration of solvent molecules from the pure solvent side of the membrane to the solution side. This phenomena is called osmosis and is illustrated in Figure 10-9. The net result of osmosis is the transfer of solvent across the membrane to the side having the lowest concentration of solvent. The mechanism of osmosis is not completely understood, but it appears to occur between aqueous solutions separated by an osmotic membrane no matter what nonvolatile solute is involved. Osmosis can be stopped if a certain opposing pressure or force per unit area is exerted on the solution side of the membrane, as illustrated in Figure 10-9. This means that osmosis results in a pressure being exerted on the solution of higher concentration of solute. This pressure is called the osmotic pressure. The osmotic pressure can be quite great and sometimes reaches hundreds of atmospheres. Osmotic pressure is one of the factors involved in the uptake of water by trees and plants.

10-13   Waters of the Hydrosphere

Recall from Section 10-3 that in the water cycle water moves from the oceans to the land by evaporation and precipitation. Then it returns to the oceans passing through streams, rivers, and lakes. Water passing through this cycle comes into intimate contact with the atmosphere and the rocks, minerals, and soils of the lithosphere. As a result of dissolving processes these natural waters become solutions containing various ions and molecules. Let us consider the compositions of some natural waters.

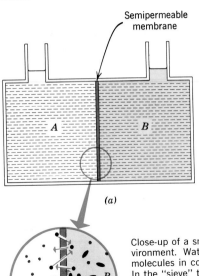

Osmosis: Enlarged view of two fluid compartments separated by a semipermeable membrane.

(a)

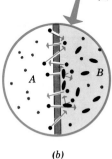

Close-up of a small section of membrane and its immediate environment. Water molecules are represented by dots; sugar molecules in compartment B are represented by shaded ovals. In the "sieve" theory of osmosis, the membrane is said to have pores large enough to permit passage of water molecules but small enough to stop solute molecules (or ions). As drawn, of every five molecules of water that get from A to B, only three return. Two others are shown colliding with sugar molecules. The result is a net flow of water from A to B, and level B rises. Parts a and b from Elements of General and Biological Chemistry, 3rd ed., by John Holum, John Wiley & Sons, 1972, 138.

(b)

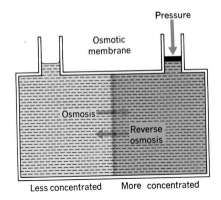

Osmotic pressure: In the osmosis process, solvent molecules pass from the less concentrated solution to the more concentrated solution. The process of osmosis can be reversed by applying pressure to the more concentrated side. The amount of pressure needed just to counteract the osmosis process is called the osmotic pressure. Pressures in excess of the osmotic pressure can cause reverse osmosis.

(c)

Figure 10-9   Osmosis and osmotic pressure.

Ocean Water. The ocean is the habitat of tremendous amounts of animal and plant life. The ocean is actually a vast solution of ions and other substances in which these plants and animals exist. Studies of seawater from many sources have shown that there are apparently only small variations in the relative concentrations of these ions. The constituents of typical sea water in grams per kilogram of seawater are shown in Table 10-2.

Since the ocean waters are in contact with the atmosphere and the biological processes of photosynthesis and decomposition of organic matter are continuously occurring in the ocean, seawater contains some dissolved gases. The most important dissolved gases in seawater are nitrogen, N_2, oxygen, O_2, and carbon dioxide, CO_2.

Table 10-2   The Composition of Seawater

<table>
<tr><td colspan="4" align="center">Major Constituents of Seawater</td></tr>
<tr><td>Positive ions</td><td>grams/kg
sea water</td><td>Negative ions</td><td>grams/kg
sea water</td></tr>
<tr><td>Sodium ion, Na^+</td><td>10.76</td><td>Chloride ion, Cl^-</td><td>19.353</td></tr>
<tr><td>Magnesium ion, Mg^{2+}</td><td>1.294</td><td>Sulfate ion, SO_4^{2-}</td><td>2.712</td></tr>
<tr><td>Calcium ion, Ca^{2+}</td><td>0.413</td><td>Hydrogen carbonate
ion, HCO_3^-</td><td>0.142</td></tr>
<tr><td>Potassium ion, K^+</td><td>0.387</td><td>Bromide ion, Br^-</td><td>0.067</td></tr>
<tr><td>Strontium ion, Sr^{2+}</td><td>0.008</td><td>Fluoride ion, F^-</td><td>0.0014</td></tr>
<tr><td>Barium ion, Ba^{2+}</td><td>0.00005</td><td>Iodide ion, I^-</td><td>0.00006</td></tr>
</table>

Minor Constituents of Seawater in grams per metric ton or parts per million (most elements occur in some ionic form and a few as organic compounds)

Boron	5.0	Cesium	4×10^{-4}
Silicon	3.0	Cerium	4×10^{-4}
Nitrogen	0.52	Yttrium	2×10^{-4}
Lithium	0.17	Silver	2×10^{-4}
Rubidium	0.12	Lanthanum	2×10^{-4}
Phosphorus	7.2×10^{-2}	Cadmium	1×10^{-4}
Indium	2.0×10^{-2}	Tungsten	1×10^{-4}
Zinc	1.0×10^{-2}	Germanium	6×10^{-5}
Iron	1.0×10^{-2}	Chromium	4×10^{-5}
Aluminum	1.0×10^{-2}	Thorium	4×10^{-5}
Molybdenum	1.0×10^{-2}	Scandium	4×10^{-5}
Selenium	4.1×10^{-3}	Lead	2×10^{-5}
Tin	3.0×10^{-3}	Mercury	2×10^{-5}
Copper	3.0×10^{-3}	Gallium	2×10^{-5}
Arsenic	3.0×10^{-3}	Bismuth	2×10^{-5}
Uranium	3.0×10^{-3}	Niobium	1×10^{-5}
Nickel	1.9×10^{-3}	Thallium	1×10^{-5}
Vanadium	1.9×10^{-3}	Gold	4×10^{-6}
Manganese	1.9×10^{-3}	Dissolved gases are quite variable	
Titanium	1×10^{-3}	Nitrogen, N_2	10
Antimony	4×10^{-4}	Oxygen, O_2	7
Cobalt	4×10^{-4}	Carbon dioxide, CO_2	600

Ocean water contains about 3.5% by-mass dissolved ions. As can be seen in Table 10-2, sodium ion and chloride are the predominant ions. It is obvious why ocean water tastes salty. When a sample of seawater is evaporated, a mixture of ionic compounds remains most of which is sodium chloride.

The **salinity** of seawater as defined by oceanographers is the mass in grams of the solids in 1 kg of seawater evaporated to a constant mass at 480°C. Since the relative concentrations of dissolved substances is somewhat invariant, the salinity of seawater can be directly related to the chlorinity of the seawater. **Chlorinity** is defined as the number of grams of chloride ion, bromide ion, and iodide ion contained in a kilogram of seawater. In the experimental determination of chlorinity, the bromide ions and iodide ions are assumed to be replaced by chloride ions for calculation purposes. The experimentally observed relationship between salinity and chlorinity is

$$salinity = 1.805 \; chlorinity + 0.030$$

Some of the major constituents found in seawater are extracted commercially. Millions of tons of salt are obtained from the ocean every year by solar evaporation. Over 100,000 tons of bromine are obtained from the ocean annually, using chlorine to convert the bromide ion to bromine.

$$Cl_2 + 2Br^- \rightarrow 2Cl^- + Br_2 \; (bromine)$$

Magnesium metal is obtained from the magnesium ion in ocean water by the following series of reactions.

$$Mg^{2+} + 2OH^- \rightarrow Mg(OH)_2$$

$$Mg(OH)_2 + 2H_3O^+ + 2Cl^- \rightarrow MgCl_2 + 4H_2O$$

$$MgCl_2 \xrightarrow{\text{electrolysis}} Mg + Cl_2$$

Lake and River Water. Ocean water has a relatively constant salinity, but river and lake waters have variable composition. This is to be expected since the rivers and lakes often contain water that has come into contact with various geological formations. This water may have flowed long or short distances over the land, dissolving minerals and substances of

Table 10-3   USPHS Drinking Water Standards

Chemical Species	Maximum Allowable Concentration in Milligrams per Liter (parts per million)
Arsenic (ionic)	0.05
Barium ion	1.0
Cadmium ion	0.01
Chloride ion	250
Chromium (ionic)	0.05
Copper (ionic)	1
Cyanide ion	0.2
Fluoride ion	about 2.0
Iron (ionic)	0.3
Lead (ionic)	0.05
Linear alkyl sulfonate (detergent)	0.5
Manganese (ionic)	0.05
Nitrate and nitrite ion	10 (as N)
Selenium (ionic)	0.01
Silver ion	0.05
Sulfate ion	250
Synthetic organic chemicals (carbon-chloroform extract)	0.2
Total dissolved solids	500
Uranyl ion	5
Zinc ion	5

decaying plant life along the way. In addition, the waters may contain materials contributed by humans.

Some land-locked lakes are known as salt lakes since they have accumulated vast amounts of dissolved mineral salts. Except for the waters of salt lakes, the natural waters of lakes and rivers are not salty and are called **fresh water.** Of course, this water is usually neither fresh nor pure but is a solution of dissolved ions and molecules. The most common positive ions in such water are calcium ion, Ca^{2+}, magnesium ion, Mg^{2+}, and sodium ion, Na^+. The most common negative ions are hydrogen carbonate (bicarbonate) ion, HCO_3^-, and sulfate ion, SO_4^{2-}. There are many other substances present in lake and river water.

Since surface water, along with well water, is for public use and consumption, national chemical standards have been established for drinking water. The U.S. Public Health Service **chemical standards for drinking water** are listed in Table 10-3. These standards along with bacteriological standards serve as guides to suppliers of public water to maintain health safety, color, taste, appearance, and odor of drinking water.

10-14   Water Use

An industrial society uses tremendous amounts of water. In the United States, an estimated 1.5 billion cubic meters of water are used each day. Actually, very little of this water is consumed in a manner in which it is chemically converted to other substances. Most of the water is used in crop irrigation, as a medium in industrial processes, and to carry away our domestic and industrial wastes. Around 40% of the water is used in agricultural irrigation. Over 50% is used by industry, including steam-generating electric power plants that account for about three-fifths of industrial use. Approximately 10% is used for municipal public water supplies.

Unlike water used in industry, agricultural water is used only once before it evaporates or reenters the environmental waters. The industrial uses of water other than steam-electric plants are shown in Figure 10-11. The Federal Water Pollution Control Act Amendments of 1972 call for more strict controls of the contents of industrial waste waters. Consequently, some industries are designing processes in which water can be recycled numerous times within a plant before being cleaned up and released to the environment. In petroleum and coal industries water is used on the average more than five times before being discarded. Many municipalities and industries are attempting to recycle water and clean up waste water effluents under the incentive of the Federal Clean Water Act.

As can be seen in Figure 10-12, it is expected that the water use in the United States will exceed the useable surface supply by the year 2000. This does not mean we will run out of water, but it does mean that we will have to reuse and recycle water and attempt to preserve the quality of the water. The reuse of water has been carried out for many years. In certain river basins in the United States water is reused 10 to 20 times before it reaches the ocean. The problem of reuse is that the water must be purified before it can be used again. Some users purify

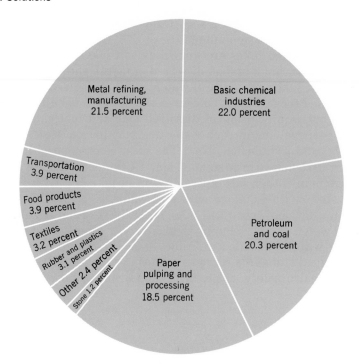

Figure 10-10   Various industrial uses of water. From *Understanding Chemistry: From Atoms to Attitudes* by T. R. Dickson, Wiley, N.Y., 1974.

their waste waters before returning them to the source. Others do not attempt much purification, which tends to degrade the quality of the source so that much effort in purification must be expended by those desiring to use the waters again. It is apparent that the users of water, whether industrial or municipal, must be more concerned with the purity of the waste waters that are returned to the environment. The purification of waste water involves much chemistry and the expenditure of energy. The methods of sewage treatment will be discussed in Section 11-12.

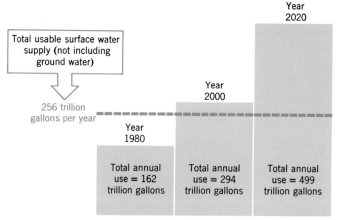

Figure 10-11   Estimated future water use and demand compared to usable surface water supply. Based on U.S. Water Resources Council Data. From *Understanding Chemistry: From Atoms to Attitudes* by T. R. Dickson, Wiley, N.Y., 1974.

Questions and Problems

1. Describe the regions of the environment in which the waters of the hydro-sphere are found and indicate the percentage of the earth's water found in these regions.

2. Explain the processes involved in the water cycle of the earth.

3. Describe ground water including the terms water table and aquifer.

4. Define the following terms:
   (a) solution
   (b) solvent
   (c) solute
   (d) concentration

5. Give a definition of molarity and describe how to calculate the molarity of a solution given the grams of solute and the volume of solution.

6. A sample of salt water is found to contain 25.4 g of sodium chloride, NaCl, in 1.000 ℓ. What is the molarity of sodium chloride in the water?

7. A vinegar solution is found to contain 25.0 g of acetic acid in 525 ml of the solution. What is the molarity of the acetic acid in the solution?

8. A cleaning solution is 9.50 M in ammonia, NH_3. How many grams of am-monia are contained in one quart of the solution? (1 qt = 0.946 ℓ)

9. A 100.0 ml sample of ocean water contains 1.076 g of sodium ion, Na^+, and 1.935 g of chloride ion, Cl^-. What is the molarity of sodium ion in the ocean water? What is the molarity of chloride ion in the ocean water?

10. What is a standard solution and how is a standard solution prepared?

11. Determine the number of grams of solute needed to prepare the following solutions.
    (a) 700 ml of a 1.000 M NaOH solution
    (b) 1.000 ℓ of a 0.600 M $HC_2H_3O_2$ (acetic acid) solution
    (c) 50.0 ml of a 0.200 M C_2H_5OH (ethyl alcohol) solution
    (d) 350 ml of a 0.1075 M Na_2SO_4 solution

12. Complete the following table involving aqua ammonia, $NH_3(aq)$ solutions.

Molarity	Volume	Number of Moles
5.00 M	200 ml	
	3.50 ℓ	7.00 moles NH_3
0.1000 M		0.0200 moles NH_3

13. If a sample of a salt solution contains 8.00 moles of NaCl and is 0.950 M in NaCl, what is the volume of the sample?

14. A solution is prepared by dissolving 51.0 ml of glacial acetic acid, $HC_2H_3O_2$, in enough water to produce 0.500 ℓ of solution. If the density of glacial acetic acid is 1.049 g/ml, determine the molarity of the acetic acid solution. Glacial acetic acid is pure acetic acid.

15. A 30.0 g sample of sodium phosphate is dissolved in water to produce 250 ml of solution. What is the molarity of sodium phosphate in the solution?

16. How many milliliters of a 0.500 M hydrochloric acid solution, $H_3O^+(aq)$ + $Cl^-(aq)$, contain 2.73 moles of hydronium ion, H_3O^+?

17. How many milliliters of a 2.00 M aqua ammonia, $NH_3(aq)$, solution contain 0.500 moles of ammonia, NH_3?

18. How many moles of acetic acid, $HC_2H_3O_2$, are contained in 25.0 ml of a 3.00 M acetic acid solution?

19. How many moles of dissolved NH_3 are contained in 10.0 ml of a 15.0 M NH_3 solution?

20. How many milliliters of an aqueous 0.100 M H_2S solution are needed so that the volume contains 5.00×10^{-3} moles of dissolved H_2S?

21. How many grams of potassium permanganate, $KMnO_4$, are needed to prepare 500 ml of a 0.100 M solution of $KMnO_4$?

22. How many grams of 3-cadmium sulfate 8-water, $3CdSO_4$ $8H_2O$, are needed to prepare 1.00 ℓ of a solution that is 2.00×10^{-3} molar in terms of $CdSO_4$? Note that there are three moles of $CdSO_4$ per each mole of hydrate.

23. If you wanted to prepare 250 ml of a 2.00 M sodium chloride, $NaCl$, solution, how would you do it? How many grams of sodium chloride are needed?

24. A 50.0 ml portion of a potassium sulfate, K_2SO_4, solution is evaporated to remove the water and is found to contain 0.250 g of potassium sulfate. What is the molarity of potassium sulfate in the solution?

25. To what volume should 50.0 ml of a 4.27 M solution of sodium chloride be diluted to produce a 3.00 M solution?

26. How many milliliters of a 3.00 M sulfuric acid solution are needed to prepare 1.00 ℓ of a 0.100 M sulfuric acid solution by dilution?

27. How many milliliters of 6.00 M hydrochloric acid are needed to prepare 100 ml of 0.100 M hydrochloric acid by dilution?

28. How many milliliters of 18 M sulfuric acid are required to prepare 1.50 ℓ of 1 M sulfuric acid by dilution with water? (In practice, acid is always added to water.)

29. To what volume should 3.00 ml of 15 M aqua ammonia be diluted to give 0.5 M aqua ammonia?

30. A solution is prepared by dissolving enough solute in 76.3 g of solvent to produce 100.0 g of solution. What is the percentage by mass of the solute in the solution? (How can you find the mass of the solute?)

31. A solution is prepared by dissolving 0.500 g of sodium chloride in 50.0 g of water at 25°C. What is the percent by mass $NaCl$ in the solution?

32. A solution was prepared by dissolving 12.20 g of potassium chlorate in 200 g of water. What is the percent by mass $KClO_3$ in the solution?

33. A solution was prepared by dissolving 16.0 g of CH_3OH (methyl alcohol) in 100.0 g of water. What is the molality of the methanol solution?

34. A solution was prepared by dissolving 50.0 g of C_2H_5OH (ethyl alcohol) in 200 g of water. What is the molality of the ethanol solution?

35. Concentrated hydrochloric acid contains 36.0% by mass dissolved HCl and has a density of 1.180 g/ml at 20°C. Calculate the molarity of concentrated hydrochloric acid.

36. Concentrated nitric acid has 71.0% by mass HNO_3 and a density of 1.420 g/ml at 20°C. Calculate the molarity of concentrated nitric acid.

37. A H_2SO_4 solution of density 1.802 g/ml contains 88.0% H_2SO_4 by mass (the rest is water). What is the molarity of the H_2SO_4 in the solution?

38. Ocean water contains 6.7×10^{-2} g of bromide ion, Br^-, per kilogram of ocean water. What is the parts per million concentration of bromide ion?

39. A 2.0 ℓ sample of drinking water is found to contain 3.0 mg of barium ion, Ba^{2+}. What is the parts per million concentration of barium ion? Does the drinking water meet the U.S. Public Health Service standards for barium ion?

40. How many milliliters of 2.00 M hydrochloric acid are needed to react with 0.600 moles of ammonia, NH_3, if the reaction is

$$H_3O^+(aq) + NH_3(aq) \rightarrow NH_4^+(aq) + H_2O$$

41. If a solution contains 5.0 g of chloride ion, Cl^-, how many milliliters of 1.00 M silver nitrate, $AgNO_3$, solution are needed to react with the chloride ion if the reaction is

$$Ag^+(aq) + Cl^-(aq) \rightarrow AgCl(s)$$

42. If we want to dissolve 2.00 g of aluminum metal in acid, how many milliliters of 6.00 M hydrochloric acid are needed if the reaction is

$$2Al(s) + 6H_3O^+(aq) \rightarrow 2Al^{3+}(aq) + 6H_2O + 3H_2(g)$$

43. How many grams of magnesium metal can react with 10.0 ml of 3.00 M hydrochloric acid if the reaction is

$$Mg(s) + 2H_3O^+(aq) \rightarrow Mg^{2+}(aq) + 2H_2O + H_2(g)$$

44. How many liters of carbon dioxide gas measured at 25°C and 758 torr pressure are formed when 25.0 ml of 6.00 M hydrochloric acid react with calcium carbonate according to the equation

$$CaCO_3(s) + 2H_3O^+(aq) \rightarrow Ca^{2+}(aq) + 3H_2O + CO_2(g)$$

45. How are the freezing point and boiling point of a solvent affected by the presence of a nonvolatile solute?

46. Describe the phenomenon of osmosis.

47. What are the three major uses of water in the United States?

48. Give definitions for the terms chlorinity and salinity of ocean water.

49. If 1.5 billion (1.5×10^9) cubic meters of water are utilized each day in the United States, how many gallons of water are used each day? (1 gal = 3.79 ℓ). How many gallons are used each year?

11

Chapter 11: Solution Dynamics

Objectives

The student should be able to:
1 Describe the dissolving process.
2 Describe how a solution containing ions conducts electricity.
3 Describe dynamic equilibrium and chemical equilibrium.
4 Apply Le Chatelier's principle to a chemical equilibrium.
5 Predict how a solute should be represented in aqueous solution based on how well the solution conducts electrical current.
6 Describe the degradation of substances in natural waters.
7 Describe the steps involved in sewage treatment.

Terms to Know

Solubility
Saturated solution
Supersaturated solution
Electrolyte
Nonelectrolyte
Chemical equilibrium
Reversible reaction
Le Chatelier's principle
Water pollution
Biochemical oxygen demand
Primary treatment
Secondary treatment
Tertiary treatment

11

Solution Dynamics

11-1   The Dissolving Process

When a solute dissolves in a solvent, the solute and solvent particles interact in a manner that depends on the structural nature of the particles. For example, water can readily dissolve many ionic substances since strong interactions between the ions and the polar water molecules can occur. These interactions are strong enough to pull the ions from the crystal lattice of the solid phase and allow them to enter the solution phase. This dissolving process is illustrated in Figure 11-1. Many substances made up of polar molecules will also readily dissolve in water because of the strong dipole-dipole interaction between solute and solvent molecules. The dissolving process with water as a solvent may involve the separation of the cations and anions in the crystal lattice of an ionic substance, or the dissolving process may just involve the mixing of the solvent and solute molecules. On the other hand, some substances actually react with water chemically when they are dissolved in water. Such reactions may be quite extensive or only small amounts of products may be formed. Specific examples of these dissolving processes are discussed in the following sections.

When a solid is dissolved in a liquid, the dissolving process involves interactions between the particles in the solid and liquid phases. Thus, as we would expect, the greater the amount of contact between the two phases the faster the dissolving process will occur. In other words, the rate of dissolution (dissolving) can be increased by increasing the state of subdivision of the solid phase and by thoroughly mixing the solid and liquid phases. For example, finely divided sugar crystals will dissolve more quickly than large crystals. Since another important factor influencing the dissolving process is the kinetic motion of the intermingling particles, we also expect the rate of dissolution, in most cases, to be increased at higher temperatures. Many substances will dissolve faster at a higher temperature. Certain energy changes occur when dissolving takes place. These changes result in heat being evolved or ab-

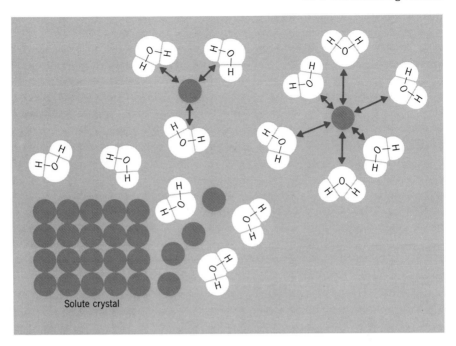

Solute crystal

Figure 11-1 The dissolving of a solute in water. Water dissolves a solute by attracting the solute particles from the crystalline solute. The nature of the interaction between the polar water molecules and the solute particles depends on whether the particles are positive ions, negative ions (ionic solids), or molecules.

sorbed. The heat change associated with the dissolving process is called the heat of solution or the enthalpy of solution. The dissolving process may be exothermic or endothermic. Of course, the nature of the heat change that accompanies the dissolving process depends on the nature of the solvent and solute involved.

When a solute is first placed in contact with the solvent, the constituent particles readily enter the solution. The dissolved solute and solvent particles are constantly moving about in the solution. As a result of this motion, some of the solute particles migrate to a position between the solution and the undissolved solute and reenter the undissolved solute phase. This reformation of the solute phase is called **recrystallization.** Dissolving and recrystallization are mutually competing processes and, as long as sufficient solute is present, a point is reached in the dissolving process when the rate of dissolving equals the rate of recrystallization. This situation is illustrated in Figure 11-2. The competition between the two processes and the eventual equality of the rates of the two processes illustrate a phenomenon called **dynamic equilibrium.** Such a state of equilibrium is not static (the two processes are taking place), but, because of the equality of the rates, no net change in the amount of solute in the solution occurs once equilibrium has been reached. Equilibrium can only be reached if the temperature is constant, since a change in temperature may change the rate of dissolving and before a new state of equilibrium is established at the new temperature, the amount of dissolved solute will have been changed.

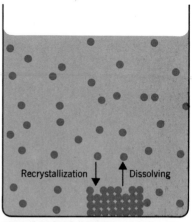

When the rate of dissolving equals the rate of recrystallization, a state of dynamic equilibrium exists between the dissolved and undissolved solute.

Figure 11-2 Dynamic equilibrium between dissolved and undissolved solute.

11-2   Solubility

The amount of solute that can be dissolved in a given amount of solvent before the state of dynamic equilibrium is reached depends on the natures of the solvent and solute and the temperature. The amount of a substance that can be dissolved in a given amount of solvent at a specific temperature is called the **solubility** of the solute. Each substance will have a characteristic solubility in a given solvent. Sometimes solubilities are expressed in terms of the number of grams of solute that can be dissolved in 100 g of solvent at a specific temperature. For example, the solubility of sodium chloride in water at 0°C is 35.7 g per 100.0 g of water.

A solution in which the dissolved solute is in dynamic equilibrium with the undissolved solute is called a **saturated solution.** Such a solution contains the maximum amount of solute that can be dissolved in the given amount of solvent at that temperature. When we want to measure the solubility of a solute in a solvent, it is necessary to prepare a saturated solution of the solute, and then the amount of solute in a specific amount of solvent can be experimentally determined.

When a solution contains less solute than could be contained in solution at saturated conditions, it is referred to as an **unsaturated solution.** The solubilities of many solid substances increase with an increase in temperature. Incidentally, it is interesting to note that gases behave in the opposite manner, that is, the solubilities of gases decrease with an increase in temperature. It is possible to prepare solutions of solid substances that contain fairly large amounts of solute by elevating the temperature of the solution so that more solute will dissolve. Usually, when such solutions are cooled down, the excess solute will recrystallize into the solid phase. Sometimes, however, when such a solution is cooled, a solution is formed that contains more solute than a saturated solution would contain at the same temperature. These solutions that contain more solute than would be expected are called **supersaturated solutions.** A supersaturated solution is in what is called a metastable state and can change to a stable saturated solution when the excess solute recyrstallizes. The crystallization process that involves the formation of the solid solute from the solution requires the proper arrangement of the solute particles into a form that produces the crystal lattice of the solid phase. If the solution is free of tiny solid particles, and the solution container has a smooth surface, such as glass or metal, no crystal growth can be initiated. Thus, on cooling, the solution will contain more solute than it would under normal saturated solution conditions. A supersaturated solution can be converted to a saturated solution by initiating the crystallization process. This initiation is accomplished by adding a small seed crystal to the solution or by agitating the solution.

11-3   Electrical Conductivity of Water Solutions

A rule of electrical safety is to never use an electrical appliance when you are wet or are standing in water. The purpose of the rule is to avoid an electrical shock caused by electricity flowing from the appliance through your body to the ground. You might wonder how water can conduct electricity. Actually, it is the presence of dissolved ions in water that make water an electrical conductor.

Some substances dissolve in water to form ions. Other substances dissolve so that their molecules become intimately mixed with the water molecules, but no ions are formed. How do we know when a dissolved substance forms ions in solution? Since ions are charged, they possess certain electrical properties. One of these properties is that oppositely charged entities attract one another. Suppose we had a solution containing **cations** (positive ions) and **anions** (negative ions), and we placed into this solution a piece of metal (such as platinum), which carried a negative charge. This negatively charged metal would attract the positive ions in the solution, and, thus, the cations would migrate toward the metal and ultimately form a layer of ions around the metal. If a positively charged piece of metal were placed in the solution, the negative ions would be attracted to the metal and would form a layer of ions around the metal. How would it be possible to obtain the two charged pieces of metal? This could be accomplished by connecting the two pieces of metal through metal wires to the two terminals of a **battery** or a **generator** as shown in Figure 11-3. The battery or generator serves as a device for pumping electrons from one piece of metal to the other. This ease of movement of electrons arises from the nature of the structure of metals. Since one piece of metal connected to the battery has an excess of electrons, it will be netatively charged, and the piece that is deficient in electrons will be positively charged. The electron pump (battery) serves to maintain what is called a **potential dif-**

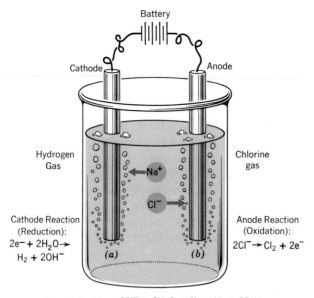

Battery

Cathode   Anode

Hydrogen   Chlorine
Gas   gas

Na$^+$

Cl$^-$

Cathode Reaction   Anode Reaction
(Reduction):   (Oxidation):
$2e^- + 2H_2O \rightarrow$ $2Cl^- \rightarrow Cl_2 + 2e^-$
$H_2 + 2OH^-$
(a)   (b)

Overall Reaction: $2Cl^- + 2H_2O \rightarrow Cl_2 + H_2 + 2OH^-$

Figure 11-3 The electrolysis of a sodium chloride solution.

ference between the two pieces of metal. If these two pieces of metal connected to the electron pump are immersed in a solution containing ions, the cations are attracted toward the negative piece of metal and the anions toward the positive metal. In this situation, the pieces of metal are called **electrodes.** The electrode that attracts the cations is called the **cathode,** and the electrode that attracts the anions is called the **anode.** An interesting phenomenon occurs if the potential difference between the electrodes is great enough. The battery or generator provides a driving force that can result in a chemical reaction that involves the gaining of electrons by some species at the cathode surface and the loss of electrons by another species at the surface of the anode. That is, since the battery or generator tends to pump electrons, a situation arises in which some species in solution lose electrons at the anode and others gain electrons at the cathode. In this manner a complete electrical circuit is set up, in which electrons are pumped through the metal to the cathode where electrons are gained by some species in solution, and, simultaneously, some species in solution lose electrons to the anode that provides more electrons to be pumped to the cathode. The fact that, under the correct conditions, ions can migrate in the solution, and electrons are lost and gained at the electrodes, means that a solution containing ions will conduct electricity. However, electrical conductivity in a solution is not the same as that which occurs in a metal. It is not possible for electrons to flow through the solution, but electrical conductivity in solution occurs as a result of the movement of ions and the electrode reactions involving the loss and gain of electrons. The species that react at the cathode and anode depend on the nature of the species present in solution. Some species react readily at the electrodes while others do not react. The process of subjecting a solution to the conditions that will produce electrode reactions is called **electrolysis.**

11-4   Electrolysis of Sodium Chloride Solutions

As an example of an electrolytic process that illustrates electrical conductivity by a solution, let us consider what happens when a rather concentrated sodium chloride solution is placed in an electrolytic cell. When the electrodes are connected to the external source of electricity (electron pump), the chloride ions are attracted to the anode where they can lose electrons and form molecular chlorine. This reaction can be represented as

$$2Cl^-(aq) \rightarrow Cl_2(g) + 2e^- (anode)$$

Since molecular chlorine is diatomic, two chloride ions must react to produce one chlorine molecule. A reaction in which a species loses or gains electrons at an electrode and is converted to a new species is called an **electrode reaction.** An electrode reaction, such as that given above, only occurs when a simultaneous reaction involving the gain of electrons occurs at the other electrode, which is part of the system. In this example, we would expect the sodium ions to migrate toward the cathode. However, sodium ions cannot gain electrons in aqueous solution to form sodium metal because sodium metal cannot exist in water. Some other species must react at the cathode. The only other species present is water that, in this case, will gain electrons according to the electrode reaction

$$2e^- (cathode) + 2H_2O \rightarrow H_2(g) + 2OH^-(aq)$$

These two electrode reactions occur simultaneously when the solution of sodium chloride is subjected to electrolysis. For every two electrons gained at the cathode, two electrons are lost at the anode. These electrode reactions are actually chemical reactions involving electron transfer. The overall result of the electron transfer process occuring during the electrolysis of the sodium chloride solution can be represented by the equation

$$2Na^+(aq) + 2Cl^-(aq) + 2H_2O \xrightarrow{\text{electrolysis}} Cl_2(g) + H_2(g) + 2Na^+(aq) + 2OH^-(aq)$$

Thus, when a rather concentrated solution of sodium chloride is subjected to electrolysis, chlorine gas is produced at the anode and hydrogen gas is produced at the cathode. Furthermore, for every two chloride ions that react, two hydroxide ions are produced. See Figure 11-3 for an illustration of this electrolysis process.

11-5   Electrolytes and Nonelectrolytes

The conduction of electricity by a solution involves the movement of ions and certain electrode reactions. Consequently, observing whether a solution will conduct electricity will be an indication of whether ions are present in the solution. A substance that forms an aqueous solution that conducts electricity is called an **electrolyte.** Of course, an electrolyte must form ions in solution when it dissolves. Soluble ionic substances are electrolytes as are some molecular substances. A substance that forms an aqueous solution that does not conduct is called a **nonelectrolyte.** Many molecular substances (e.g., sugar and alcohol) are nonelectrolytes. The fact that a substance is a nonelectrolyte indicates that it does not form ions when it dissolves. Whether or not a substance is an electrolyte can be determined by preparing an aqueous solution of the substance and then experimentally observing whether the solution conducts electricity. When this is done, it is found that solutions of some substances conduct, and solutions of other substances do not. We classify such substances as electrolytes or nonelectrolytes, respectively. When conductivity experiments are carried out, it is also found that solutions of some electrolytes are strong conductors while solutions of others are quite weak conductors. Thus, some electrolytes are called **strong electrolytes,** and others are called **weak electrolytes.** The difference is the result of the extent to which substances form ions when they are dissolved in water. Substances that dissolve in water to form no products other than cations and anions are strong electrolytes. Recall that dissolving may involve merely the dissolution of ions from the crystal lattice of an ionic substance or the reaction of a molecular substance with water. Strong electrolytes can be ionic or molecular substances. Weak electrolytes are substances that dissolve in water but only react with water to a slight extent to produce relatively few ions. The fact that only a few ions are present accounts for the weak conductivity of electricity. In any case, by observing the conductivity of a solution of a substance, we can classify the substance as strong electrolyte, weak electrolyte, or a nonelectrolyte.

11-6   Chemical Equilibrium

Suppose we had a water solution containing a certain concentration of hydrogen chloride (HCl) and another solution of the same concentration of acetic acid ($HC_2H_3O_2$). When we test the conductivity of these solutions, they both are found to conduct electricity so we classify hydrogen chloride and acetic acid as electrolytes. However, the hydrogen chloride solution is found to be a much stronger conductor of electricity than the acetic acid solution. This difference in conductivity indicates that the acetic acid solution has fewer ions than the hydrogen chloride solution. Since they are both electrolytes, and we are comparing solutions of the same concentration, we might expect the solutions to contain the same number of ions, but they do not. The difference in the number of ions is a result of the extent to which these two substances enter into chemical reaction with water to form ions.

   When some molecular substances, such as hydrogen chloride, dissolve in water, they enter into a chemical reaction that consumes the substance to form ions. The reaction between hydrogen chloride and water is

$$HCl + H_2O \rightarrow H_3O^+ + Cl^-$$

On the other hand, other molecular substances dissolve in water and enter into a chemical reaction that produces relatively few ions. For example, when acetic acid is dissolved in water, the reaction is

$$HC_2H_3O_2 + H_2O \rightleftarrows H_3O^+ + C_2H_3O_2^-$$

This reaction is an example of a **reversible chemical reaction** involving chemical equilibrium (note the double arrow). In previous discussions we gave chemical equations showing the reactants and products separated by a single arrow. Since many reactions actually involve chemical equilibrium, let us consider the nature of chemical equilibrium.

Figure 11-4 illustrates **chemical equilibrium** using an analogy involving fish in tanks. In Figure 11-4a the fish in the left-hand tank represent the initial reactants. When the barrier between tanks is removed, some of the fish make their way to the right-hand tank and become products. As they accumulate on the right, some will swim back to the left-hand tank. After a certain time they will distribute between the two tanks so that there is a certain "concentration" of reactants and products. But the fish will continuously swim back and forth between the tanks changing places with other fish. The idea is that even though they are swimming back and forth, the relative "concentrations" in both tanks remain constant. When this occurs a state of **dynamic equilibrium** is said to exist. The term dynamic means that some of the "reactants" are forming "products" and some of the "products" are forming "reactants." The term equilibrium means that the rates of the two changes are the same, and no net change in the "concentrations" of the "reactants" and "products" is apparent. That is, the relative numbers of fish in the two tanks remain constant even though the fish are continuously moving back and forth between tanks.

Now let us consider the reversible reaction between acetic acid and water in terms of chemical equilibrium. (See Figure 11-4b.) The $HC_2H_3O_2$ enters into reaction with H_2O to form H_3O^+ (hydronium ion) and $C_2H_3O_2^-$ (acetate ion). When these two ions begin to accumulate, they enter into a reaction with one another that reforms the acetic acid and water. A point is reached when the rate of the reaction forming the ions equals the rate of the reaction of the ions, and a state of dynamic chemical equilibrium exists. That is, acetic acid and water are continuously reacting to give hydronium ion and acetate ion, and the ions are continuously reacting to give the original reactants; but, since the rates of the two competing reactions are the same, the relative concentrations of the reactants and products becomes fixed at certain levels. In fact, the equilibrium reaction in the case of acetic acid produces low concentrations of ions. Thus, a solution of acetic acid contains a mixture of acetic acid molecules, water molecules, and some hydronium ion and acetate ion. Of course, chemical species do not swim as fish do, but they do move about in solution colliding with one another. A solution in which a state of equilibrium exist can be viewed as consisting of a collection of chemical species constantly moving about within the solution volume. A collision between reactant particles may result in a forward reaction while a collision between product particles may result in the reverse reaction. Even though the collisions and reactions are occurring continuously, the rates of the forward and reverse reactions are

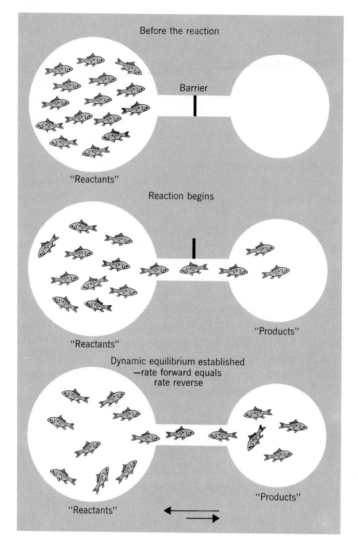

Fish anology for dynamic equilibrium. Note that once dynamic equilibrium is established there is no net change in the "concentration" of the reactants or products, but there is a continuous interchange between the two sides.

Figure 11-4a   An idealized view of dynamic equilibrium.

equal at equilibrium. Thus the relative concentrations of the species involved remain constant.

A state of dynamic equilibrium in a reaction is denoted by use of a double arrow in the equation for the reaction. The relative sizes of the arrows used indicate which species exist at greater concentrations at equilibrium. The larger arrow points in the direction of the species present at greater concentrations. The equilibrium is said to favor this direction. For example, the reversible reaction between acetic acid and water (see the above equation and note arrows) favors the formation of water and acetic acid. Another way of stating this is that water and

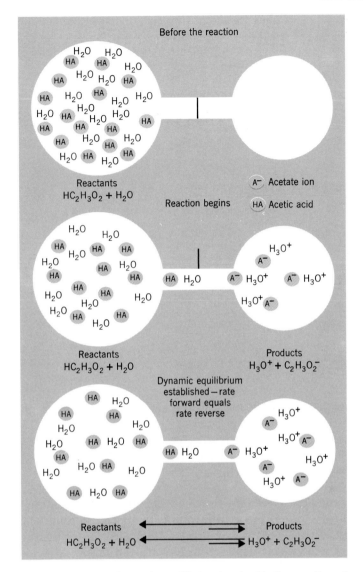

Figure 11-4b   Dynamic equilibrium involved in the reaction of acetic acid with water to form acetate ion and hydronium ion.

acetic acid enter into a slight chemical reaction to form hydronium ion and acetate ion.

Reversible chemical reactions involving chemical equilibrium are quite common, and many of the reactions we have discussed are reversible reactions. Usually, it is only necessary to refer to the equilibrium to explain certain chemical phenomena. A few examples of equilibrium reactions are given in Table 11-1.

When we want to emphasize the reversibility or state of equilibrium of a reaction, we include the double arrow in the equation. When we are not concerned with denoting equilibrium we use a single arrow between

Table 11-1   Some Reactions Involving Chemical Equilibrium

The formation of dinitrogen tetroxide from nitrogen dioxide

$$2NO_2 \rightleftharpoons N_2O_4$$

The formation of ammonia from nitrogen and hydrogen

$$N_2 + 3H_2 \rightleftharpoons 2NH_3$$

The formation of ozone from oxygen

$$3O_2 \rightleftharpoons 2O_3$$

The reaction of ammonia and water

$$NH_3 + H_2O \rightleftharpoons NH_4^+(aq) + OH^-(aq)$$

The reaction of hydrogen fluoride and water

$$HF + H_2O \rightleftharpoons H_3O^+(aq) + F^-(aq)$$

The dissolving of silver chloride

$$AgCl(s) \rightleftharpoons Ag^+(aq) + Cl^-(aq)$$

The dissolving of calcium carbonate

$$CaCO_3(s) \rightleftharpoons Ca^{2+}(aq) + CO_3^{2-}(aq)$$

the reactants and products in the same way that we represent a non-equilibrium reaction.

11-7   Le Chatelier's Principle

When a chemical system is at equilibrium, it will remain in this state indefinitely unless the equilibrium is disturbed in some manner. It is of interest to consider what happens when some factor upsets the equilibrium. A general principle regarding this situation is called **Le Chatelier's principle,** which can be stated as follows.

When a chemical equilibrium is upset by a change in any factor affecting the equilibrium, the equilibrium will shift in a manner that tends to counteract the change.

To illustrate this principle the formation of ammonia from hydrogen and nitrogen can be used.

$$N_2(g) + 3H_2(g) \rightleftharpoons 2NH_3(g) + \text{energy}$$

Let us consider how the equilibrium is affected by changes in pressure, temperature, and concentrations of reactants or products. First, what happens when the pressure of the equilibrium system is increased? Note that in the reaction 1 mole of nitrogen and 3 moles of hydrogen give 2 moles of ammonia. Fewer molecules would tend to reduce the pressure and counteract the pressure increase. Thus, using Le Chatelier's principle, we would predict that more ammonia would form; that is, the equilibrium would shift to the ammonia side since this would tend to decrease the pressure.

What happens to the above equilibrium when the temperature is changed? Since the reaction is exothermic, an increase in the temperature adds energy to the system. By Le Chatelier's principle, an increase in the energy of the system results in a shift of the equilibrium toward the nitrogen-hydrogen side. On the other hand, a decrease in temperature decreases the energy of the system resulting in a shift in the equilibrium towards the ammonia side.

When a system is at equilibrium, the concentrations of each species is constant. The equilibrium can be shifted by increasing or decreasing the concentration of a reactant or product. The equilibrium shifts in the direction that tends to diminish an increase in concentration or replenish a decrease in concentration. For example, if more hydrogen and nitrogen are added to the above equilibrium, the equilibrium will shift toward the ammonia side. If the concentration of ammonia is decreased by removal of ammonia, the equilibrium will shift toward the ammonia side.

According to Le Chatelier's principle, a chemical equilibrium in solution can be affected by increasing the concentration of one species in the solution. For instance, if extra acetate ion (from sodium acetate) is added to a solution of acetic acid

$$HC_2H_3O_2 + H_2O \rightleftharpoons H_3O^+(aq) + C_2H_3O_2^-(aq)$$

the equilibrium will shift to the acetic acid side decreasing the amount of H_3O^+ and increasing the amount of $HC_2H_3O_2$. If extra OH^- is added to a solution of ammonia

$$NH_3 + H_2O \rightleftharpoons NH_4^+(aq) + OH^-(aq)$$

the equilibrium will shift increasing the concentration of ammonia. When extra silver ion (from $AgNO_3$) is added to a saturated solution of silver chloride

$$AgCl(s) \rightleftharpoons Ag^+(aq) + Cl^-(aq)$$

the equilibrium will shift precipitating some silver chloride and decreasing the amount of chloride ion.

11-8 Substances in Solution

When we are working with solutions, a question arises concerning what symbols we should use to represent the dissolved solutes. This concern is especially significant in the case of weak electrolytes. That is, we know that a weak electrolyte reacts somewhat with water to produce some ions. Ammonia reacts slightly with water according to the equation

$$NH_3(aq) + H_2O \rightleftharpoons NH_4^+(aq) + OH^-(aq)$$

It is not convenient to write down all of the species when we want to represent a solution of ammonia. Since NH_3 is the species present in the greatest concentration and the ions are present in small concentrations, a solution of ammonia can be represented by the formula $NH_3(aq)$. Therefore, whenever we want to refer to a solution of ammonia, we use $NH_3(aq)$. It is standard practice to represent a solution of a weak electrolyte by the formula of the most predominant species in solution and to follow this formula by the parenthetical *aq*, to indicate a water solution. Keep in mind that water is always present in an aqueous solution, and its presence is implied or indicated by the parenthetical *aq*. In the case of a weak electrolyte the most predominant species in solution is the molecular form of the dissolved substance.

Example 11-1   An aqueous solution of acetic acid, $HC_2H_3O_2$, is found to be a weak conductor of electricity. How should an aqueous solution of acetic acid be represented? Since acetic acid is a weak electrolyte, it is best to represent a solution of acetic acid by this molecular formula of the acid:

$$HC_2H_3O_2(aq)$$

Since nonelectrolytes do not react with water, solutions of these substances should be represented by the formula of the substance followed by a parenthetical *aq*. For instance, an aqueous solution of the nonelectrolyte ethyl alcohol should be represented by the formula

$$C_2H_5OH(aq)$$

A substance that is a strong electrolyte is essentially completely converted to cations and anions in solution. It is important to represent solutions of strong electrolytes in the proper manner. Such solutions are best represented by the formulas of the major species in solution. Of course, the major species in solution will be the cation and anion formed when the substance dissolves. In the case of most ionic substances, it is normally a simple task to decide which ions are involved because we merely use the formulas of the cation and anion comprising the compound. However, in the case of strong electrolytes that are molecular in nature, we must know how the substance reacts with water so that we can decide which ions are formed upon dissolving.

Example 11-2   How would a solution of sodium chloride, NaCl, be represented? Since ionic substances are strong electrolytes, we predict that the best representation of such a solution is the formulas of the constituent cation and anion. So, we represent a solution of sodium chloride as

$$Na^+(aq) + Cl^-(aq)$$

Solutions of common ionic substances with which we will deal are best represented by the separate formulas of the ions comprising the compound. However, some ionic substances have ions that react chemically with water. To represent such compounds, it is necessary to know the nature of the reaction with water so that the nature of the species in solution can be deduced. As an example, consider the reaction that occurs when solid sodium peroxide, Na_2O_2, is dissolved in water:

$$2Na_2O_2(s) + 2H_2O \rightarrow 4Na^+(aq) + 4OH^-(aq) + O_2(g)$$

Example 11-3   How would a solution of potassium sulfate, K_2SO_4, be represented? To represent the solution we write down the formulas of the ions comprising the compound and include the notation that the ions are in aqueous solution. So, a solution of potassium sulfate would be represented as

$$K^+(aq) + SO_4^{2-}(aq)$$

All that is done in describing a solution in this manner is to indicate which major species are present in solution. Of course, in the case of potassium sulfate solution, there will be twice as many potassium ions present as there are sulfate ions. To emphasize this, we could represent the solution as

$$2K^+(aq) + SO_4^{2-}(aq)$$

but this is usually not necessary if we just want to indicate which species are present in solution.

As long as we know the nature of the reaction between water and a molecular substance, which is a strong electrolyte, we can represent the solution by the formulas of the ions that are produced in the reaction. Recall that hydrogen chloride gas dissolves in water and reacts according to the equation

$$HCl(g) + H_2O \rightarrow H_3O^+(aq) + Cl^-(aq)$$

Since hydronium ions and chloride ions are the species formed when HCl dissolves in water, we should represent an aqueous solution of HCl (called hydrochloric acid) as

$$H_3O^+(aq) + Cl^-(aq)$$

Some important examples of solutions of strong and weak electrolytes will be examined in the next chapter. The important things to keep in mind are that solutions of ionic substances are usually represented by the separate formulas of the constituent ions, solutions of molecular substances that are strong electrolytes are represented by the formulas of the ions formed by reaction with water, and solutions of weak electrolytes and nonelectrolytes are represented by the formula of the molecular species involved. Such representations of substances in the solution phase will greatly simplify the discussion of chemical reactions that occur in solutions. Solution chemistry is discussed in Chapter 12.

To know how to represent aqueous solutions of substances, we should keep in mind which common substances are strong electrolytes, which are weak electrolytes, and which are nonelectrolytes. Most soluble ionic compounds are strong electrolytes. These compounds include metal-non-metal and metal-polyatomic ion compounds such as sodium hydroxide, NaOH, ammonium chloride, NH_4Cl, potassium sulfate, K_2SO_4, and zinc acetate, $Zn(C_2H_3O_2)_2$. However, as we shall see in the next chapter, not all ionic compounds are soluble in water. Several molecular compounds

Table 11-2   Some Electrolytes and Nonelectrolytes

Strong Electrolytes (species dissolved to form ions)	Predominant Species in Solution
H_2SO_4	$H_3O^+(aq) + HSO_4^-(aq)$
HCl	$H_3O^+(aq) + Cl^-(aq)$
HNO_3	$H_3O^+(aq) + NO_3^-(aq)$
(soluble ionic compounds)	
NaCl	$Na^+(aq) + Cl^-(aq)$
K_2SO_4	$K^+(aq) + SO_4^{2-}(aq)$
NH_4NO_3	$NH_4^+(aq) + NO_3^-(aq)$
NaOH	$Na^+(aq) + OH^-(aq)$
KOH	$K^+(aq) + OH^-(aq)$
$Ba(OH)_2$	$Ba^{2+}(aq) + OH^-(aq)$
$NaC_2H_3O_2$	$Na^+(aq) + C_2H_3O_2^-(aq)$
$AgNO_3$	$Ag^+(aq) + NO_3^-(aq)$
Weak Electrolytes (species react slightly with water to form some ions)	
NH_3	$NH_3(aq)$
HF	$HF(aq)$
H_2S	$H_2S(aq)$
H_3PO_4	$H_3PO_4(aq)$
$HC_2H_3O_2$	$HC_2H_3O_2(aq)$
Nonelectrolytes (species do not form ions in solution)	
CH_3OH	$CH_3OH(aq)$
C_2H_5OH	$C_2H_5OH(aq)$
$C_{12}H_{22}O_{11}$	$C_{12}H_{22}O_{11}(aq)$
CH_3COCH_3	$CH_3COCH_3(aq)$

involving hydrogen and nonmetals (binary acids and oxyacids) are strong electrolytes. These compounds are called the strong acids and include hydrogen chloride, HCl, sulfuric acid, H_2SO_4, and nitric acid, HNO_3. These compounds react with water to give a hydronium ion and the corresponding negative ion. However, most other binary acids and oxyacids are weak electrolytes. These compounds include acetic acid, $HC_2H_3O_2$, hydrogen fluoride, HF, phosphoric acid, H_3PO_4, hydrogen sulfide, H_2S, and hydrogen cyanide, HCN. Many other soluble molecular compounds are nonelectrolytes. These compounds include ethyl alcohol, C_2H_5OH, methyl alcohol, CH_3OH, sucrose, $C_{12}H_{22}O_{11}$, and acetone, CH_3COCH_3. Some common electrolytes and nonelectrolytes are tabulated in Table 11-2. To decide how to represent a water solution of one of these substances, we have to decide whether the substance is a strong electrolyte, weak electrolyte, or nonelectrolyte. Once this is determined, we represent the solution by giving the formulas of the ions present or the formula of the molecular species present.

11-9   Water Pollutants

As we have seen in Chapter 10, natural waters are dilute solutions containing many chemicals. Natural water also contains a variety of suspended matter and bacteria. The presence of most of these impurities result from natural processes and cannot be avoided. The quality of fresh water can be adversely affected by the addition of other impurities by humans.

Water has a variety of uses. A body of water may be used for recreation purposes, to support fish and wildlife, for agricultural irrigation, for industrial purposes or for a public water supply. Obviously, water of varying purity is required for these uses.

What is polluted water? Actually, there are varying degrees of water pollution depending upon the intended use. In California, and, similarly other states, **water pollution** is defined legally as follows:

An impairment of the quality of water that unreasonably affects: (1) such water for beneficial uses, or (2) facilities that serve such beneficial uses.

Often the taste, odor and appearance of water indicates that it is polluted. In some cases, only precise chemical tests reveal the presence of dangerous pollutants. In any case, polluted water is water that is not fit for the intended use.

Let us now consider the nature of some **water pollutants.** Table 11-3 lists the major kinds of pollutants. Pollutants in liquid forms come from the discharge of municipal, agricultural, and industrial wastes into the waterways as well as seepage from septic tanks, animal feedlots, and sanitary landfills. These liquids contain dissolved minerals, human and animal wastes, man-made chemicals, and suspended and biological matter. Pollutants in solid forms include such material as sand, clay, soil, ashes, solid sewage, agricultural vegetable matter, grease, tars, garbage, paper, rubber, wood, metals, and plastics. Some of these solids or **physical pollutants** are of natural origin but many are synthetic

Table 11-3   Common Physical, Chemical, and Biological Pollutants

Type	Example	Type	Example
Chemical	Organic compounds	Physical	Floating solids
	Inorganic ions		Suspended material
	Radioactive material		Settleable material
Biological	Pathenogenic bacteria		Foam
	Viruses		Insoluble liquids
	Algae		Heat
	Aquatic weeds		

From *Understanding Chemistry: From Atoms to Attitudes* by T. R. Dickson, Wiley, N.Y., 1974.

substances that enter the waters by human activities. Other physical pollutants include foams, scums, oil slicks, and heat (thermal pollution). Physical pollutants affect the appearance of water and, by settling out on the bottom or floating on top of the waters, they interfere with animal life.

Chemical pollutants include dissolved or dispersed organic and inorganic substances. Inorganic pollutants come from the domestic, agricultural and industrial discharges that contain a variety of dissolved substances. These pollutants include soluble metallic chlorides, sulfates, nitrates, phosphates and carbonates. Also included are waste acids, bases and toxic dissolved gases, such as sulfur dioxide, ammonia, hydrogen sulfide, and chlorine. Acids can be deadly to aquatic life and can cause the corrosion of metals and concrete. **Organic pollutants** are carbon-containing compounds from domestic, agricultural, and industrial wastes. These include organics from human and animal wastes, food processing and slaughterhouse wastes, industrial chemicals and solvents, oils, tars, dyes, and synthetic organic chemicals such as insecticides. As we shall see in the next section, organic pollutants tend to deplete the dissolved oxygen in water. Some organics, such as oils and insecticides, can interfere with or be toxic to aquatic life.

Biological pollutants include disease causing bacteria and viruses, algae, and other aquatic plants. Certain bacteria are harmless and others are involved in the decomposition of organic compounds in water. The undesirable bacteria and viruses are those that cause diseases like typhoid, dysentery, poliomyelitis, hepatitis, and cholera. The control of viruses in water is difficult. Waterborne viral diseases must be watched closely, especially if water is to be reused.

11-10   Degradation in Water

Some pollutants are decomposed by chemical and biological processes occurring in water. These are called **degradable** or **biodegradable pollutants. Degradation** refers to breaking down into simpler substances. Most organic pollutants are degradable. However, some pesticides and detergents are nondegradable, or they decay very slowly in water. These are called **hard** or **refractory organics. Hard pesticides** include chlorinated pesticides such as DDT, chlordane, and endrin. In addition to hard organics, nondegradable chemical pollutants include nitrate ion, phosphate ion, sulfate ion, and various metal ions the least desirable of which are mercury, lead, and cadmium ions.

The decomposition of organic materials in water occurs mainly through the action of bacteria and other organisms in the water. The bacteria use the organics as foods and utilize them as sources of energy in biological oxidation processes. In such bacterial decay, dissolved oxygen is utilized and carbon dioxide, water and various nondegradable ions are produced. The following are some general reactions showing the bacterial decay of organics in the presence of oxygen. Such decay is called **aerobic (Greek: air-life) decay.**

$$CH \text{ (hydrocarbons)} + O_2 \xrightarrow{\text{bacteria}} CO_2 + H_2O$$

$$CH_2O \text{ (carbohydrates)} + O_2 \xrightarrow{\text{bacteria}} CO_2 + H_2O$$

$$\begin{array}{l}\text{organic sulfur-}\\\text{containing compounds}\end{array} + O_2 \xrightarrow{\text{bacteria}} CO_2 + H_2O + SO_4^{2-}$$

$$\begin{array}{l}\text{organic nitrogen-}\\\text{containing compounds}\end{array} + O_2 \xrightarrow{\text{bacteria}} CO_2 + H_2O + NO_3^{-}$$

$$\begin{array}{l}\text{organic phosphorus-}\\\text{containing compounds}\end{array} + O_2 \xrightarrow{\text{bacteria}} CO_2 + H_2O + PO_4^{3-}$$

All of the reactions consume dissolved oxygen from water. Organics which enter into aerobic decay are called **oxygen depleting pollutants.** When insufficient oxygen is present in water, bacterial decay of most organics can continue to occur. However, such decomposition in the absence of oxygen does not produce the same products. This **anaerobic (without oxygen) decay** is illustrated by the general equation

$$\begin{array}{l}\text{organic sulfur and}\\\text{nitrogen-containing}\\\text{compounds}\end{array} + H_2O \xrightarrow[\text{decay}]{\text{anaerobic}} CO_2 + H_2S + CH_4 + NH_4^{+}$$

Anaerobic decay produces gases that bubble from the water and contribute offensive odors to the water.

11-11 Biochemical Oxygen Demand

One of the most remarkable properties of natural water is the capacity of the water to reduce organic pollution by bacterial action. This property, along with the fact that when wastes are dumped into natural waters they are greatly diluted, has enabled us to use our waterways for disposal of vast amounts of waste. However, as the amounts of wastes increase, we begin to strain the capacity of the waters to handle the pollution load. Recall that the aerobic decay of organics utilizes dissolved oxygen. As the dissolved oxygen is used up, it is replaced by more oxygen from the atmosphere. This replacement of dissolved oxygen occurs more readily in a moving, turbulent body of water. As long as there is sufficient dissolved oxygen in the water, aerobic decay will continue. If the level of organic pollution is high, the oxygen is used up more rapidly than it is replaced. In such a situation, the dissolved oxygen in the water is soon depleted. Once the oxygen is depleted, anaerobic decay begins, and the waters become fouled by the products of this kind of decay. Furthermore, the depletion of oxygen can result in the death of fish and other aquatic life. The decay products of these animals can add to the organic level increasing the amount of anaerobic decay. Of course, anaerobic conditions do not always occur, but the chances increase as the amount of degradable organics increases. The **biochemical oxygen demand (BOD)** serves as a quantitative measure of the level of organic oxygen demanding wastes in water. In a water sample the BOD expresses the number of milligrams of dissolved oxygen per liter used up as the organic wastes are consumed by bacteria in the water. BOD is usually expressed in terms of parts per million oxygen and is determined by measuring the decrease over a period of five days of the dissolved oxygen in a water sample maintained at 20°C. Water of high BOD requires large amounts of oxygen. Thus, the BOD of water can indicate the organic pollution load of the water. **Potable** (drinkable) water normally has a BOD of 0.75 to 1.5 parts per million oxygen (ppm). Relatively unpolluted water has a BOD of around 1 to 3 parts per million. Water is considered to be polluted if the BOD exceeds around 5 ppm. Untreated municipal sewage has a BOD range of 100 to 400 ppm. Some industrial and agricultural wastes have BOD levels in the thousands of parts per million range. Of course, such BOD levels are diminished upon dilution of the wastes, but, in some cases, very drastic pollution problems are caused by these wastes. It is possible to decrease the BOD level of wastes by sewage treatment that is discussed in the next section.

11-12   Sewage Treatment

Large amounts of sewage wastes are generated by industries and municipalities each year. In addition to this, there are large amounts of agricultural wastes and irrigation water runoff. Most of the domestic and industrial wastes are treated in some way to decrease the BOD level and to remove dangerous components. A portion of the industrial sewage is fed into municipal sewage lines. However, many industries treat their own waters or dispose of wastes in special ways. Unfortunately, some industries and municipalities are responsible for adding large amounts of untreated or partially treated sewage to our waters. These wastes often have high BOD levels, and some contain toxic or refractory chemicals.

The treatment of domestic sewage is usually directly controlled by municipal or regional agencies. In our highly urbanized society well over 60% of the population is served by public sewers. Over 30% utilize septic tanks or cesspools, while less than 10% use nonwater outdoor facilities. Presently around 10% of the sewage handled by public sewers receives no treatment. This raw sewage is dumped into various bodies of water. The vast majority of domestic sewage does receive some treatment in sewage disposal plants before being released into environmental waters. Some of this treatment is adequate while some is marginal. Many domestic sewage systems are tied in with storm drain systems. This causes problems during rainy seasons when large

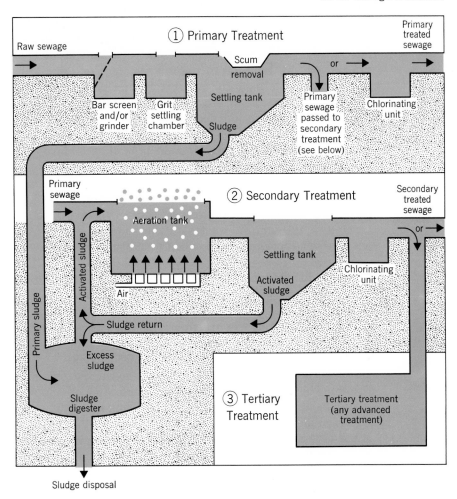

Figure 11-5   Sewage treatment steps. From *Understanding Chemistry: From Atoms to Attitudes* by T. R. Dickson, Wiley, N.Y., 1974.

amounts of storm water wash untreated sewage directly through the plant into the environment.

As shown in Figure 11-5 there are three potential steps in the treatment of domestic sewage. These three steps are called primary, secondary, and tertiary treatment. The main purpose of sewage treatment is to lower the BOD level of the waste water. Around 30% of domestic sewage receives only primary treatment while around 60% is subject to secondary treatment. Very little tertiary treatment is being applied to domestic sewage. Secondary treatment can cost twice as much as primary treatment. Tertiary treatment, at the present, is quite costly. Let us consider each of these treatment steps. See Figure 11-5.

Primary Treatment. As the waste water enters the plant it is passed through a bar screen to screen out large objects. Some plants have large grinders that pulverize the large objects so that they can pass through the treatment step. The sewage then flows slowly through a large chamber where the sand, pebbles, and other heavy material settle out. From this chamber the sewage flows into a large settling tank. The suspended solids settle to the bottom as sewage sludge, and the greases and oils float to the top as scum. The water between the sludge and scum is drained off and released to the environment or is passed to secondary treatment. Sometimes the effluent is first chlorinated to kill bacteria before it is released to the environment. Primary treatment removes around 60% of the suspended solids and around 35% of the BOD.

Secondary Treatment. The most common type of secondary treatment is **activated sludge treatment.** Sewage flows from primary treatment to an **aeration tank** where air (or pure oxygen in a few plants) is bubbled through it. The aeration of the sewage results in rapid growth of bacteria and other microorganisms. The bacteria utilize the oxygen to decompose the organic wastes in the sewage. Suspended solids mixed with bacteria form a sludge called activated sludge.

Some plants use a device called a **trickling filter** rather than the activated sludge process. In this method the sewage water is sprayed on a bed of stones about 6-ft deep. As the water trickles over and around the stones it comes into contact with bacteria that decompose the organic pollutants. The bacteria are then consumed by a variety of other organisms residing in the filter.

Treated sewage from the aeration tank or trickling filter is passed to another large settling tank where the activated sludge settles to the bottom. The activated sludge that accumulates on the bottom is transferred back to the aeration tank and mixed with incoming sewage. Excess sludge is collected, treated, and disposed of. The disposal of sewage sludge is a major problem. A large treatment plant produces appreciable amounts of sludge that must be disposed of as solid wastes. Some is used for fertilizer, but most has to be buried in a landfill or dumped in the ocean. Primary treatment followed by secondary treatment removes up to 90% of the suspended solids and around 90% of the BOD. After secondary treatment the waste water is usually subjected to chlorination before being released to the environment.

Tertiary Treatment. Any kind of treatment beyond the secondary step is termed tertiary treatment. The purpose of tertiary treatment is to remove organic pollutants, nutrients such as phosphate ion and nitrate ion, or excessive ions (mineral salts). The major objective of tertiary treatment is to make the waste water as pure as feasible before returning it to the environment. There are several kinds of tertiary treatment. Precipitation, sedimentation, and filtration can be used for nutrient removal. Carbon adsorption is used for removal of organics. Techniques such as reverse osmosis, ion exchange and electrodialysis are used for demineralization; the removal of inorganic ions.

Questions and Problems

1. Describe what happens, in terms of the interaction between water molecules and ions, when an ionic substance is dissolved in water.

2. Why does a solid substance usually dissolve faster when it is in the form of tiny crystals rather than large crystalline chunks? Why do most solid substances dissolve faster in hot water as compared to cold water?

3. Define the following terms.
   (a) solubility
   (b) saturated solution
   (c) supersaturated solution

4. Describe how a solution containing ions conducts electricity when electrodes are placed in the solution.

5. Why is it dangerous to handle electrical equipment when you are in a tub of water or standing in water?

6. How is it possible to experimentally determine the presence of ions in a solution?

7. What are electrolytes and nonelectrolytes?

8. What is the difference between strong electrolytes and weak electrolytes.

9. Give an example of a reversible chemical reaction involving chemical equilibrium and use the example to describe chemical equilibrium.

10. Give a statement of Le Chatelier's principle.

11. Using Le Chatelier's principle, predict any shift in equilibrium in each of the following:
    (a) $HF(aq) + H_2O \rightleftharpoons F^-(aq) + H_3O^+(aq)$, when extra H_3O^+ (from hydrochloric acid) is added
    (b) $PbCl_2(s) \rightleftharpoons Pb^{2+}(aq) + 2Cl^-(aq)$, when extra Cl^- (from sodium chloride) is added
    (c) $2H_2(g) + O_2(g) \rightleftharpoons 2H_2O(g) +$ energy, when the pressure is increased
    (d) $2H_2(g) + O_2(g) \rightleftharpoons 2H_2O(g) +$ energy, when the temperature is increased
    (e) $H_2(g) + I_2(g) \rightleftharpoons 2HI(g)$, when the pressure is decreased

12. The industrial manufacture of ammonia involves the equilibrium reaction

$$N_2(g) + 3H_2(g) \rightleftharpoons 2NH_3(g) + \text{energy}$$

This reaction occurs very slowly at low temperatures but the rate of the reaction increases with an increase in temperature. Industrially the reaction is run at 425°C and 500 atm pressure. Which of these conditions is consistent with Le Chatelier's principle? If the rate of the reaction increases with temperature how can Le Chatelier's principle be used to explain why the reaction is not carried out at very high temperatures.

13. In the internal combustion engine a gasoline-air mixture is compressed in the piston chamber and burned producing a high temperature. Nitrogen oxide, NO, an air pollutant is formed in the internal combustion engine by the reaction

$$\text{energy} + N_2(g) + O_2(g) \rightleftharpoons 2NO(g)$$

According to Le Chatelier's principle what condition in the internal combustion engine causes a shift in this equilibrium towards the formation of NO?

14. Hemoglobin (represented symbolically as Hb) is a complex protein found in red blood cells. Hemoglobin is involved in oxygen transport in the body according to the reaction

$$Hb + O_2 \rightleftharpoons HbO_2$$

Using Le Chatelier's principle explain how hemoglobin transports oxygen from the lungs where there is a high concentration of oxygen to the cells fed by blood capillaries where there is a low concentration of oxygen.

15. Give the formulas of the predominant species that are found in solutions of the following compounds:
    (a) glucose, $C_6H_{12}O_6$ (nonelectrolyte)
    (b) hydrogen sulfide, H_2S (weak electrolyte)
    (c) potassium chloride, KCl (strong electrolyte)
    (d) ammonium acetate, $NH_4C_2H_3O_2$ (strong electrolyte)
    (e) hydrogen bromide, HBr (strong electrolyte like HCl)
    (f) methyl alcohol, CH_3OH (nonelectrolyte)
    (g) magnesium sulfate, $MgSO_4$ (strong electrolyte)
    (h) phosphoric acid, H_3PO_4 (weak electrolyte)
    (i) nitric acid, HNO_3 (strong electrolyte like HCl)
    (j) sodium hydrogen carbonate, $NaHCO_3$ (strong electrolyte)
    (k) hydrogen fluoride, HF (weak electrolyte)

16. What is water pollution?

17. Describe the process of aerobic decay that occurs in natural waters.

18. What is biochemical oxygen demand, BOD?

19. A water sample is found to contain 25 parts per million oxygen. After a five-day BOD test the water contains 10 parts per million oxygen. What is the biochemical oxygen demand or BOD of the water?

20. Describe primary sewage treatment and secondary sewage treatment.

21. What is tertiary sewage treatment?

12

Chapter 12: Chemical Reactions (Precipitation, Acid-Base, and Oxidation-Reduction)

Objectives

The student should be able to:

1 Write the balanced net-ionic equation for a reaction given the initial reactants and the products.

2 Predict and give the balanced net-ionic equation for a precipitation reaction that can occur when two solutions containing ions are mixed based on the solubility list.

3 Give a balanced net-ionic equation for an acid-base reaction that occurs when a solution of an acid and a solution of a base are mixed given an acid-base table.

4 Write and balance the net-ionic equation for an oxidation reduction reaction given the major reactants and a redox table.

5 Calculate the number of grams or moles of a sought species using specified titration data.

6 Calculate the molarity of a solution of unknown concentration using specified titration data.

7 Calculate the pH of a solution given the hydronium ion concentration.

8 Calculate the hydronium ion concentration or pH of a buffer solution given the acid-base pair concentrations.

Terms to Know

Chemical reaction
Net-ionic equation
Spectator ion
Precipitation reaction
Acid
Base
Acid-base reaction
Strong acid
Weak acid
Acidic solution

Basic solution
Oxidation
Reduction
Redox reaction
Oxidizing agent
Reducing agent
pH
Titration
Buffer solution

12

Chemical Reactions (Precipitation, Acid-Base, and Oxidation-Reduction)

12-1 Chemical Reactions

Some common types of chemical reactions will be discussed in this chapter. First, let us consider the nature of a chemical reaction. A **chemical reaction** involves the interaction of certain chemical species that results in the formation of new chemical species. In a chemical reaction some chemical bonds are broken, and new bonds are formed. Reactions occur in a variety of ways. Here we will discuss only a few examples.

Consider the reaction that takes place when hydrogen chloride gas is mixed with water. We know that collections of chemical species are dynamic, and the particles move about colliding with one another. The hydrogen chloride molecules and water molecules collide with one another. When such a collision occurs in a manner in which the molecules are correctly oriented and the particles have high enough kinetic energy, a reaction may take place. The covalent bond between the hydrogen and chlorine breaks and the hydrogen forms a new bond with the oxygen of the water. The products are hydronium ion and chloride ion.

$$\text{Cl—H} \;+\; \text{O} \overset{\displaystyle \text{H}}{\underset{\displaystyle \text{H}}{\big\langle}} \;\longrightarrow\; \text{Cl}^- \;+\; \text{H—O}^+ \overset{}{\underset{\text{H} \quad \text{H}}{\big\langle}}$$

This reaction involves the breaking of a covalent bond and the formation of a new covalent bond.

As another example consider the reaction that takes place when a sodium chloride solution is added to a silver nitrate solution. Upon mixing, the ions move about colliding with one another and water molecules. However, sufficient attractive force exists between the silver ions and the chloride ions to result in ionic bond formation and the production of solid silver chloride that separates from the solution.

$$Na^+(aq) + Cl^-(aq) + Ag^+(aq) + NO_3^-(aq) \rightarrow$$
$$AgCl(s) + Na^+(aq) + NO_3^-(aq)$$

In this reaction ionic bonds are formed between silver ions and chloride ions to produce a solid ionic compound that is not soluble in water.

As a third example, consider the reaction between sodium metal and water. The water molecules react with the sodium atoms at the surface of the sodium metal. In the course of the reaction sodium atoms lose electrons to form sodium ions, and the water molecules are broken up to form hydrogen gas and hydroxide ions.

$$2Na(s) + 2H_2O \rightarrow 2Na^+(aq) + 2OH^-(aq) + H_2(g)$$

In this reaction, the net result is the loss of electrons by sodium atoms and the gain of electrons by water molecules. The result is the formation of ions and the breaking and forming of covalent bonds.

12-2   Observing Chemical Reactions

How do you know that a chemical reaction has occurred in a solution? To detect the occurrence of a reaction, some observable change must occur. A reaction may produce an insoluble solid that will separate from the solution (precipitate). For instance, when a solution containing silver ion, Ag^+, is added to a solution containing chloride ion, Cl^-, the white solid silver chloride will form. A reaction may produce a gas that will be evolved from the solution. For instance, when a solution containing hydronium ion, H_3O^+, is added to a solution containing carbonate ion, CO_3^{2-}, carbon dioxide gas is produced. As another example, when a piece of zinc metal (or other metals such as nickel, cobalt, iron, aluminum, and magnesium) is added to a solution containing hydronium ion, the metal ion and hydrogen gas are produced. Of course, in addition to the observation of a gas being produced, we also observe that the metal is consumed in the reaction. A reaction may be observed by the heat that is released or absorbed. That is, the temperature of solution in which the reaction occurs may increase or decrease. For instance, when a solution containing hydronium ion is added to a solution containing hydroxide ion, water is produced and the temperature of the solution increases as a result of the heat released by the reaction. However, a temperature change upon mixing two substances does not necessarily prove that a chemical reaction occurred. A reaction may be observed by the color change that occurs when a colored reactant is used up or a colored product is produced. A chemical reaction may occur in a solution, and no apparent physical change may accompany the reaction. Chemical evidence is needed to confirm such reactions. The various

Table 12-1   Types of Evidence of Chemical Reactions

Precipitation — The formation of a solid when two solutions are mixed or a substance is mixed with a solution.

$$Ca^{2+}(aq) + CO_3^{2-}(aq) \rightarrow CaCO_3(s) \text{ (solid calcium carbonate)}$$
$$Zn(s) + Cu^{2+}(aq) \rightarrow Zn^{2+}(aq) + Cu(s) \text{ (solid copper metal)}$$

Gas Formation — The formation of a gas when two solutions are mixed or a substance is mixed with a solution.

$$2H_3O^+(aq) + S^{2-}(aq) \rightarrow 2H_2O + H_2O + H_2S(g) \text{ (poisonous hydrogen sulfide gas)}$$

Consumption of a Solid — A solid is consumed as the reaction proceeds.

$$CaCO_3(s) + 2H_3O^+(aq) \rightarrow Ca^{2+}(aq) + 3H_2O + CO_2(g)$$

(calcium carbonate
is consumed)

Heat Change — The temperature of the solution changes as a result of the release or absorption of heat in the reaction.

$$H_2SO_4(\ell) + H_2O \rightarrow H_3O^+(aq) + HSO_4^-(aq) + heat$$

(*Danger:* Since large amounts of heat are released in this reaction always add sulfuric acid to water slowly. Never add water to the acid since it may boil and splatter the acid.)

Color Change — The color of a solution may change due to the consumption of a colored species or the production of a colored species or both.

$$8H_3O^+(aq) + Cr_2O_7^{2-}(aq) + 3H_2C_2O_4 \rightarrow 2Cr^{3+}(aq) + 6CO_2(g) + 15H_2O$$
           (yellow)   (green)

(Solution changes from yellow to green as the reaction proceeds.)

$$Cu^{2+}(aq) + 4NH_3(aq) \rightarrow Cu(NH_3)_4^{2+}(aq)$$
   (pale blue)   (dark blue)

(Solution changes from pale blue to dark blue.)

types of physical evidence for chemical reactions are summarized in Table 12-1.

12-3 Net-Ionic Equations

Reactions in solutions often involve ions as reactants and products. In this section we shall consider how to write equations for these reactions. One important thing to keep in mind is how to represent water solutions of substances. By determining whether the substance is a strong electrolyte, a weak electrolyte, or a nonelectrolyte, we shall know how to represent it in solution. Turn back to Section 11-8 (Table 11-2) to review which substances are strong or weak electrolytes or nonelectrolytes.

To write an equation for a reaction in solution, first write down the species present in the two solutions that we mix together. Then, write down the product or products of the reaction. Of course, the products of the reaction must be known. This information can be obtained by experimental observation and sometimes by logical prediction. Finally, balance the equation. As examples, let us consider a few reactions.

Example 12-1   A solution of the ionic compound $CaCl_2$ is added to a solution of the ionic compound Na_2CO_3, and the product solid $CaCO_3$ is formed. Write an equation for the reaction.

First, since each of the ionic compounds are electrolytes, we represent their solutions as the separated ions.

$$Ca^{2+}(aq) + Cl^-(aq) + Na^+(aq) + CO_3^{2-}(aq)$$

The product of the reaction is $CaCO_3$, so we could write the equation as

$$Ca^{2+}(aq) + 2Na^+(aq) + CO_3^{2-}(aq) + 2Cl^-(aq) \rightarrow CaCO_3(s) + 2Cl^-(aq) + 2Na^+(aq)$$

Notice that it turned out that the sodium ions, $Na^+(aq)$, and the $Cl^-(aq)$ ions did not react in any way. That is, these ions appear on either side of the equation, and they are not changed by the reaction. They did not enter into any chemical reaction. This often happens in reactions occurring in solutions. Certain species in solution do not undergo reactions while other species do react. Ions that are present in solution during a reaction but do not actually react are called **spectator ions.** You might wonder why these ions are present in the first place. The fact is that all solutions are electrically neutral, and these ions must be present to keep the solution neutral. So, spectator ions are ions that are not involved in a reaction, but they must be present to maintain the same amount of negative and positive charge in the solution. When we write equations representing reactions in solutions, we often do not include the spectator ions. Such equations are called **net-ionic equations** or simply ionic equations. Net-ionic equations show which species react and which are produced but do not include the spectator ions. The net-ionic equation for the above reaction would be

$$Ca^{2+}(aq) + CO_3^{2-}(aq) \rightarrow CaCO_3(s)$$

This reaction indicates that when a solution containing $Ca^{2+}(aq)$ is added to a solution containing $CO_3^{2-}(aq)$, a precipitation occurs to form $CaCO_3(s)$. Remember that spectator ions from the two solutions are present but are not included in the equation.

Example 12-2   A solution of the ionic compound sodium hydroxide, NaOH, is added to a solution of the weak electrolyte acetic acid, $HC_2H_3O_2$. Write an equation for the reaction that produces acetate ion, $C_2H_3O_2^-$, and water.

First, write down the species present in the two solutions.

$$Na^+(aq) + OH^-(aq) + HC_2H_3O_2(aq)$$

The solution of NaOH contains sodium ions and hydroxide ions, and the solution of acetic acid, a weak electrolyte, is represented by the molecular acid formula.

Second, write down the products and balance the equation.

$$Na^+(aq) + OH^-(aq) + HC_2H_3O_2(aq) \rightarrow Na^+(aq) + C_2H_3O_2^-(aq) + H_2O$$

Third, delete the spectator ions to form the net-ionic equation

$$OH^-(aq) + HC_2H_3O_2(aq) \rightarrow C_2H_3O_2^-(aq) + H_2O$$

Example 12-3   A solution of chlorine dissolved in water, $Cl_2(aq)$, is mixed with a solution of the ionic compound iron(II) sulfate. A reaction occurs in which iron(III) ion and

chloride ion are produced. Write an equation for the reaction.
First, write down the species present in the two solutions.

$$Cl_2(aq) + Fe^{2+}(aq) + SO_4^{2-}(aq)$$

Chlorine is represented by the molecular formula Cl_2. Iron(II) sulfate is a soluble ionic compound, so the solution is represented by the ions. Second, write down the products and balance the equation.

$$Cl_2(aq) + Fe^{2+}(aq) + SO_4^{2-}(aq) \rightarrow Fe^{3+}(aq) + SO_4^{2-}(aq) + 2Cl^-(aq)$$

Third, delete the spectator ions.

$$Cl_2(aq) + Fe^{2+}(aq) \rightarrow Fe^{3+}(aq) + 2Cl^-(aq)$$

Now, is the equation balanced? Well, it is balanced chemically — two chlorines and one iron on each side. But look a little closer at the charges. A net-ionic equation must be both chemically balanced and balanced with respect to charge. That is, no extra positive or negative charges can be produced in the reaction. This means that the net number of positive and negative charges must be the same on each side of the equation. Positive and negative charges on the same side of the equation have a neutralizing effect. The net charge on a given side of the arrow will be the sum of the number of positive and negative charges. Referring to our last equation, we see that the charge is not balanced.

$$Cl_2(aq) + Fe^{2+}(aq) \rightarrow Fe^{3+}(aq) + 2Cl^-(aq)$$
$$\text{net charge } (2+) \quad (3+) + (2-) = (1+)$$

The charge can be balanced by adjusting the coefficients. If we place a coefficient of 2 in front of the iron(II) ion and the iron(III) ion, the equation will be both chemically and electrically (charge) balanced.

$$Cl_2(aq) + 2Fe^{2+}(aq) \rightarrow 2Fe^{3+}(aq) + 2Cl^-(aq)$$
$$\text{net charge } 2(2+) = (4+) = (6+) + (2-) = (4+)$$

The reason it is necessary to balance the charge in a net-ionic equation is that, since all solutions are electrically neutral, no extra positively or negatively charged species will be produced in a reaction. Let us summarize the approach to writing net-ionic equations.

1. Insoluble substances, gases, weak electrolytes, and nonelectrolytes are represented by their formulas.
2. Strong electrolytes (soluble ionic substances and strong acids) are represented in solution by the ions that they form.
3. Write down the species present in the two solutions that are mixed.
4. Write down the products.
5. Delete the spectator ions and balance the equation both chemically and electrically (charge balance).

12-4   Precipitation Reactions and Solubility

As we know, many soluble ionic substances exist in solution in the form of the cations and anions that comprise the compound. These ions in the solution phase are in continuous motion and, thus, are moving about momentarily interacting with one another. As long as the substance corresponding to the cation and anion is soluble and enough water is present so that it is not a supersaturated solution, the interaction between ions is not sufficient to cause the aggregation of the ions into a crystalline solid. If we mix two solutions containing dissolved ionic substances, all of the ions will intermingle and mutually interact. Now, if any of the ions that are mixed constitute the ions of an insoluble ionic substance, the ions will combine into aggregations that will result in the solid phase separating from the solution. This solid phase is, of course, the insoluble ionic substance. The process involving the combination of the ions in a solution to form the solid phase is called **precipitation.** The solid phase that separates from the solution is called the **precipitate.** This process generally occurs quite quickly, and the precipitate forms the instant the solutions are mixed and slowly settles from the solution. Precipitates take on various appearances. Some are white and others are colored (e.g., $Zn(OH)_2$ is white, PbS is black, $Cu(OH)_2$ is green, and Ag_2CrO_4 is orange.). Precipitates may be flocculent (fluffy), gelatinous (jellylike), coagulated (curdy or lumpy) or crystalline (powdery or silky). How can a precipitation be predicted when two solutions are mixed? This can often be done by knowing which ionic substances are soluble and which are insoluble. If we know that a given substance is insoluble, then, when the constituent ions of the substance are brought together in the solution phase, precipitation of the substance will usually result. The solubilities of many common ionic substances in water can be summarized in the form of a solubility list based on experimental observations. Recall that ionic substances involve metal ions in combination with nonmetal ions or polyatomic anions. For our purposes, the following **solubility list** will be useful. Refer to this list any time you want to check the solubility of a substance or to Table 12-2.

1. Almost all ionic compounds containing sodium ion, Na^+, potassium ion, K^+, or ammonium ion, NH_4^+, are **soluble.**

2. Almost all ionic compounds containing nitrate ion, NO_3^-, or acetate ion, $C_2H_3O_2^-$, are **soluble.**

3. All ionic compounds containing sulfate ion, SO_4^{2-}, are **soluble** except

> calcium sulfate, $CaSO_4$
> strontium sulfate, $SrSO_4$
> barium sulfate, $BaSO_4$
> lead(II) sulfate, $PbSO_4$

4. All ionic compounds containing chloride ion, Cl^-, bromide ion, Br^-, iodide ion, I^-, and fluoride ion, F^- are **soluble** except

> silver, mercury(I), lead(II) chloride: $AgCl$, Hg_2Cl_2, $PbCl_2$
> silver, mercury(I), lead(II) bromide: $AgBr$, Hg_2Br_2, $PbBr_2$
> silver, mercury(I), lead(II) iodide: AgI, Hg_2I_2, PbI_2
> magnesium, calcium, strontium, barium, lead(II) fluoride:
> MgF_2, CaF_2, SrF_2, BaF_2, PbF_2

5. All ionic compounds containing hydroxide ion, OH^-, are **insoluble** except

sodium hydroxide, NaOH
potassium hydroxide, KOH
barium hydroxide, Ba(OH)$_2$

6. Almost all ionic compounds containing carbonate ion, CO$_3^{2-}$, chromate ion, CrO$_4^{2-}$, and phosphate ion, PO$_4^{3-}$, are **insoluble** except those containing sodium, potassium or ammonium ions. This solubility list is summarized in Table 12-2.

12-5   Prediction of Precipitation Reactions

It is possible to predict whether a precipitation reaction occurs when two solutions are mixed by writing down the ions involved and then deciding if any cation and anion can combine to form an insoluble substance. Of course, we can use the solubility list to decide which substances are insoluble. Furthermore, as long as the concentration of ions is assumed to be great enough, insoluble substances can be precipitated from the solution phase. Remember, whenever we are dealing with a solution of an ionic substance we represent the substance in solution by writing the cation and anion as separate species.

Example 12-4   Write the balanced equation for any reaction that occurs when a solution of silver nitrate is added to a solution of calcium chloride.
Since both solutions involve ionic substances, we first write down the ions involved. Each cation and anion is written separately:

$$Ag^+(aq) + NO_3^-(aq) + Ca^{2+}(aq) + Cl^-(aq)$$

Now we decide whether any of the cations will react with any anions to form an insoluble substance. According to solubility rule 1, all nitrates are soluble so calcium nitrate will not precipitate. However, according to solubility rule 4, silver chloride is insoluble. Therefore, we would predict the combination of the silver ions and chloride ions to form a silver chloride precipitate. The equation representing the reaction is

$$Ag^+(aq) + Cl^-(aq) \rightarrow AgCl(s)$$

Since both ions carry only a single charge, one silver ion reacts with one chloride ion to give silver chloride. Remember that it is always necessary, in an equation involving ions, to balance the charge as well as the number of atoms of each element.

Table 12-2   A Solubility List of Common Ionic Compounds

Anion	Cations that Form Precipitate	Cations that Do Not Form Precipitate
NO$_3^-$	None	All other common
C$_2$H$_3$O$_2^-$	None	All other common
SO$_4^{2-}$	Ca^{2+}, Sr^{2+}, Ba^{2+}, Pb^{2+}	Most other common
Cl$^-$, Br$^-$, I$^-$	Ag$^+$, Hg$_2^{2+}$, Pb^{2+}	Most other common
OH$^-$	Most	Na$^+$, K$^+$, Ba^{2+}
F$^-$	Mg^{2+}, Ca^{2+}, Sr^{2+}, Ba^{2+}, Pb^{2-}	Most other common
CO$_3^{2-}$, CrO$_4^{2-}$, PO$_4^{3-}$	Most	Na$^+$, K$^+$, NH$_4^+$
Almost all ionic compounds of Na$^+$, K$^+$ and NH$_4^+$ are soluble		

Example 12-5   Write the balanced equation for any reaction that occurs when a solution of aluminum sulfate is added to a solution of barium hydroxide.
First, we write down the ions present in these two solutions:

$$Al^{3+}(aq) + SO_4{}^{2-}(aq) + Ba^{2+}(aq) + OH^-(aq)$$

Now we decide whether any cation and anion combinations can occur. According to solubility rule 5, aluminum hydroxide is insoluble. So, we would predict the reaction

$$Al^{3+}(aq) + 3OH^-(aq) \rightarrow Al(OH)_3(s)$$

Of course, to balance the equation, three OH^- are needed to react with each Al^{3+}. According to rule 3, barium sulfate is insoluble. Thus, we would predict the additional reaction

$$Ba^{2+}(aq) + SO_4{}^{2-}(aq) \rightarrow BaSO_4(s)$$

Since both the barium and the sulfate ion carry double charges, they will react on a one-to-one basis.

The above example involved two precipitation reactions occurring at the same time. Usually this is not the case, but keep in mind that when solutions are mixed it is possible that no reaction may occur, one precipitation reaction may occur, or two precipitation reactions may occur.

Example 12-6   Write the balanced equation for any reaction that occurs when a solution of potassium acetate is added to a solution of sodium chloride.
The ions that would be involved are the cations and anions that are present in the solutions of the two substances. These ions are

$$K^+(aq) + C_2H_3O_2{}^-(aq) + Na^+(aq) + Cl^-(aq)$$

Now we decide on any possible reactions. Sodium ion will not react with acetate ion according to rules 1 and 2. Potassium ion will not react with chloride ion according to rules 1 and 4. Thus, we would predict no precipitation reactions in this case and the mixing of the two solutions would merely produce a solution consisting of all four ions.

12-6   Acids and Bases

As you probably know, vinegar has a distinct sour taste. However, you may not know that when you drop a piece of chalk in a vinegar solution, carbon dioxide will bubble off as the chalk dissolves, or that a piece of freshly sandpapered zinc, when placed in vinegar, will slowly dissolve accompanied by the evolution of hydrogen gas. Vinegar is a water solution of an acid called acetic acid.

It is possible to classify certain compounds according to similarities in chemical properties, that is, similarities in the kinds of chemical reactions the substances undergo. Chemists noted long ago that certain substances, which became known as **acids,** were characterized as having sour tastes, being able to dissolve certain metals, changing the color of the vegetable dye called litmus from blue to red and reacting with chemicals called bases. Those substances that became known as **bases** have bitter tastes, feel slippery to the touch, change the color of red litmus to blue and react chemically with acids. As a deeper understanding of the nature of chemicals developed, it became apparent that there were many substances that could be classified as acids or bases. It is now possible to give more definite chemical definitions for these classes of compounds.

A Swedish chemist, Svante Arrhenius, in 1884, proposed the first significant definitions. Several are possible. He defined an acid as a substance that forms hydrogen ions (H^+) in water solution and a base as a substance that forms hydroxide ions (OH^-) in water solution. The Arrhenius theory was significant, since it provided a basis for the description of ions in aqueous solutions. However, for acids and bases in aqueous solutions, the best definitions come from the **Brønsted-Lowry theory** of acids and bases. According to this theory, the terms acid and base should be defined as follows:

Acid. A chemical species that can donate hydrogen ions or protons (H^+) in a chemical reaction. A proton donor.

Base. A chemical species that can accept hydrogen ions or protons in a chemical reaction. A proton acceptor.

According to these definitions, an acid is any hydrogen containing species in which the covalent bond holding the hydrogen can be broken so that the hydrogen ion can be lost. A base is a species that is capable of forming a new covalent bond with a proton donated by an acid. Of course, when an acid loses a proton a base must be present to accept it. Thus, an acid can react with a base in a **proton transfer** or **acid-base reaction.** Such a reaction can be generally represented as (see Figure 12-1)

$$\text{H—A} \quad \text{B} \rightarrow \text{H—B} + \text{A}$$
$$\text{acid} \qquad \text{base}$$

As an example of an acid-base reaction, consider the reaction between acetic acid and the hydroxide ion present in a sodium hydroxide solution.

$$HC_2H_3O_2(aq) + OH^-(aq) \rightleftharpoons H_2O + C_2H_3O_2^-(aq)$$

An acid loses a proton in a chemical reaction, and a base gains a proton. An acid loses a proton to a species that is capable of bonding to the proton more strongly than the original acid. Thus, we say that an acid, H—A, can lose a proton to leave a weaker base, A, and a base, B, can gain a proton to form a weaker acid, H—B.

H—A	+	B	→	H—B	+	A
weaker proton binder		proton acceptor		stronger proton binder		(the weaker base)
(the stronger acid)		(the stronger base)		(the weaker acid)		

Each acid in the Brønsted-Lowry sense will have a corresponding base,

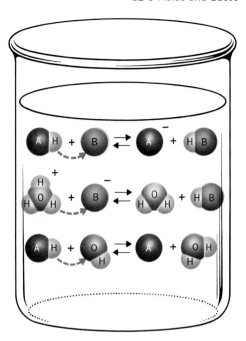

Figure 12-1   Acid-base reactions. When a solution of an acid and a solution containing a base are mixed, a reaction may occur in which protons are transferred from the acid to the base. This is called an acid-base reaction.

and each base will have a corresponding acid. An acid and a base that are related by a transfer of a proton are called a **conjugate acid-base pair.** When an acid loses a proton, it must be gained by a base. That is, an acid can only lose a proton to a base and, in this manner, be converted to its corresponding conjugate base. The base that gains the proton will in turn be converted to its corresponding conjugate acid. A reaction in which a proton is transferred from an acid to a base and the corresponding conjugate acid and base are formed is called an acid-base reaction. See Figure 12-1. Acid-base reactions are common reactions in solution. Let us consider the previous example of an acid-base reaction. When acetic acid and hydroxide ion react a proton is transferred from $HC_2H_3O_2$ to OH^- to form the conjugate acid H_2O corresponding to OH^-. In any Brønsted-Lowry acid-base reaction we should be able to label each acid and base. The acetic acid-hydroxide ion reaction can be labeled as

$$
\begin{array}{cccc}
\text{acid 1} & \text{base 2} & \text{acid 2} & \text{base 1} \\
HC_2H_3O_2(aq) + & OH^-(aq) & \rightleftharpoons \quad H_2O & + \ C_2H_3O_2^-(aq) \\
\text{(the stronger} & \text{(the stronger} & \text{(the weaker} & \text{(the weaker} \\
\text{acid} & \text{base)} & \text{acid)} & \text{base)}
\end{array}
$$

The stronger acid, acetic acid, reacts with the stronger base, hydroxide ion, to form the weaker acid, water, and the weaker base, acetate ion. In any acid-base reaction, an acid can be recognized as a proton donor or a species formed by a base that gains a proton. A base can be recognized as a proton acceptor or a species formed by an acid that loses a proton. When an acid and a base are brought together in a solution, a proton transfer reaction can occur. Generally, acid-base reactions are equilibrium reactions and a double arrow is used to denote the equilibrium.

Since an acid-base reaction will favor the formation of the weaker acid and the weaker base, the longer arrow should be drawn in this direction and the shorter arrow in the other direction.

12-7   Common Acids and Bases

Some common acids and bases are given in Table 12-3. Notice that some of the acids listed in Table 12-3 are the same acids that were listed as strong electrolytes in Section 11-8. These acids, hydrogen chloride, HCl, sulfuric acid, H_2SO_4, and nitric acid, HNO_3, are called the **strong acids.** This term is used because when they are mixed with water, they react completely with water in an acid-base reaction to form hydronium ion and the corresponding anion. For example,

$$H_2SO_4 + H_2O \rightarrow H_3O^+(aq) + HSO_4^-(aq)$$

Recall from Section 6-5 that water solutions of these compounds are named as acids. Now that we know the nature of the solutions of these strong acids, we can represent such solutions as consisting of hydronium ion and the corresponding anion formed upon reaction with water. That is, water solutions of hydrochloric acid, sulfuric acid, and nitric acid contain hydronium ion and the corresponding conjugate base of the acid. For instance, a sulfuric acid solution is represented as

$$H_3O^+(aq) + HSO_4^-(aq)$$

The other acids listed in Table 12-3 are **weak acids.** These acids do not react extensively with water when they are mixed with water.

Table 12-3   Common Brønsted-Lowry Acids and Bases

	Acids	Bases	
Strong acids (react completely with water to form H_3O^+ and conjugate base)	H_2SO_4 HCl HNO_3	HSO_4^- Cl^- NO_3^-	Very weak bases (Will not react with H_3O^+ to form conjugate acid)
Weak acids (do not react extensively with water) (will react with OH^- to given conjugate base and water) Acids are listed according to decreasing strength	H_3O^+ HSO_4^- H_3PO_4 HF $HC_2H_3O_2$ H_2S $H_2PO_4^-$ HSO_3^- NH_4^+ HCN HCO_3^- HPO_4^{2-} HS^- H_2O	H_2O SO_4^{2-} $H_2PO_4^-$ F^- $C_2H_3O_2^-$ HS^- HPO_4^{2-} SO_3^{2-} NH_3 CN^- CO_3^{2-} PO_4^{3-} S^{2-} OH^-	Bases (will react with H_3O^+ to form conjugate acid and H_2O) Bases are listed according to increasing strength

Consequently, we represent water solutions of these acids, which are called the weak acids, by the formula of the acid. For example, a water solution of hydrogen fluoride, HF, is represented as

$$HF(aq)$$

and a water solution of hydrogen sulfide, H_2S, is represented as

$$H_2S(aq)$$

Table 12-3 also lists the common **bases.** Notice that these are the conjugate bases of the strong and weak acids. Of course, we recognize these bases as some of the common monoatomic and polyatomic ions. As ions, these bases occur in many ionic compounds. In fact, when soluble ionic compounds containing these ions are dissolved in water, the resulting solution is represented by the ions comprising the compound. For example, a solution of sodium sulfide, Na_2S, is represented as

$$Na^+(aq) + S^{2-}(aq)$$

and a solution of potassium hydrogen sulfate, $KHSO_4$, is represented as

$$K^+(aq) + HSO_4{}^-(aq)$$

Acids and bases vary in their tendencies to lose and gain protons. Thus, we say that some acids are stronger than other acids and that some bases are stronger than other bases. (See Table 12-3). The hydroxide ion, OH^-, is the strongest base that can exist in water. Solutions containing hydroxide ion can be prepared by dissolving an ionic compound containing hydroxide ion. The common soluble hydroxide ion containing compounds called **hydroxides** are sodium hydroxide, NaOH, potassium hydroxide, KOH, and barium hydroxide, $Ba(OH)_2$. In aqueous solution these compounds are represented as hydroxide ion and the cation. For instance, a water solution of NaOH is represented as

$$Na^+(aq) + OH^-(aq)$$

and a water solution of $Ba(OH)_2$ is represented as

$$Ba^{2+}(aq) + OH^-(aq)$$

12-8 Acidic and Basic Solutions

As another example of an acid-base reaction, consider the reversible reaction of ammonia with water.

$$NH_3 + H_2O \rightleftharpoons NH_4{}^+ + OH^-$$

base   acid ammonium hydroxide
ion   ion

Ammonia is quite soluble in water but only reacts to a slight extent. Nevertheless, ammonia is reacting as a base (proton acceptor) and water is reacting as an acid (proton donor). A water solution of ammonia is called **aqua ammonia.** Note that water is an acid in this reaction, but behaved as a base in the reaction with sulfuric acid shown in Section

12-7. Water can be an acid in some cases and a base in other cases. A species that can act as an acid or a base is called **amphiprotic.** Since water is amphiprotic, a proton transfer can occur between water molecules.

$$H_2O + H_2O \rightleftarrows H_3O^+ + OH^-$$

This reversible reaction does not occur to a great extent but does indicate that in pure water equal but small concentrations of hydronium ion and hydroxide ion are present. Pure water or a water solution that contains neither an acid or base will have equal concentrations of hydronium and hydroxide ions and are called **neutral solutions.** When an acid is mixed with water it reacts with water to form more hydronium ion. A water solution that has a concentration of hydronium ion that is greater than pure water is called an **acidic solution.** When an acid is dissolved in water, an acidic solution results. When a base is mixed with water it reacts to form more hydroxide ion. A water solution that has a concentration of hydroxide ion that is greater than pure water is called a **basic solution.** When a base dissolves in water a basic solution results.

12-9 Acid-Base Reactions

When we mix a solution containing an acid with a solution containing a base, an acid-base reaction can occur. These reactions result in a transfer of protons from the acid to the base. We can use Table 12-3 to predict an acid-base reaction. The general rule is that an acid will react with any base that is below it in the acid-base table. The reason for this is that the acids and bases are listed according to relative strengths, and an acid-base reaction will favor the formation of the weaker acid and weaker base. For example, consider mixing a solution of hydrogen fluoride, HF(aq), with a sodium hydroxide solution. First, write down the species which are mixed.

$$HF(aq) + Na^+(aq) + OH^-(aq)$$

The hydrogen fluoride is an acid and the hydroxide ion is a base. Referring to Table 12-3 we note that the OH^- is below HF in the table. Thus, we predict a reaction to form the corresponding conjugate base and acid.

HF $\longrightarrow$ F$^-$

H$_2$O $\longleftarrow$ OH$^-$

The balanced equation for the reaction is

$$HF(aq) + OH^-(aq) \rightarrow H_2O + F^-(aq)$$

A summary of this approach to writing acid-base reactions is

1. Write down the formulas of the species that are mixed.
2. Pick out any acid and base.
3. Note the positions of the acid and base in the acid-base table.

4. If the acid is above the base in the table write the equation for the acid-base reaction.

Example 12-7   Give the equation for any reaction that occurs when a solution of hydrochloric acid is mixed with a solution of sodium hydroxide. First, write down the species present in the two solutions.

$$H_3O^+(aq) + Cl^-(aq) + Na^+(aq) + OH^-(aq)$$

Second, pick out the acid and base.

$$H_3O^+(aq) + OH^-(aq)$$

Next, note that the H_3O^+ is above the OH^- in the acid-base table. Thus, the acid-base reaction which occurs is

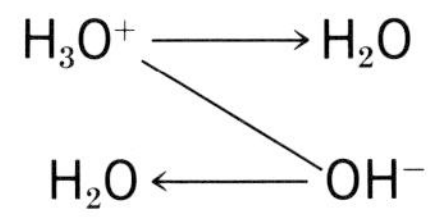

$$H_3O^+ \longrightarrow H_2O$$
$$H_2O \longleftarrow OH^-$$

The equation for the reaction is

$$H_3O^+(aq) + OH^-(aq) \rightarrow 2H_2O$$

The reaction between a solution of a hydroxide and a solution of a strong acid always produces water. When the solutions are added in the correct amounts so that enough H_3O^+ is present to react with the OH^-, the solution is neither acidic nor basic but is neutral. Consequently, such a reaction involving a solution of a strong acid and a solution of a hydroxide is called a **neutralization** reaction.

Example 12-8   Give the equation for any reaction that occurs when a solution of aqua ammonia is added to a solution of hydrochloric acid.
First, write down the species present in the two solutions.

$$NH_3(aq) + H_3O^+(aq) + Cl^-(aq)$$

Second, pick out the acid and base.

$$NH_3(aq) + H_3O^+(aq)$$

Next, note that the H_3O^+ is above the NH_3 in the acid-base table. Thus, the acid-base reaction that occurs is

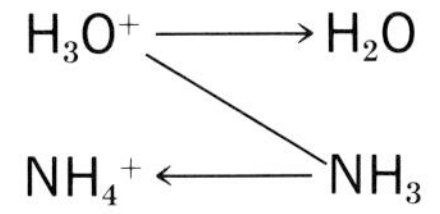

$$H_3O^+ \longrightarrow H_2O$$
$$NH_4^+ \longleftarrow NH_3$$

The equation for the reaction is

$$H_3O^+(aq) + NH_3(aq) \rightarrow NH_4^+(aq) + H_2O$$

Occasionally, the product formed when a solution of a strong acid is added to a solution of a weak base is only slightly soluble. In such cases, the slightly soluble compound is given off as a gas. An example of this is the reaction that occurs between a strong acid solution and a solution containing carbonate ion. For instance, when excess hydrochloric acid is added to a solution of sodium carbonate, the following reaction occurs.

$$2H_3O^+(aq) + CO_3^{2-}(aq) \rightarrow 3H_2O + CO_2(g)$$

The carbon dioxide gas bubbles off from the solution. The product formed when CO_3^{2-} gains two protons is theoretically carbonic acid, H_2CO_3. However, this compound decomposes rapidly to form CO_2 and H_2O ($H_2CO_3 \rightarrow H_2O + CO_2$). Carbon dioxide is not very soluble in water and leaves the solution as a gas.

Example 12-9   Give the equation for any reaction that occurs when a solution of sodium hydroxide, NaOH, is mixed with a solution of acetic acid, $HC_2H_3O_2$.
First, write down the species present in the solution.

$$HC_2H_3O_2(aq) + Na^+(aq) + OH^-(aq)$$

Second, pick out the acid and the base.

$$HC_2H_3O_2(aq) + OH^-(aq)$$

Next, note that $HC_2H_3O_2$ is above OH^- in the acid-base table. Thus, the acid-base reaction which occurs is

$$HC_2H_3O_2 \longrightarrow C_2H_3O_2^-$$
$$H_2O \longleftarrow OH^-$$

The equation for the reaction is

$$HC_2H_3O_2(aq) + OH^-(aq) \rightarrow C_2H_3O_2^-(aq) + H_2O$$

Some of the weak acids have more than one proton to lose. Those with two protons to lose are called diprotic acids, and those with three are called triprotic acids. If enough base is added to solutions of these acids, they will lose all of the exchangeable protons.

Example 12-10   Excess potassium hydroxide, KOH, solution is added to a solution of phosphoric acid, H_3PO_4. Give the equation for any reaction.
First, write down the species present in the solutions.

$$H_3PO_4(aq) + K^+(aq) + OH^-(aq)$$

Second, pick out the acid and base.

$$H_3PO_4(aq) + OH^-(aq)$$

Next, note that the H_3PO_4 is above the OH^- in the acid-base table. However, also note that it requires $3OH^-$ to react with all three protons of the H_3PO_4. The reaction is

$$H_3PO_4 \longrightarrow H_2PO_4^-$$
$$H_2PO_4^- \longrightarrow HPO_4^{2-}$$
$$HPO_4^{2-} \longrightarrow PO_4^{3-}$$
$$H_2O \longleftarrow OH^-$$

The equation for the reaction as long as sufficient OH^- is available is

$$H_3PO_4(aq) + 3OH^-(aq) \rightarrow 3H_2O + PO_4^{3-}(aq)$$

12-10   Oxidation and Reduction

When a piece of zinc metal is added to a solution containing copper(II) ions, the following reaction occurs spontaneously:

$$Zn(s) + Cu^{2+}(aq) \rightarrow Zn^{2+}(aq) + Cu(s)$$

It is possible to carry out this reaction in another manner. A piece of zinc (zinc electrode) is placed in a solution containing a low concentration of zinc ions, and a piece of copper (copper electrode) is placed in a solution containing a high concentration of copper(II) ions. The two solutions are connected by a conducting bridge as shown below.

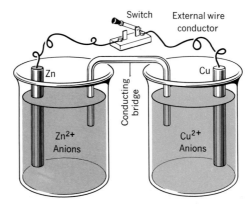

Since the two solutions are not in direct contact, no reaction occurs. However, if the two electrodes are connected by a wire, an electron transfer can occur by the flow of electrons (electricity) through the wire. The reaction at the zinc electrode or anode is

$$Zn(s) \rightarrow Zn^{2+}(aq) + 2e^{-}$$

The electrons that are lost can flow in the external wire which results in the following reaction at the copper electrode or cathode

$$Cu^{2+}(aq) + 2e^{-} \rightarrow Cu(s)$$

The reaction will occur spontaneously, and current (electrons) will flow through the wire. The current flow can be a source of useful electrical energy. An apparatus in which such reactions are separated by a conducting bridge and the electrodes can be connected by an external conductor to obtain useful electrical work is called a **voltaic cell, galvanic cell** or battery.

A battery is an interesting chemical device. Inside the battery case are chemicals that enter into reactions when the battery terminals are connected by wires. The reactions involve the loss and gain of electrons. The electrons flow through the circuit and can be used to do electrical work. The reactions involved in batteries are typical of a class of solution reactions which are neither precipitation or acid-base reactions. These reactions, which generally involve the increase in the oxidation number of one element and the simultaneous decrease in the oxidation number of another element, are called **oxidation-reduction reactions**

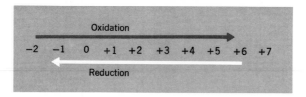

Figure 12-2   Oxidation is an increase in oxidation number. Reduction is a decrease in oxidation number.

or **redox reactions.** As an example of a redox reaction, consider the reaction when sodium metal is mixed with water:

$$2\text{Na}(s) + 2\text{H}_2\text{O} \rightarrow 2\text{Na}^+(aq) + 2\text{OH}^-(aq) + \text{H}_2(g)$$

In this reaction the oxidation number of sodium has changed from zero to +1, and the oxidation number of hydrogen has changed from +1 in water to zero in hydrogen gas. We say that the sodium has been oxidized and the hydrogen in the water has been reduced. The definitions of these terms are:

Oxidation. The increase in oxidation number of an element; loss of electrons.

Reduction. The decrease in oxidation number of an element; gain of electrons.

Oxidation-reduction reactions can be recognized by noting whether any changes have occurred in the oxidation numbers of any elements involved in the reaction (See Figure 12-2). Since oxidation and reduction must occur simultaneously, a redox reaction will involve an increase in the oxidation number of one element and the decrease of another element.

Most redox reactions can be symbolically separated into a process involving the loss of electrons by one species and a process involving the gain of electrons by another species. For example, the reaction between sodium metal and water can be represented in the following manner.

$$\text{Na}(s) \rightarrow \text{Na}^+(aq) + \text{e}^- \qquad \text{(Oxidation)}$$
$$2\text{e}^- + 2\text{H}_2\text{O} \rightarrow \text{H}_2(g) + 2\text{OH}^-(aq) \qquad \text{(Reduction)}$$

Such representations of a redox reaction are called **half reactions,** since they represent the **oxidation half** of the reaction and the **reduction half** of the reaction. A redox reaction can usually be represented by two half reactions. The complete redox equation can be obtained from the half reaction equations by multiplying either or both of the half reactions by an appropriate factor that will make the number of electrons lost equal the number of electrons gained. After the coefficients of the half reactions have been adjusted, the two half reactions can be added to obtain the completely balanced redox reaction. For example, if the above oxidation half is multiplied by 2 so that two electrons are lost for every two that are gained, and the two half reactions are added, the following balanced equation is obtained.

$$2Na(s) + 2H_2O \rightarrow 2Na^+(aq) + 2OH^-(aq) + H_2(g)$$

Species that gain electrons and are reduced are called oxidizing agents and species that lose electrons and are oxidized are called reducing agents. Be careful to use these terms correctly. Since the oxidation and reduction processes occur simultaneously, the **oxidizing agent** oxidizes the other species and in the process become reduced; and the **reducing agent** reduces the other species and in the process become oxidized. That is, the oxidizing agent oxidizes the reducing agent and the reducing agent reduces the oxidizing agent. In the overall reaction

$$2Na(s) + 2H_2O \rightarrow 2Na^+(aq) + 2OH^-(aq) + H_2(g)$$

the sodium is the reducing agent and is oxidized to Na^+ and the H_2O is the oxidizing agent and is reduced to H_2.

Oxidation-reduction reactions are quite common and useful. Redox reactions are involved in metal plating, batteries, fuel cells, the refining of metals, and even biochemical respiration.

12-11   Redox Reactions

How is it possible to predict a redox reaction that may occur when an oxidizing agent is mixed with a reducing agent? The reduced form of an oxidizing agent is a potential reducing agent since it could react to form the original oxidizing agent. In other words, an oxidizing agent and the corresponding reduced form are capable of reacting from the oxidized to the reduced form or from the reduced form to the oxidized form. For example, chlorine, Cl_2, can act as an oxidizing agent, and the chloride ion, Cl^-, can act as a reducing agent. How they react depends on the other substance with which they are mixed. In fact, oxidizing and reducing agents vary in their ability to react with other agents. A strong oxidizing agent has a great tendency to oxidize reducing agents. The reduced form of a strong oxidizing agent is a weak reducing agent. A strong reducing agent has a great tendency to reduce oxidizing agents. The oxidized form of a strong reducing agent is a weak oxidizing agent. By carrying out numerous oxidation-reduction reactions, it is possible to determine which oxidizing agents will react with which reducing agents. In this manner it is possible to prepare a table arranged according to relative ability to react. In such a table the stronger oxidizing agents would be at the top and the strengths would decrease down the table. This means that the reducing agents will be ordered according to increasing strength down the table. There are hundreds of oxidizing and reducing agents. For convenience, a redox table can involve only species of interest. A possible redox table is shown in Table 12-4. Note that some oxidizing agents require the presence of hydronium ion (an acidic solution) to behave as oxidizing agents.

This table can be used to predict redox reactions since, generally, any of the oxidizing agents will react with any reducing agent lying below it in the table. Two oxidizing agents will not react with one another nor

Table 12-4   A Redox Table Including Some Common Oxidizing and Reducing Agents

Oxidizing Agents		Reducing Agents

H_2O_2 $\quad + \quad 2H_3O^+(aq) + 2e^- \rightleftharpoons \quad\quad\quad 4H_2O$

$MnO_4^- \quad + \quad 8H_3O^+(aq) + 5e^- \rightleftharpoons 12H_2O + Mn^{2+}(aq)$

$Cl_2(g) \quad\quad\quad\quad\quad\quad\quad\quad + 2e^- \rightleftharpoons \quad\quad\quad 2Cl^-(aq)$

$Cr_2O_7^-(aq) + 14H_3O^+(aq) + 6e^- \rightleftharpoons 21H_2O + 2Cr^{3+}(aq)$

$Br_2(l) \quad\quad\quad\quad\quad\quad\quad\quad + 2e^- \rightleftharpoons \quad\quad\quad 2Br^-(aq)$

$NO_3^-(aq) \quad + \quad 4H_3O^+(aq) + 3e^- \rightleftharpoons \quad 6H_2O + NO(g)$

$Fe^{3+}(aq) \quad\quad\quad\quad\quad\quad\quad\quad + 1e^- \rightleftharpoons \quad\quad\quad\quad Fe^{2+}(aq)$

$I_2(s) \quad\quad\quad\quad\quad\quad\quad\quad\quad + 2e^- \rightleftharpoons \quad\quad\quad\quad 2I^-(aq)$

$2CO_2(g) \quad + \quad 2H_3O^+(aq) + 2e^- \rightleftharpoons \quad 2H_2O + H_2C_2O_4(aq)$

will two reducing agents. Furthermore, no oxidizing agent will react with a reducing agent lying above it in the table. When a reaction is predicted from the table, the oxidizing agent will react with the reducing agent to produce the reduced form of the oxidizing agent and the oxidized form of the reducing agent. To write the equation, the reaction of the oxidizing agent (reduction half reaction) is taken from the table as written. The reaction of the reducing agent (oxidation half reaction) is reversed. These half reactions are then multiplied by the proper coefficient to give the same number of electrons lost and gained and then added to give the final equation. For example, the predicted reaction when a solution containing iron(III) ion, Fe^{3+}, is mixed with a solution containing iodide ion, I^-, is found from Table 12-4 to be

$$Fe^{3+} + e^- \longrightarrow Fe^{2+}$$

$$I_2 + 2e^- \longleftarrow 2I^-$$

The half reactions are

$Fe^{3+} + e^- \rightarrow Fe^{2+}$ $\quad\quad$ reduction

$2I^- \rightarrow I_2 + 2e^-$ $\quad\quad$ oxidation $\quad\quad$ (Note that this is the reverse of the reaction in the table.)

Multiplying the reduction reaction by two and adding the half reactions gives the final equation

$$2Fe^{3+} + 2e^- \rightarrow 2Fe^{2+}$$
$$\underline{2I^- \rightarrow I_2 + 2e^-}$$
$$2Fe^{3+} + 2I^- \rightarrow 2Fe^{2+} + I_2$$

What reaction would be predicted when an acidic solution containing dichromate ion, $Cr_2O_7^{2-}$, is mixed with a solution containing bromide ion, Br^-? From Table 12-3 we would predict the reaction

$$14H_3O^+ + Cr_2O_7^{2-} + 6e^- \longrightarrow 2Cr^{3+} + 21H_2O$$

$$Br_2 + 2e^- \longleftarrow 2Br^-$$

The half reactions are

$$14H_3O^+ + Cr_2O_7^{2-} + 6e^- \rightarrow 2Cr^{3+} + 21H_2O \qquad \text{(Reduction)}$$
$$2Br^- \rightarrow Br_2 + 2e^- \qquad \text{(Oxidation)}$$

Multiplying the oxidation half reaction by three and adding the two half reactions gives the final equation

$$14H_3O^+ + Cr_2O_7^{2-} + 6e^- \rightarrow 2Cr^{3+} + 21H_2O$$
$$\underline{6Br^- \rightarrow 3Br_2 + 6e^-}$$
$$14H_3O^+ + Cr_2O_7^{2-} + 6Br^- \rightarrow 3Br_2 + 2Cr^{3+} + 21H_2O$$

What reaction would be predicted when an acidic solution containing dichromate ion, $Cr_2O_7^{2-}$, is mixed with a solution containing chloride, Cl^-? Noting the positions of these species in Table 12-4 we would predict no reaction since Cl^- is above $Cr_2O_7^{2-}$ in the table.

$$Cl_2 + 2e^- \longrightarrow 2Cl^-$$
$$14H_3O^+ + Cr_2O_7^{2-} + 6e^- \longrightarrow 2Cr^{3+} + 21H_2O$$

Another redox table involving some common metals, water and hydronium ion, H_3O^+, is shown in Table 12-5. This table is sometimes called the **electromotive** or **activity series** of the metals. The most active metals are the stronger reducing agents at the bottom of the table. This table can be used to predict reactions in the same manner as the previous table. That is, any oxidizing agent will react with any reducing agent below it in the table. For instance, when any of the metals below

Table 12-5   A Redox Table Including Common Metals (the electromotive series of metals)

Oxidizing Agents		Reducing Agents	
$Hg^{2+}(aq)$	$+ 2e^-$	$\rightleftharpoons Hg(l)$	
$Ag^+(aq)$	$+ 1e^-$	$\rightleftharpoons Ag(s)$	
$Fe^{3+}(aq)$	$+ 1e^-$	$\rightleftharpoons Fe^{2+}(aq)$	
$Cu^{2+}(aq)$	$+ 2e^-$	$\rightleftharpoons Cu(s)$	
$2H_3O^+(aq)$	$+ 2e^-$	$\rightleftharpoons H_2(g) + 2H_2O$	
$Pb^{2+}(aq)$	$+ 2e^-$	$\rightleftharpoons Pb(s)$	
$Ni^{2+}(aq)$	$+ 2e^-$	$\rightleftharpoons Ni(s)$	These metals dissolve in acid solutions
$Co^{2+}(aq)$	$+ 2e^-$	$\rightleftharpoons Co(s)$	
$Cd^{2+}(aq)$	$+ 2e^-$	$\rightleftharpoons Cd(s)$	
$Fe^{2+}(aq)$	$+ 2e^-$	$\rightleftharpoons Fe(s)$	
$Zn^{2+}(aq)$	$+ 2e^-$	$\rightleftharpoons Zn(s)$	
$2H_2O$	$+ 2e^-$	$\rightleftharpoons H_2(g) + 2OH^-(aq)$	
$Al^{3+}(aq)$	$+ 3e^-$	$\rightleftharpoons Al(s)$	
$Mg^{2+}(aq)$	$+ 2e^-$	$\rightleftharpoons Mg(s)$	
$Na^+(aq)$	$+ 1e^-$	$\rightleftharpoons Na(s)$	These metals react spontaneously in water
$Ca^{2+}(aq)$	$+ 2e^-$	$\rightleftharpoons Ca(s)$	
$K^+(aq)$	$+ 1e^-$	$\rightleftharpoons K(s)$	

H_3O^+ in the table are added to an acidic solution, a reaction will occur producing H_2 gas and the ion of the metal. However lead metal does not readily react with acids. (DANGER: The metals at the bottom of the table may react so violently in acid solution that an explosion may occur.) For example, when some zinc metal is added to a solution of hydrochloric acid (strong acid containing H_3O^+), we would predict the following reaction.

$$2H_3O^+(aq) + Zn(s) \rightarrow Zn^{2+}(aq) + H_2(g) + 2H_2O$$

Notice that H_2O appears in the table as an oxidizing agent. We would predict that any metal below H_2O in the table would react with H_2O to form H_2 and the metal ion. However, Mg only reacts in hot water, and Al does not react at normal temperatures. The other metals will react vigorously with H_2O. For instance, when potassium metal is placed in water, the following predicted reaction occurs.

$$2K(s) + 2H_2O \rightarrow 2K^+(aq) + 2OH^-(aq) + H_2(g)$$

The table also provides for the prediction of a reaction that may occur when a metal is added to a solution containing ions of another metal. Any metal ion will react with any of the metals that appear below it in the table. Such a reaction will produce the metallic form of the original metal ion and the ion of the original metal. For example, when a piece of nickel is added to a solution containing Ag^+, the following reaction is predicted.

$$2Ag^+(aq) + Ni(s) \rightarrow 2Ag(s) + Ni^{2+}(aq)$$

In summary, it is possible to predict reactions from redox tables by locating the oxidizing agent and reducing agent in the table. If the reducing agent occurs below the oxidizing agent in the table, a reaction is predicted. The equation for the reaction is obtained from the half reactions in the table.

12-12   Titrations

In the laboratory it is sometimes necessary to determine the molarity of a solution or the amount of some species contained in a sample of a solution. One way to do this is to carry out a solution phase reaction involving the sought species and some other species. By knowing the equation for the reaction and by measuring the amount of other species needed to react with the sought species, it is possible to determine the amount of sought species by stoichiometric calculations. As an example, consider how it would be possible to determine the number of grams of chloride ion in a sample of solution by adding a solution containing silver ion. The reaction that occurs would be

$$Ag^+(aq) + Cl^-(aq) \rightarrow AgCl(s)$$

By measuring the amount of silver ion containing solution of known concentration needed to provide enough silver ion to react with the chloride ion in the sample, it is possible to determine the number of grams of chloride ion in the sample. Assume that it requires 25.0 ml of a 1.00 M solution of silver nitrate to provide enough silver ion to react with the chloride. First, from the volume and concentration of the silver nitrate solution, it is possible to find the number of moles of silver ion needed to react with the chloride ion. The number of moles of silver ion is found by multiplying the volume by the molarity.

$$25.0 \text{ ml} \left(\frac{1 \text{ liter}}{10^3 \text{ ml}}\right) \left(\frac{1.00 \text{ mole Ag}^+}{1 \text{ liter}}\right) = 0.0250 \text{ mole Ag}^+$$

From the balanced equation it can be seen that the molar ratio involving silver ion and chloride is

$$\left(\frac{1 \text{ mole Cl}^-}{1 \text{ mole Ag}^+}\right)$$

Using this factor, the number of moles of chloride ion that reacted with the silver ion can be found.

$$0.0250 \text{ mole Ag}^+ \left(\frac{1 \text{ mole Cl}^-}{1 \text{ mole Ag}^+}\right) = 0.0250 \text{ mole Cl}^-$$

Finally, the number of grams of chloride ion in the sample can be found by converting the number of moles to grams.

$$0.0250 \text{ mole Cl}^- \left(\frac{35.4 \text{ g Cl}^-}{1 \text{ mole Cl}^-}\right) = 0.885 \text{ g Cl}^-$$

The process of adding a measured amount of a solution of known concentration to a sample of another solution for purposes of determining the concentration of the solution or the amount of some species in the solution is called **titration.** Titration is a very important process for the analysis of solutions. It involves adding to a sample of a solution, containing an unknown amount of some species, the correct amount of reacting species to react with the sought species. The addition and measurement of the volume of the solution of known concentration is carried out by use of a **buret,** pictured in Figure 12-3. A titration is usually performed by placing a sample of the unknown solution in a flask, filling a buret with the known solution (called the titrant), and then slowly delivering the known solution to the flask until the necessary amount has been mixed with the unknown solution. The point at which the necessary amount has been added is called the **equivalence point** of the titration.

The equivalence point of a titration is often detected by placing in the reaction flask a small amount of a substance called an **indicator.** The indicator is chosen so that it will react with the titrant when the equivalence point is reached. The reaction of the indicator produces a colored product, the appearance of which is called the end point of the titration. Some indicators are colored to begin with and produce a different colored product, so that the change indicates the end point. In any case, once the end point has been found, and it is sufficiently close to the equivalence point, the volume of titrant used to reach the end point can be determined from the buret. See Figure 12-3 for an illustration of the titration process. Using the volume and concentration of the titrant, the

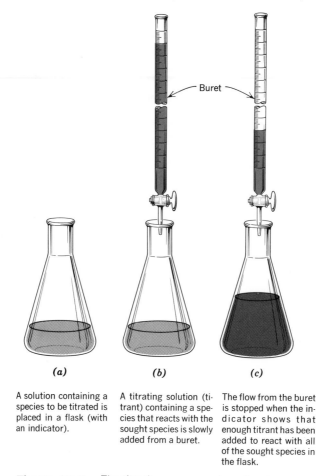

(a)	*(b)*	*(c)*
A solution containing a species to be titrated is placed in a flask (with an indicator).	A titrating solution (titrant) containing a species that reacts with the sought species is slowly added from a buret.	The flow from the buret is stopped when the indicator shows that enough titrant has been added to react with all of the sought species in the flask.

Figure 12-3   The titration process using a buret.

number of moles of reactive species required can be found. Then the number of moles or number of grams of the sought species can be found using the balanced equation involving the species. If the molarity of the unknown solution is to be calculated, it is necessary to measure the volume of the original unknown solution before titrating. Then, the molarity can be found by dividing the calculated number of moles of sought species in the solution sample by the volume of the sample.

Example 12-11   Calculate the molarity of a solution of acetic acid if a titration revealed that 30.0 ml of a 0.100 M solution of NaOH were required to react with the acetic acid in a 25.0 ml sample of the acid. The reaction involved is

$$OH^-(aq) + HC_2H_3O_2(aq) \rightleftharpoons H_2O + C_2H_3O_2^-(aq)$$

First, organize the data given.

volume OH$^-$ solution (titrant)	molarity OH$^-$ solution	volume HC$_2$H$_3$O$_2$ solution
30.0 ml	$\left(\dfrac{0.100 \text{ mole OH}^-}{1 \text{ liter}}\right)$	25.0 ml

From the data it is possible to determine the number of moles OH$^-$ used by multiplying the volume by the molarity.

$$30.0 \text{ ml} \left(\frac{1 \text{ liter}}{10^3 \text{ ml}}\right) \left(\frac{0.100 \text{ mole OH}^-}{1 \text{ liter}}\right) = 0.00300 \text{ mole OH}^-$$

The number of moles of $HC_2H_3O_2$ that reacted with the OH$^-$ can be found using the molar ratio from the balanced equation.

$$0.00300 \text{ mole OH}^- \left(\frac{1 \text{ mole } HC_2H_3O_2}{1 \text{ mole OH}^-}\right) = 0.00300 \text{ mole } HC_2H_3O_2$$

Finally, the molarity of the acid solution can be found by dividing the number of moles of $HC_2H_3O_2$ by the volume of the sample (in liters).

$$\left(\frac{0.00300 \text{ mole } HC_2H_3O_2}{25.0 \text{ ml}}\right) \left(\frac{10^3 \text{ ml}}{1 \text{ liter}}\right) = \left(\frac{0.120 \text{ mole } HC_2H_3O_2}{1 \text{ liter}}\right) \text{ or } 0.120 \text{ M } HC_2H_3O_2$$

Generally, this calculation could be accomplished as shown below.

$$M_u = \left(\frac{V_t}{V_u}\right) (M_t) \left(\frac{\text{moles } u}{\text{moles } t}\right)$$

where M_u is the molarity of the unknown solution
M_t is the molarity of the titrant
V_u is the volume of the unknown solution
V_t is the volume of the titrant
$\dfrac{\text{moles } u}{\text{moles } t}$ is the molar ratio of the titrated species to the titrant species

When the molar ratio is $1:1$, the relation reduces to

$$M_u = \frac{V_t}{V_u} M_t$$

Applying this relation to the previous example, the molarity of the acetic acid solution is

$$M_u = \left(\frac{30.0 \text{ ml}}{25.0 \text{ ml}}\right) 0.100 = 0.120 \text{ M } HC_2H_3O_2$$

12-13   pH

Recall from Section 12-8 that the acidity of a solution depends on the concentration of hydronium ion in the solution. The presence of an acid or base in a water solution effects the equilibrium between water, hydronium ion, and hydroxide ion.

$$H_2O + H_2O \rightleftharpoons H_3O^+ + OH^-$$

In pure water or a neutral solution not containing an acid or a base, the concentrations of hydronium ion and hydroxide ion are equal. In fact, in a neutral solution at 25°C the concentration of these two ions is 10^{-7} molar. That is,

$$[H_3O^+] = [OH^-] = 10^{-7} \text{ M}$$

where the square bracket, [ ], denotes molarity of the species enclosed in the bracket.

When an acid is in solution it reacts with water to increase the concentration of hydronium ion. The increased hydronium ion concentration causes the water equilibrium to shift resulting in a lower concentration of hydroxide ion. See Le Chatelier's principle in Section 11-7. An acidic solution is characterized by a concentration of hydronium ion which is greater than 10^{-7} M. When a base is in solution it increases the concentration of hydroxide ion. The increased hydroxide ion concentration causes the water equilibrium to shift resulting in a lower concentration of hydronium ion. A basic solution is characterized by a concentration of hydroxide ion that is greater than 10^{-7} M and a concentration of hydronium ion that is less than 10^{-7} M.

The concentration of hydronium ion in solutions can vary greatly. A 1 M hydrochloric acid solution has a hydronium ion concentration of 1 M while a 1 M sodium hydroxide solution has a hydronium ion concentration of 10^{-14} M. In neutral solutions the concentration of hydronium ion is 10^{-7} M. Since these concentrations of hydronium ion include a very wide range, a special scale has been devised to express such concentrations in aqueous solutions. This is called the **pH** (read: pee-ach) scale and is based on the following definition: **pH is the negative logarithm of the hydronium ion concentration.**

$$pH = -\log\ [H_3O^+]$$

The $[H_3O^+]$ represents the concentration of hydronium ion in moles per liter. The pH allows for the expression of concentrations of hydronium ion as a simple number, as shown below.

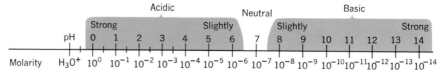

Note that there is a 10-fold difference between subsequent pH units. For instance, pH 6 is 10 times more acid than pH 7, and pH 5 is 100 times more acid than pH 7, and so on. For solutions with $[H_3O^+]$ ranging from 1 M to 10^{-14} M, the pH ranges from 0 to 14. Solutions with pH below 7 are acidic while solutions with pH above 7 are basic (sometimes called alkaline). A neutral solution at 25°C has a pH of 7. The pH is merely used as a means of expressing the hydronium-ion concentration in a solution. If the $[H_3O^+]$ is an integral power of 10, the pH is easily determined as

$$[H_3O^+] = 10^{-pH}$$

That is, the pH is the numerical value of the power of 10. Thus, a solution with a $[H_3O^+]$ of 10^{-4} M has a pH of 4, and a solution with a $[H_3O^+]$ of 10^{-9} M has a pH of 9. When the pH is not an integral power of 10, the logarithm of the concentration must be taken. For example, a solution with $[H_3O^+]$ of 2×10^{-3} M has a pH of

$$pH = -\log(2 \times 10^{-3})$$
$$pH = -\log 2 - \log 10^{-3}$$
$$(\log 2 = 0.30 \text{ and } \log 10^{-3} = -3)$$
$$pH = -0.30 - (-3) = -0.30 + 3 = 2.7$$

Table 12-6   The pH Values of Typical Solutions

pH	Solution	pH	Solution
1	Volcanic waters	7	Pure water
	Gastric juice		Blood
2	Lemon juice	8	Sea water
	Vinegar		
3	Orange juice	9	Alkaline lakes
	Wine		
4	Tomatoes	10	Soap solutions
5	Coffee, black	11	Household ammonia
6	Urine	12	

The log of 10 raised to a whole number power is the power ($\log 10^{-3}$ is -3). The log of a number between 1 and 10 can be found in a table of logarithms or from the relationship shown below.

Number	1	2	3	4	5	6	7	8	9	10
Logarithm	0	.18 .30	.40 .48	.59 .60	.65 .70	.74 .79	.81 .84	.86 .90	.93 .95 .98	1

Hydronium ion is present in many common solutions and mixtures. Gastric juices, citric fruit juices, soft drinks, vinegar, urine, milk, and some natural fresh water are acidic in nature. Pure water and saliva are neutral, and blood is slightly basic. Table 12-6 lists the pH of some common solutions.

In certain biological processes that occur in solution, the pH must be maintained within certain limits. Human blood normally has a pH of about 7.4 at room temperature. Any deviation in the pH of blood above 7.9 or below 7.0 will lead fairly quickly to death.

A special instrument called a pH meter can be used to measure the pH of an aqueous solution. Actually, the pH meter measures an electrical property of the solution that is proportional to the pH. The approximate pH of a solution can be found with acid-base indicators. The nature and use of acid-base indicators is shown in Figure 12-4.

12-14   Buffer Solutions

Pure water has a pH of 7. When a small amount of acid is added to water, the pH is lowered; the solution becomes acidic. When a small amount of base is added to pure water, the pH increases; the solution becomes basic. Blood serum, the fluid base of blood, behaves differently than water. Blood serum has a pH of 7.4. When a small amount of an acid or base is added to serum, the pH changes only a slight amount if at all. The difference between serum and water is that serum is a buffer solution. A **buffer solution** is a solution that resists a change in pH on addition of small amounts of an acid or base to the solution. A buffer solution contains a mixture of an acid and the conjugate base of

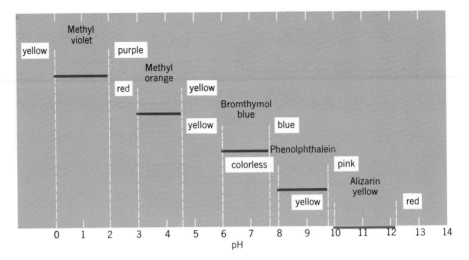

Acid-base indicators are chemicals that take on colors that depend on the pH of a solution. Indicators are actually weak acids that react with water as shown by the general equation:

$$HIN + H_2O \rightleftharpoons H_3O^+ + IN^-$$

acid	base
color	color

When an indicator is placed in a solution the hydronium-ion concentration will dictate which colored form the indicator will favor. If the pH is low the solution takes on the color of the acid form of the indicator. At higher pH the solution takes on the color of the base form. As shown various indicators turn characteristic colors in specific pH ranges. For example methyl orange turns red below pH 3 and yellow above pH 4.4. Between pH 3 and 4.4 the color is orange—a mixture of red and yellow. The pH of a solution can be approximated by testing portions of the solution with small amounts of indicators. For instance if a solution turns methyl orange red and methyl violet purple the pH of the solution is in the range of 2 to 3.

Figure 12-4   Acid base indicators.

the acid at comparable concentrations. For example, one buffer system in blood involves dihydrogen phosphate ion, $H_2PO_4^-$, and its conjugate base, hydrogen phosphate ion, HPO_4^{2-}, existing in equilibrium.

$$H_2PO_4^- + H_2O \rightleftharpoons HPO_4^{2-} + H_3O^+$$

This equilibrium serves to maintain a fairly constant pH. This is a result of the tendency of the equilibrium concentrations of the species involved to be maintained by a shift in the equilibrium.

If some OH^- is added to the solution it will react with the H_3O^+ and be neutralized. The decrease in H_3O^+ concentration results in a shift in the equilibrium to replenish the hydronium ion that was used up. If some H_3O^+ is added to the solution, a reaction will occur between the H_3O^+ and the HPO_4^{2-} to produce additional H_2O and $H_2PO_4^-$. This reaction will serve to deplete the system of the excess H_3O^+. Thus, because of the presence of HPO_4^{2-} and the $H_2PO_4^-$, the H_3O^+ concentration and pH will be maintained at a fairly constant level even when small amounts of acid or base are added to the system. Of course, such a buffer system has a limited capacity to resist changes in H_3O^+ and, if too much acid or base is added, the capacity of the buffer can be exceeded. Nevertheless, the dihydrogen phosphate ion-hydrogen phosphate ion buffer system in blood helps to maintain the hydronium ion concentration of blood within critical limits.

Buffers are used in chemistry when we want a solution that will have a specific pH and will resist change in pH upon the addition of small amounts of acid or base. Buffer solutions can be prepared by adding the appropriate acid and conjugate base of the acid in proper proportions. To see how the pH of a buffer solution is determined, we must first consider calculations involving chemical equilibrium.

As we know, chemical equilibrium is denoted by a double arrow between reactants and products. The general equilibrium reaction can be represented as

$$aA + bB + \ldots \rightleftharpoons cC + dD \ldots$$

where a, b, c, and d are the coefficients in the balanced equation. A few examples of equilibrium reactions are:

$$HC_2H_3O_2(aq) + H_2O \rightleftharpoons H_3O^+(aq) + C_2H_3O_2^-(aq)$$

$$NH_3(aq) + H_2O \rightleftharpoons NH_4^+(aq) + OH^-(aq)$$

$$AgCl(s) \overset{water}{\rightleftharpoons} Ag^+(aq) + Cl^-(aq)$$

The fact that the concentrations of reactants and products are constant when a reaction is at equilibrium (at a constant temperature and pressure) can be expressed as an **equilibrium constant** for the reaction. The equilibrium constant for a reversible reaction has the form

$$K_{eq} = \frac{[C]^c[D]^d}{[A]^a[B]^b} = \frac{(products)}{(reactants)}$$

where the [ ] are used to denote molar concentrations of the species and each concentration is raised to a power corresponding to the coefficient used in the balanced equations. The numerical value of the equilibrium constant of a reversible reaction carried out under specific conditions can be experimentally determined. A few examples of equilibrium constants involving the reactions given above are

$$\left.\begin{array}{l}\text{acetic acid reacting}\\\text{with water}\end{array}\right\} K_a(\text{acid}) = \frac{[H_3O^+][C_2H_3O_2^-]}{[HC_2H_3O_2]} = 1.75 \times 10^{-5}$$

Water is not included in the equilibrium expression, since in an aqueous solution the concentration of water is essentially constant

$$\left.\begin{array}{l}\text{ammonia reacting}\\\text{with water}\end{array}\right\} K_b(\text{base}) = \frac{[NH_4^+][OH^-]}{[NH_3]} = 1.8 \times 10^{-5}$$

Water is not included for the reason mentioned above.

$$\left.\begin{array}{l}\text{saturated solution}\\\text{of AgCl}\end{array}\right\} K_{SP}(\text{solubility product}) = [Ag^+][Cl^-] = 1.78 \times 10^{-10}$$

The solid AgCl is not included in the expression because if solid is present the amount does not effect the equilibrium. Equilibrium constants for many reactions are tabulated and can be used in calculations concerning the reactions. The magnitude of the K_a for a weak acid gives an indication of its strength relative to other acids. The larger the K_a, the stronger the acid. Why? The reaction of a weak acid with water can be represented as $HA + H_2O \rightleftharpoons H_3O^+ + A^-$ and the equilibrium constant as

$$K_a = \frac{[H_3O^+][A^-]}{[HA]}$$

The equilibrium expression for an acid can be used to determine the hydronium ion concentration of a buffer solution that contains a known concentration of acid and conjugate base of the acid. For example, the hydronium ion concentration of a buffer solution that is 1.0 molar in acetic acid, and 1.0 molar in acetate ion is found as follows.

$$HC_2H_3O_2 + H_2O \rightleftharpoons H_3O^+ + C_2H_3O_2^-$$

$$K_a = \frac{[H_3O^+][C_2H_3O_2^-]}{[HC_2H_3O_2]} = 1.75 \times 10^{-5}$$

solving for $[H_3O^+]$ gives

$$[H_3O^+] = \frac{K_a[HC_2H_3O_2]}{[C_2H_3O_2^-]} = 1.75 \times 10^{-5}\frac{(1.0\ M)}{(1.0\ M)}$$

therefore

$$[H_3O^+] = 1.75 \times 10^{-5}$$

The hydronium ion concentration for any buffer solution of a weak acid and the conjugate base

$$HA + H_2O \rightleftharpoons H_3O^+ + A^-$$

is given by the expression

$$H_3O^+ = K_a\frac{[HA]}{[A^-]}$$

Where K_a is the acid constant, [HA] is the concentration of the acid, and $[A^-]$ is the concentration of the conjugate base of the acid.

When the concentration of the acid and conjugate base are equal, the hydronium ion concentration equals the equilibrium constant, K_a. A buffer can be prepared by mixing an ionic compound containing the conjugate base (called the acid salt) with a solution of the acid.

Questions and Problems

1. What is a chemical reaction?

2. Give the formulas for the predominant ions that are present in aqueous solutions of the following ionic compounds.

(a) sodium chloride, NaCl
(b) potassium sulfate, K_2SO_4
(c) silver nitrate, $AgNO_3$
(d) potassium permanganate, $KMnO_4$
(e) sodium dichromate, $Na_2Cr_2O_7$
(f) sodium hydroxide, NaOH
(g) ammonium bromide, NH_4Br
(h) potassium hydrogen sulfate, $KHSO_4$

3. What is a net-ionic equation? Write balanced net-ionic equations for the following reactions.

(a) $Cl_2(aq)$ reacts with a solution containing $Br^-(aq)$ to give $Cl^-(aq)$ and Br_2.
(b) Some Pb metal is placed in a solution containing $Ag^+(aq)$, and a reaction producing Ag metal and $Pb^{2+}(aq)$ occurs.
(c) A solution of Na_2CrO_4 is added to a solution of $Pb(NO_3)_2$, and a precipitate of $PbCrO_4$ is formed.
(d) A solution of $Ni(NO_3)_2$ is added to a solution of NaOH and a precipitate of $Ni(OH)_2$ is formed.
(e) A solution of KOH is added to a solution of H_2S producing H_2O and S^{2-}.
(f) Some solid $CaCO_3$ is added to a solution of hydrochloric acid producing CO_2, H_2O and Ca^{2+}. (Hint: $CaCO_3$ is one reactant.)

4. Describe how a precipitation reaction can occur when two solutions containing ions are mixed.

5. Write balanced net-ionic equations for any precipitation reactions that occur when the following solutions are mixed. Use the general approach to predicting precipitation reactions and refer to the solubility list given in Section 12-4.

(a) A solution of lead(II) nitrate is added to a solution of potassium chloride.
(b) A solution of sodium fluoride is added to a solution of calcium nitrate.
(c) A solution of ammonium chloride is added to a solution of sodium chloride.
(d) A solution of sodium hydroxide is added to a solution of magnesium chloride.
(e) A solution of barium nitrate is added to a solution of sodium sulfate.
(f) A solution of sodium chloride is added to a solution of zinc chloride.

6. Predict and give balanced net-ionic equations for any precipitation reactions that occur when the following solutions are mixed. Base the predictions on the solubility list given in Section 12-4.

(a) A solution of $Hg_2(NO_3)_2$ is added to a solution of NaCl.
(b) A solution of KCl is added to a solution of $NaNO_3$.
(c) A solution of KOH is added to a solution of $CaCl_2$.
(d) A solution of $Pb(NO_3)_2$ is added to a solution of Na_2SO_4.
(e) A solution of NaOH is added to a solution of $Cd(NO_3)_2$.
(f) A solution of NH_4Br is added to a solution of $AgNO_3$.
(g) A solution of $CaCl_2$ is added to a solution of Na_2CO_3.
(h) A solution of NaOH is added to a solution of KCl.
(i) A solution of $Ba(OH)_2$ is added to a solution of $ZnSO_4$.

(j)   A solution of $MgCl_2$ is added to a solution of NaF.
(k)   A solution of $AgNO_3$ is added to a solution of KI.
(l)   A solution of $Pb(NO_3)_2$ is added to a solution of KBr.
(m) A solution of $CaCl_2$ is added to a solution of NaF.
(n)   A solution of $CaCl_2$ is added to a solution of K_2SO_4.
(o)   A solution of $FeCl_3$ is added to a solution of KOH.
(p)   A solution of $Pb(NO_3)_2$ is added to a solution of Na_2CrO_4.
(q)   A solution of $Pb(NO_3)_2$ is added to a solution of NaF.
(r)   A solution of NaCl is added to a solution of K_2SO_4.
(s)   A solution of $Al(NO_3)_3$ is added to a solution of NaOH.
(t)   A solution of $MgCl_2$ is added to a solution of $Ba(OH)_2$.
(u)   A solution of $Zn(NO_3)_2$ is added to a solution of KCl.
(v)   A solution of $FeCl_3$ is added to a solution of $Pb(NO_3)_2$.

7. State the Brønsted-Lowry definitions for an acid and a base.

8. What is an acid-base reaction?

9. Explain the difference between a weak acid and a strong acid.

10. What is a conjugate acid-base pair? For each of the following acid-base reactions pick out the acids and the bases and indicate the conjugate acid-base pairs.
(a) $HNO_3(\ell) + H_2O \rightarrow H_3O^+(aq) + NO_3^-(aq)$
(b) $OH^-(aq) + H_3PO_4(aq) \leftrightharpoons H_2O + H_2PO_4^-(aq)$
(c) $OH^-(aq) + NH_4^+(aq) \leftrightharpoons NH_3(aq) + H_2O$
(d) $H_3O^+(aq) + OH^-(aq) \leftrightharpoons H_2O + H_2O$
(e) $H_3O^+(aq) + F^-(aq) \leftrightharpoons HF(aq) + H_2O$

11. What is an acidic solution as compared to a neutral solution?

12. What is a basic solution as compared to a neutral solution?

13. What is a neutralization reaction? Give the net-ionic equation for the neutralization reaction that occurs when a solution of nitric acid is added to a solution of potassium hydroxide.

14. Predict and give balanced net-ionic equations for any acid-base reactions that occur when the following solutions are mixed. Base the predictions on the general method of writing acid-base reactions and refer to the acid-base table (Table 12-3).
(a)   A sodium hydroxide, NaOH, solution is mixed with a sodium hydrogen sulfate, $NaHSO_4$, solution.
(b)   Excess sodium hydroxide, NaOH, solution is mixed with a solution of H_2S.
(c)   A solution of hydrochloric acid is added to a solution of NaHS.
(d)   A solution of nitric acid is added to a solution of aqua ammonia.
(e)   A solution of hydrochloric acid is added to a solution of $NaNO_3$.
(f)   A solution of hydrochloric acid is added to a solution of KOH.
(g)   Excess nitric acid solution is added to a solution of Na_3PO_4.
(h)   A solution of NaOH is added to a solution of HF.
(i)   A solution of KOH is added to a solution of NaHS.
(j)   A solution of NaOH is added to a Na_3PO_4 solution.
(k)   Excess NaOH solution is added to a solution of H_2SO_4.
(l)   A solution of hydrochloric acid is added to a solution of NaF.

(m) An excess amount of a solution of KOH is added to a solution of NaH_2PO_4.

(n) A solution of NaOH is added to a solution of NH_4Cl.

(o) A solution of hydrochloric acid is added to a solution of NaOH.

(p) A solution of nitric acid is added to a solution of $KC_2H_3O_2$.

(q) A solution of hydrochloric acid is added to a solution of K_2CO_3.

(r) Some H_2S is added to water.

(s) Some HNO_3 is mixed with water.

15. Give definitions for the terms oxidation and reduction.

16. What is an oxidation-reduction or redox reaction?

17. What is an oxidizing agent? What is a reducing agent? In each of the following redox reactions pick out the oxidizing agent and the reducing agent.

(a) $2MnO_4^-(aq) + 10Cl^-(aq) + 16H_3O^+(aq) \rightarrow$
$$2Mn^{2+}(aq) + 5Cl_2(g) + 24H_2O(aq)$$

(b) $H_2O_2 + 2Fe^{2+}(aq) + 2H_3O^+(aq) \rightarrow 4H_2O + 2Fe^{3+}(aq)$

(c) $MnO_2 + 2Cl^-(aq) + 4H_3O^+(aq) \rightarrow Mn^{2+}(aq) + Cl_2(g) + 6H_2O$

18. Using Table 12-4 give the balanced net-ionic equation for any redox reaction that occurs when the following solutions are mixed. The term acidified means that hydronium ion is present in the solution.

(a) A NaCl solution is added to an acidified hydrogen peroxide, H_2O_2, solution.

(b) A NaI solution is added to an acidified solution of $KMnO_4$.

(c) An $FeCl_2$ solution is added to an acidified solution of $Na_2Cr_2O_7$.

(d) Bromine, Br_2, is added to a NaI solution.

(e) An oxalic acid solution, $H_2C_2O_4$, is added to an acidified hydrogen peroxide, H_2O_2, solution.

19. Using Table 12-5 give the balanced net-ionic equation for any redox reaction that occurs when the following are mixed.

(a) A piece of sodium metal is added to water.

(b) A piece of zinc metal is added to a $Hg(NO_3)_2$ solution.

(c) A piece of iron metal is added to a solution of hydrochloric acid.

(d) A piece of nickel metal is added to a sodium chloride solution.

(e) A small piece of calcium metal is added to a hydrochloric acid solution.

(f) A piece of aluminum metal is added to a $AgNO_3$ solution.

(g) A piece of copper metal is added to a hydrochloric acid solution.

20. What is a titration?

21. A sample of a substance containing SO_4^{2-} is dissolved in water and titrated with a 0.500 M $BaCl_2$ solution. It is found that 25.0 ml of the titrant are needed to titrate the SO_4^{2-} in the sample. If the reaction involved is

$$Ba^{2+}(aq) + SO_4^{2-}(aq) \rightarrow BaSO_4(s)$$

Determine the number of grams of SO_4^{2-} in the sample.

22. A sample of a substance containing Cl^- is dissolved in water and titrated with a 0.500 M solution of silver nitrate. It is found that 30.2 ml of the silver nitrate solution are needed to titrate the Cl^- in the sample. The reaction involved is

$$Ag^+(aq) + Cl^-(aq) \rightarrow AgCl(s)$$

Determine the number of grams of Cl^- in the sample.

23. Calculate the molarity of a solution of hydrochloric acid if 25.0 ml of a 0.200 M solution of NaOH are needed to titrate a 30.0 ml sample of the solution. The reaction involved is

$$H_3O^+(aq) + OH^-(aq) \rightarrow 2H_2O$$

24. A sample of a solution of oxalic acid, $H_2C_2O_4$, is titrated with a 0.500 M solution of NaOH. It is found that 32.0 ml of the OH^- solution are needed to titrate the acid. If the reaction is

$$2OH^-(aq) + H_2C_2O_4(aq) \rightarrow 2H_2O + C_2O_4{}^{2-}(aq)$$

Calculate the number of grams of $H_2C_2O_4$ in the original sample.

25. Calculate the molarity of an acetic acid solution if 30.0 ml of a 0.300 M KOH solution are needed to titrate a 25.0 ml sample of the acid solution. The reaction involved is

$$HC_2H_3O_2(aq) + OH^-(aq) \rightarrow H_2O + C_2H_3O_2{}^-(aq)$$

26. Give a definition of pH.

27. Determine the pH of the following solutions and indicate whether the solution is acidic or basic.
    (a) An ammonia solution that has 1.0×10^{-9} M H_3O^+.
    (b) A sample of saliva that has 3.0×10^{-7} M H_3O^+.
    (c) A sample of urine that has 2.0×10^{-6} M H_3O^+.
    (d) A citric acid solution that has 6.5×10^{-3} M H_3O^+.
    (e) A 1.0 M NaOH solution that has 1.0×10^{-14} M H_3O^+.

28. What is a buffer solution?

29. Calculate the hydronium ion concentration of a buffer solution containing 1.0 M acetic acid, $HC_2H_3O_2$, and 0.50 M sodium acetate, $NaC_2H_3O_2$. ($K_a = 1.75 \times 10^{-5}$ for $HC_2H_3O_2$)

30. Calculate the hydronium ion concentration of a buffer solution containing 2.0 M hydrofluoric acid, HF, and 1.5 M potassium fluoride, KF. ($K_a = 3.53 \times 10^{-4}$ for HF)

31. Calculate the pH of a buffer solution containing 1.0 M sodium dihydrogen phosphate, NaH_2PO_4, and 1.0 M sodium hydrogen phosphate, Na_2HPO_4. The equilibrium is

$$H_2PO_4{}^-(aq) + H_2O \rightleftharpoons H_3O^+(aq) + HPO_4{}^{2-}(aq)$$

and $K_a = 6.23 \times 10^{-8}$ for $H_2PO_4^-$ which is the acid in the buffer. (Hint: find the hydronium ion concentration and convert it to pH.)

32. Describe a battery in terms of an oxidation-reduction reaction.

13

Chapter 13: Organic Chemistry

Objectives

The student should be able to:

1. Describe and give examples of the group of organic compounds called the alkanes.
2. Recognize and name the simple alkyl groups.
3. Write the structural formula of a simple alkane given the name.
4. Describe and give examples of an alkene.
5. Describe the polymerization reaction of an alkene.
6. Write an equation representing the polymerization of a specific alkene.
7. Recognize an alcohol, an aldehyde, a carboxylic acid, and an ester from the formulas.
8. Recognize a phosphate ester from the formula.
9. Recognize an amine, an amide, and an amino acid from the formulas.
10. Recognize a cyclic compound from the formula.
11. Recognize a benzene (aromatic) compound from the formula.
12. Recognize a heterocyclic compound from the formula.

Terms to Know

Organic chemistry
Structural formulas
Double bond
Triple bond
Hydrocarbons
Saturated hydrocarbons
Alkanes
Isomers
Alkyl groups
Unsaturated hydrocarbons
Alkenes or olefins
Polymer
Monomer

Alcohols
Fermentation
Aldehydes
Carboxylic acids
Esters
Amines
Amides
Amino acids
Benzene
Cyclic compounds
Aromatic compounds
Heterocyclic compounds

13

Organic Chemistry

13-1   Synthetic Organic Chemicals

For centuries chemists have isolated and investigated chemical compounds found in nature. They classified these compounds into two groups. Inorganic compounds were those isolated from mineral sources. Organic compounds were those isolated from plant and animal sources. A wide variety of compounds were found to be manufactured by plants and animals during life processes. Upon investigation, it was found that the one thing that organic compounds had in common was that they all were carbon-containing compounds. Today, organic compounds are considered as a special group of synthetic and natural carbon-containing compounds. Inorganic compounds are composed of a wide variety of the other elements, and, except for the carbonates, most inorganic compounds do not contain carbon.

Much effort was expended in determining the composition of the variety of compounds found in nature. Chemists began investigating organic compounds and the ability of carbon to form bonds. They found that they were able to synthesize some chemical compounds that were chemically equivalent to naturally occurring compounds. Moreover, they learned that it was possible to synthesize numerous chemical compounds that were not found in nature. That is, they were able to create new combinations of elements which were not known to exist. These achievements contributed to the development of a vast technology in which usable synthetic chemicals are produced.

Several decades ago **synthetic organic** substances were not common in consumer products. Today these human-made substances surround us and are incorporated into many of the things we use. Synthetic organic chemicals are prepared from naturally occurring molecules

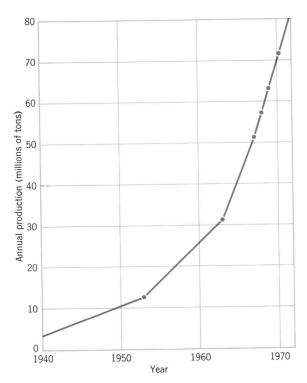

Figure 13-1   Synthetic organic chemical production.

(usually derived from petroleum) by industrial chemical processes that convert the natural products to more useful molecules.

The polyester shirt or blouse you might be wearing is made of a synthetic fiber, and the plastic buttons and colored dyes used on the garment are synthetics. The vinyl tiles or carpet on the floor are synthetic plastics, and the rubber heels on your shoes are synthetic rubber. The aspirin or vitamin tablets you may have taken today contain synthetic pharmaceuticals. The bread you ate for breakfast came from a plastic-wrapped container, and the bread might have contained synthetic food additives as preservatives. The apple you had for lunch may have carried a small amount of synthetic pesticide into your body.

The increase in the annual amounts of synthetic organic chemicals produced in the United States has been phenomenal. As shown in Figure 13-1, annual synthetics production increased from 5 million tons in 1943 to 80 million tons in 1973. These synthetics include plastics, plasticizers, paints, pesticides, preservatives, and pharmaceuticals, among others.

13-2   Carbon Chemistry

A branch of chemistry called **organic chemistry** is devoted to the study of the compounds of carbon. The most important property of carbon is the ability of carbon atoms to form chemical bonds with one another and

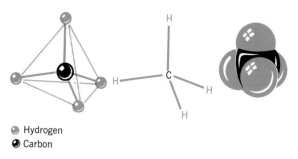

● Hydrogen
◐ Carbon

Figure 13-2   Three representations of a methane molecule.

with a variety of other elements. Carbon forms four covalent bonds in organic compounds. Thus, carbon is said to be **tetravalent.** This tetravalency or tendency to form four bonds correlates with the fact that carbon has four valence electrons.

$$\cdot \overset{\displaystyle \cdot}{\underset{\displaystyle \cdot}{C}} \cdot$$

When a carbon atom forms covalent bonds with four other atoms, the four pairs of electrons are **tetrahedrally** distributed about the carbon atom as shown in Figure 13-2. For writing convenience, the four possible covalent bonds of carbon may be represented in a plane as

$$-\overset{\displaystyle |}{\underset{\displaystyle |}{C}}-$$

Of course, other atoms are bonded to carbon through these bonds.

One of the reasons there are numerous organic compounds is that carbon atoms can form strong bonds with other carbon atoms while at the same time forming strong bonds with other nonmetals. This means that it is possible to have chains of carbon atoms bonded to one another and to other types of atoms. It is possible for two carbon atoms to be linked by a single covalent bond.

$$-\overset{\displaystyle |}{\underset{\displaystyle |}{C}}-\overset{\displaystyle |}{\underset{\displaystyle |}{C}}-$$

A given carbon atom can form more than one single bond with other carbon atoms. This gives rise to a very large number of possible sequences of carbon atoms bonded to one another.

As illustrated in Figure 13-3, carbon atoms can bond together in a variety of ways much like the links of a chain. In fact, the bonded carbon sequence in an organic compound is sometimes called the **carbon chain.** There are literally billions of possible sequences of bonded carbon atoms. This accounts for the fact that there are millions of known organic compounds. Fortunately, most of these compounds can be studied as groups of compounds having similar structures. Organic chemistry involves the study of the various groups of compounds. In this chapter, 12 groups of the most common types of organic compounds will be described.

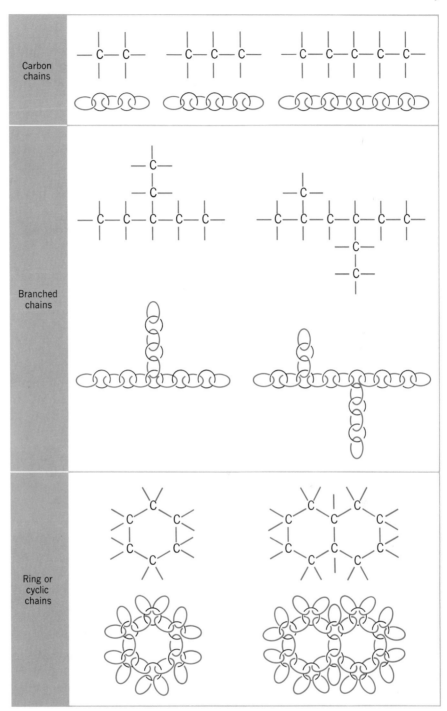

Figure 13-3   The ability of carbon atoms to link together in a variety of ways is similar to the formation of chains from individual links. Bonded carbon sequences in molecules are sometimes called carbon chains. When molecules include carbon sequences linked to the main chain they are sometimes called branched chains, and the side chains are called branches. Carbons sometimes link together in a cyclic fashion in which the end carbon bonds to the first carbon. These are called cyclic or ring molecules. From *Understanding Chemistry: From Atoms to Attitudes* by T. R. Dickson, Wiley, N.Y., 1974.

13-3   Bonding in Organic Compounds

Organic compounds involve bonded carbon sequences in which some of the carbon bonds are used in bonding to atoms of other elements. Hydrogen is the element most often found bonded to carbon in organic compounds. A hydrogen atom can share one pair of electrons with a carbon atom.

$$-\overset{|}{\underset{|}{C}}-H$$

The formula above does not represent a compound but just indicates how a carbon-hydrogen bond can be represented. In a compound, all of the bonds of carbon must be used. The compound methane, CH_4, can be represented as (also see Figure 13-2)

$$H-\overset{\overset{\displaystyle H}{|}}{\underset{\underset{\displaystyle H}{|}}{C}}-H$$

This formula, which indicates the atoms that are bonded, is called a **structural formula.** The formula, CH_4, which only indicates the type and number of atoms comprising a molecule, is called a **molecular formula.** Structural formulas are used to indicate which atoms are bonded to which atoms in the molecule. On the other hand, molecular formulas are used to indicate the composition of the molecules of the compounds.

Butane, C_4H_{10}, has the following structural formula:

$$H-\overset{\overset{\displaystyle H}{|}}{\underset{\underset{\displaystyle H}{|}}{C}}-\overset{\overset{\displaystyle H}{|}}{\underset{\underset{\displaystyle H}{|}}{C}}-\overset{\overset{\displaystyle H}{|}}{\underset{\underset{\displaystyle H}{|}}{C}}-\overset{\overset{\displaystyle H}{|}}{\underset{\underset{\displaystyle H}{|}}{C}}-H$$

Instead of writing the complete structural formula of a compound like butane, it is possible to represent the compound by a **condensed structural formula** that indicates the bonding sequence without showing all of the bonds. For example, the condensed structural formula for butane is

$$CH_3CH_2CH_2CH_3$$

Such a formula should be interpreted as indicating that the carbons are bonded to one another in sequence, and each carbon is bonded to the hydrogens or other atoms that are next to it in the formula. Condensed formulas are more convenient to write than full structural formulas.

Two carbon atoms are capable of sharing two pair of electrons with one another to form a **double covalent bond (double bond).**

$$\overset{\diagdown}{\diagup}C=C\overset{\diagup}{\diagdown}$$

The double bond uses two bonds for each carbon leaving two other bond positions on each carbon that are involved in bonding to other atoms. For example, the compound ethylene, C_2H_4, involves a double bond.

$$\overset{H}{\underset{H}{\diagdown}} C = C \overset{H}{\underset{H}{\diagup}} \quad \text{or} \quad CH_2{=}CH_2$$

Sometimes two carbon atoms actually share three pairs of electrons to form a **triple covalent bond (triple bond),**

$$-C{\equiv}C-$$

When two carbon atoms are bonded by a triple bond, each carbon can form one other bond with a different atom. For example, acetylene, C_2H_2, involves a triple bond.

$$H{-}C{\equiv}C{-}H \text{ or } HC{\equiv}CH$$

Oxygen can bond to carbon in two ways. Oxygen can form two covalent bonds so that it is possible for oxygen to bond to a carbon by a double bond.

$$\overset{O}{\underset{}{\underset{\displaystyle C}{\|}}}\diagdown$$

This leaves two bond positions on the carbon that are used for bonding to other atoms. The structural formula of formaldehyde, CH_2O, is

$$\overset{O}{\underset{H \quad H}{\underset{\displaystyle C}{\|}}}$$

Carbon and oxygen can be bonded by a single bond.

$$-\overset{|}{\underset{|}{C}}{-}O{-}$$

This leaves one bond position on the oxygen and three on the carbon. The structural formula of ethyl alcohol, C_2H_5OH, is

$$H{-}\overset{H}{\underset{H}{\overset{|}{\underset{|}{C}}}}{-}\overset{H}{\underset{H}{\overset{|}{\underset{|}{C}}}}{-}O{-}H \text{ or } CH_3CH_2OH$$

Usually nitrogen and carbon are bonded by a single bond.

$$-\overset{|}{\underset{|}{C}}{-}\overset{}{\underset{|}{N}}{-}$$

This leaves two bond positions on the nitrogen and three on the carbon.

The structural formula of methylamine, CH_3NH_2, is

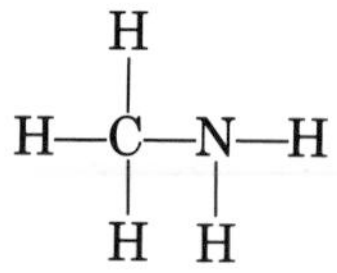

$$
\begin{array}{c}
\quad\ H \\
\quad\ | \\
H-C-N-H \\
\quad | \quad | \\
\quad H \ \ H
\end{array}
$$

However carbon-nitrogen double and triple bonds occur in some organic compounds. Carbon can form single bonds with the halogens (F, Cl, Br, I).

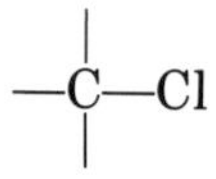

$$
\begin{array}{c}
| \\
-C-Cl \\
|
\end{array}
$$

The structural formula of chloroform, $CHCl_3$, is

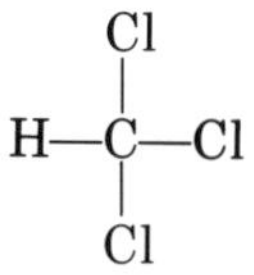

$$
\begin{array}{c}
\quad\ Cl \\
\quad\ | \\
H-C-Cl \\
\quad\ | \\
\quad\ Cl
\end{array}
$$

The knowledge of the manner in which carbon forms bonds with atoms serves as a foundation for the discussion of organic compounds.

13-4   Alkanes and Isomerism

Organic compounds containing only carbon and hydrogen are called **hydrocarbons.** Hydrocarbons that involve only single bonded carbons (no double or triple bonds) are called **saturated hydrocarbons.** It is important to examine some of these hydrocarbons since they serve as a basis for the nomenclature and structural formulas of a large number of organic compounds. There are numerous possible hydrocarbons. The simplest hydrocarbon is methane, CH_4
Second is ethane, C_2H_6.

$$CH_3CH_3$$

Third is propane, C_3H_8.

$$CH_3CH_2CH_3$$

Notice that these compounds differ by $-CH_2-$ units.

$$CH_3-CH_2-CH_3$$

The structural formulas of a large number of hydrocarbons can be written that differ by $-CH_2-$ units. Any group of compounds in which the members differ in this manner is called an **homologous series.** The saturated hydrocarbons comprise an homologous series corresponding to the general formula C_nH_{2n+2} (CH_4 n = 1, C_2H_6 n = 2, C_3H_8 n = 3). These compounds are called the **alkanes.** Eight of the most common alkanes are listed in Table 13-1.

Let us consider the five-carbon alkane, C_5H_{12}. In such a compound,

Table 13-1   Some Alkanes

Name	Molecular Formula	Structural Formula	Condensed Structural Formula
Methane	CH_4	H—C—H (with H above and H below)	
Ethane	C_2H_6	H—C—C—H	CH_3CH_3
Propane	C_3H_8	H—C—C—C—H	$CH_3CH_2CH_3$
Butane	C_4H_{10}	H—C—C—C—C—H	$CH_3CH_2CH_2CH_3$ or $CH_3(CH_2)_2CH_3$ [a]
Pentane	C_5H_{12}	H—C—C—C—C—C—H	$CH_3CH_2CH_2CH_2CH_3$ or $CH_3(CH_2)_3CH_3$ [a]
Hexane	C_6H_{14}	H—C—C—C—C—C—C—H	$CH_3CH_2CH_2CH_2CH_2CH_3$ or $CH_3(CH_2)_4CH_3$ [a]
Heptane	C_7H_{16}	H—C—C—C—C—C—C—C—H	$CH_3CH_2CH_2CH_2CH_2CH_2CH_3$ or $CH_3(CH_2)_5CH_3$ [a]
Octane	C_8H_{18}	H—C—C—C—C—C—C—C—C—H	$CH_3CH_2CH_2CH_2CH_2CH_2CH_2CH_3$ or $CH_3(CH_2)_6CH_3$ [a]

[a] The $(CH_2)_n$ notation refers to n CH_2 units in a row.

the sequence of carbons can be

H—C—C—C—C—C—H (pentane chain with H atoms above and below each carbon)

But notice that it is possible to have other branched carbon sequences with the same molecular formula.

$$
\begin{array}{ccccc}
 & & \overset{\displaystyle H}{\underset{\displaystyle |}{H-C-H}} & & \\
 & H\;H & | & H & \\
H-\overset{|}{\underset{|}{C}}-\overset{|}{\underset{|}{C}} & & \overset{|}{\underset{|}{C}} & \overset{|}{\underset{|}{C}}-H & \\
 & H\;H & H & H &
\end{array}
$$

These compounds do not have the same structure as the first, but they do have the same molecular formula, C_5H_{12}. Compounds with the same molecular formula but different structural formulas are called **isomers.** Isomerism is quite common in organic compounds and greatly increases the number of possible compounds. Some additional examples of structural isomers are

$$CH_3CH_2CH_2CH_3$$
butane

$$\overset{\displaystyle CH_3}{\underset{\displaystyle |}{CH_3CHCH_3}}$$
isobutane

$$CH_3CH_2OCH_2CH_3$$
ethyl ether

$$CH_3CH_2CH_2CH_2OH$$
butyl alcohol

13-5   Alkyl Groups

A large number of organic compounds can be considered to be derived from the substitution of one or more hydrogens of alkanes by another atom or group of atoms. For instance, the compound

$$
\overset{\displaystyle H}{\underset{\displaystyle H}{H-\overset{|}{\underset{|}{C}}-OH}}
$$

can be viewed as being derived from the substitution of an H of methane with a OH group. Another view of this compound is that it is composed of a H group bonded to a —OH.

$$
H-\overset{\displaystyle |}{\underset{\displaystyle |}{\underset{\displaystyle H}{C}}}-
$$

Groups corresponding to the alkanes can be considered to be formed by removing one hydrogen to leave a bonding position. Such groups

Table 13-2 Some Alkyl Groups

Formula	Name
CH_3—	Methyl group
CH_3CH_2—	Ethyl group
$CH_3CH_2CH_2$—	Propyl group
$\begin{array}{c} CH_3 \\ \mid \\ CH_3CH \end{array}$— or $(CH_3)_2CH$—	Isopropyl group
$CH_3CH_2CH_2CH_2$—	Butyl group
$\begin{array}{c} CH_3 \\ \mid \\ CH_3CHCH_2 \end{array}$— or $(CH_3)_2CHCH_2$—	Isobutyl group
$\begin{array}{c} CH_3 \\ \mid \\ CH_3{-}C{-} \\ \mid \\ CH_3 \end{array}$ or $(CH_3)_3C$—	Tertiary butyl group
R—	General symbol for any alkyl group

formed from the alkanes are called **alkyl groups.** They are named by using the name of the alkane from which they are derived with the -ane ending changed to -yl. For example,

$$\begin{array}{c} H \\ \mid \\ H{-}C{-} \\ \mid \\ H \end{array} \text{ is methyl group } \text{ and } \begin{array}{c} H \quad H \\ \mid \quad \mid \\ H{-}C{-}C{-} \\ \mid \quad \mid \\ H \quad H \end{array} \text{ is ethyl group}$$

These groups can also be shown in a condensed, more easily written formulas as CH_3— (methyl) and CH_3CH_2— (ethyl). Table 13-2 lists some common alkyl groups. A special symbol, R—, is used to represent any alkyl group. A few examples of substituted compounds containing alkyl groups are:

CH_3OH methyl alcohol
CH_3CH_2Cl ethyl chloride
$(CH_3)_2CHNH_2$ isopropyl amine

13-6 Alkenes

Hydrocarbons that have a double bond between any two carbons are called **unsaturated hydrocarbons** or **olefins.** The simplest olefin is ethylene:

$$\begin{array}{ccc} H & & H \\ \diagdown & & \diagup \\ & C{=}C & \\ \diagup & & \diagdown \\ H & & H \end{array}$$

The compounds formed from this compound by adding longer carbon sequences are members of the homologous series called the **alkenes.** The general formula for alkenes is C_nH_{2n} (CH_2=CH_2 $n = 2$, CH_3CH=CH_2 $n = 3$, etc.).

Some common alkenes and substituted alkenes are listed below.

CH_2=CH_2 ethylene

CH_3—CH=CH_2 or CH_3CH=CH_2 propylene

$$CH_2{=}C\begin{smallmatrix} H \\ \\ Cl \end{smallmatrix} \quad \text{or} \quad CH_2{=}CHCl$$ vinyl chloride

$$\begin{smallmatrix} F \\ \\ F \end{smallmatrix}C{=}C\begin{smallmatrix} F \\ \\ F \end{smallmatrix} \quad \text{or} \quad F_2C{=}CF_2$$ tetrafluoroethylene

Whenever a double bond occurs in a sequence of carbons, it is called an **unsaturated linkage.** Molecules of **polyunsaturated vegetable oils** involve carbon sequences with unsaturated linkages. However, vegetable oils are not classified as alkenes since they contain oxygen as well as carbon and hydrogen. Vegetable oils are classified as lipids and are discussed in Section 14-3. Compounds with double bonds are generally quite chemically reactive in comparison with alkanes. The typical reactions involving alkenes are **addition reactions** in which the double bond is broken and other atoms become bonded to the carbons of the original double bonds. Two examples of addition reactions are

$$\begin{smallmatrix} H \\ \\ H \end{smallmatrix}C{=}C\begin{smallmatrix} H \\ \\ H \end{smallmatrix} + HCl \rightarrow H{-}\underset{\underset{H}{|}}{\overset{\overset{H}{|}}{C}}{-}\underset{\underset{H}{|}}{\overset{\overset{H}{|}}{C}}{-}Cl$$ ethyl chloride

$$CH_3{-}CH{=}CH_2 + Cl_2 \rightarrow CH_3\underset{\underset{}{|}}{\overset{\overset{Cl}{|}}{C}}H{-}\underset{\underset{}{|}}{\overset{\overset{Cl}{|}}{C}}H_2$$ propylene chloride

13-7   Polymerization and Plastics

Some alkenes can react under specific conditions in the presence of a catalyst so that the individual alkene molecules add to one another. In this reaction, the double bonds are broken and hundreds or thousands of molecules link together to form very large molecules called **polymers.** The original alkene used to prepare the polymer is called the **monomer.** These reactions are called **polymerization reactions.** (See Figure 13-4.) The polymer molecules form solids that are known as **plastics** and **elastomers,** which are used to make many of the materials and objects that we use in everyday life. Table 13-3 lists some of the products which are made of polymers. Many substances that occur in nature contain polymers. Cotton, wool, silk, and natural rubber are polymers. In

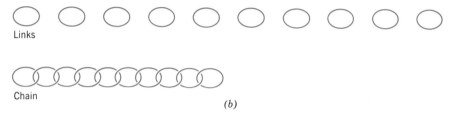

Figure 13-4   Polymerization. (a) In a polymerization reaction, monomer molecules are linked together to form large molecules called polymers. (b) The formation of a polymer can be likened to the forming of a chain. Monomer units are linked together one after another to form the chain-like polymer molecule. Polymerization reactions are sometimes called chain reactions and the bonded atoms comprising the polymer are sometimes called the polymer chain. From *Understanding Chemistry: From Atoms to Attitudes* by T. R. Dickson, Wiley, N.Y., 1974.

fact, as we shall see in Sections 14-2 and 14-4, many important biological molecules are polymers.

In an equation for a polymerization reaction, it is not possible to give the exact formula of the polymer, since its individual molecules, made up of hundreds or thousands of monomer units, vary in chain length. We could represent the polymerization of ethylene as

$$n\ CH_2{=}CH_2 \xrightarrow[\text{heat pressure}]{\text{catalyst}} H{-}\underset{\underset{H}{|}}{\overset{\overset{H}{|}}{C}}{-}\underset{\underset{H}{|}}{\overset{\overset{H}{|}}{C}}{-}\underset{\underset{H}{|}}{\overset{\overset{H}{|}}{C}}{-}\underset{\underset{H}{|}}{\overset{\overset{H}{|}}{C}}{-}\underset{\underset{H}{|}}{\overset{\overset{H}{|}}{C}}{-}\underset{\underset{H}{|}}{\overset{\overset{H}{|}}{C}}{-}\underset{\underset{H}{|}}{\overset{\overset{H}{|}}{C}}{-}\ .\,.\,.$$

but this is not convenient. However, notice that the polymer could be represented by the repeating sequence

$$-\underset{\underset{H}{|}}{\overset{\overset{H}{|}}{C}}{-}\underset{\underset{H}{|}}{\overset{\overset{H}{|}}{C}}{-}$$

This sequence repeats along the polymer except at the ends, but the ends are a very small part of the entire polymer. Polymers are often represented by a characteristic repeating sequence. Therefore, the polymer of ethylene could be represented as

Table 13-3   Polymers—Plastics, Resins, and Rubbers

Name	Use
Polyethylene	Electrical insulation, packaging material (sandwich bags, plastic wrap), molded toys and utensils, milk carton coatings
Polypropylene	Molded containers, bottles, hospital utensils (sterilizable), washing machine parts, automobile interior parts
Polyvinyl chloride	Electrical insulation, toys, garden hoses, automobile seat covers, washable wallpaper, packaging material, bottles (shampoo), "patent leather," vinyl flooring, water pipes
Polytetrafluoroethylene	Teflon, electrical insulator, chemically inert material, nonstick coatings (pots, pans and tools)
Polyvinylacetate	Paints, adhesives for textiles, paper and wood, sizing for textiles
Polyvinyl alcohol	Emulsifiers in cosmetics, water soluble packaging materials
Polymethylmethacrylate (Acrylic)	Plexiglas, Lucite, paints, light fixtures, signs, airplane windows, helicopter bubbles, dentures.
Polystyrene	Styrofoam, packaging material, bottle caps, refrigerator interiors, toys, containers, kitchen utensils, foam insulation
Polyvinyldiene chloride	Saran, packing material, coating material
Nylon (Polyamides)	Machine parts (gears, cams), boil-in-the bag containers, filaments used in brushes, surgical sutures, fishing lines, carpets, women's stockings
Cellulose acetate	Photographic film, toothbrushes, combs
Phenol-formaldehyde Resins	Bakelite, electrical insulators,
Melamine-formaldehyde Resins	Dishes, buttons, Melmac, Formica
Polyester Resins	Molding compounds, fiberglass resins, Mylar film, synthetic fibers, paints, permanent-press clothing
Epoxide Resins	Epoxy glues, coatings
Polyurethanes	Foams for packaging, insulation, furniture
Silicones	Polishes, lubricants, high temperature rubber
cis-Polyisoprene	Natural rubber
Styrene-butadiene Rubber	Synthetic rubber
Polychloroprene	Neoprene synthetic rubber

$$\left(\begin{matrix} H & H \\ | & | \\ -C - C - \\ | & | \\ H & H \end{matrix} \right)_n$$

where n is some large number corresponding to the number of monomer units comprising the polymer. Now the equation for the polymerization of ethylene to form polyethylene can be written as

$$n\ CH_2{=}CH_2 \xrightarrow[\text{heat pressure}]{\text{catalyst}} \left(-\underset{\underset{H}{|}}{\overset{\overset{H}{|}}{C}}-\underset{\underset{H}{|}}{\overset{\overset{H}{|}}{C}}-\right)_n \text{(polymer)}$$

(monomer)

The polymer of ethylene is called polyethylene and is used to make plactic bottles, toys, and other products. A few additional polymerization reactions are given below.

$$n\ CH_2{=}CHCH_3 \rightarrow \left(-\underset{\underset{H}{|}}{\overset{\overset{H}{|}}{C}}-\underset{\underset{H}{|}}{\overset{\overset{CH_3}{|}}{C}}-\right)_n$$

(propylene)   (polypropylene)

Polypropylene is used to make synthetic fibers, films, and other products.

$$n\ CF_2{=}CF_2 \longrightarrow \left(-\underset{\underset{F}{|}}{\overset{\overset{F}{|}}{C}}-\underset{\underset{F}{|}}{\overset{\overset{F}{|}}{C}}-\right)_n$$

(tetrafluoroethylene)   (polytetrafluoroethylene or Teflon)

Teflon is a chemically inert plastic used for nonstick coatings on machines, tools, and cooking utensils.

Large quantities of polymers are produced each year in the United States in the forms of molded plastics, films, and fibers and a variety of synthetic rubbers. The many uses of plastics and rubbers are listed in Table 13-3. Synthetic fibers are in wide use in carpets, clothing and other fabrics. In fact, as shown in Figure 13-5, synthetic fibers are now used in greater quantities than natural cotton and wool fibers.

13-8   Alcohols

In this and the following sections a few important groups of organic compounds are discussed. Compounds are grouped according to similarities. The alkanes are saturated hydrocarbons, and the alkenes are hydrocarbons with a double bond. Compounds can be grouped according to some structural characteristic that sets them apart from other compounds.

The group of compounds involving a —OH (hydroxy group) bonded to an alkyl group (R—) are called **alcohols** (R—OH). The **hydroxy group** is characteristic of alcohols, some of which have familiar common names. A few typical alcohols are described below.

CH_3OH methyl alcohol or methanol
(wood alcohol)

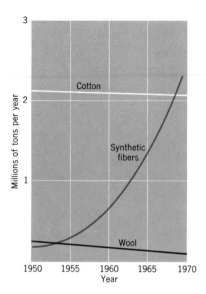

Figure 13-5　Annual textile fiber use in the United States.

Around 2.5 million tons of methyl alcohol are manufactured ($CO + 2H_2$ $\longrightarrow$ CH_3OH) in the United States each year. It is used in the manufacture of numerous chemical products such as formaldehyde, jet fuel, and antifreeze. (DANGER: Methyl alcohol or wood alcohol is very poisonous.)

$$\overset{\displaystyle OH}{\underset{\displaystyle |}{CH_3CHCH_3}} \quad or \quad (CH_3)_2CHOH \quad \text{isopropyl alcohol}$$

Isopropyl alcohol is used as a disinfectant and in rubbing alcohol

$$CH_3CH_2OH \ or \ C_2H_5OH \quad \text{ethyl alcohol or ethanol}$$

Millions of tons of ethyl alcohol are produced annually in the United States. It is used in numerous manufacturing processes and in the preparation of alcoholic beverages. Ethanol can be produced by the fermentation of sugars.

Fermentation is a chemical process in which complex organic molecules are broken down into simpler compounds like ethanol. This process is catalyzed by certain enzymes, which are complex chemical catalysts produced by living cells. The sugars used for fermentation are often formed by enzymatic decomposition of starches from corn, potatoes, rice, or grain. The fermentation reactions that produce ethyl alcohol are shown in the following equations.

$$\text{starch} \xrightarrow[\substack{\text{from malt}}]{\substack{\text{enzyme} \\ \text{diastase}}} n\ C_{12}H_{22}O_{11} \quad \text{(maltose)}$$

$$\underset{\text{maltose}}{C_{12}H_{22}O_{11}} \xrightarrow[\substack{\text{from yeast}}]{\substack{\text{enzyme} \\ \text{maltase}}} 2C_6H_{12}O_6 \quad \text{(glucose)}$$

several enzymes

$$C_6H_{12}O_6 \xrightarrow[\text{yeast}]{\text{also in}} 2CO_2 + 2C_2H_5OH \qquad \text{(ethyl alcohol)}$$

glucose

In addition to fermentation, over 1 million tons of ethyl alcohol are manufactured each year by the following industrial method.

$$CH_2{=}CH_2 + H_2O \xrightarrow{H_2SO_4} CH_3CH_2OH$$

ethylene

Alcohols that have more than one hydroxy group attached to a carbon sequence are called polyhydroxy alcohols. Two important polyhydroxy alcohols are described below.

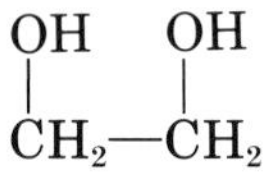

$$\begin{array}{cc} OH & OH \\ | & | \\ CH_2 & CH_2 \end{array} \quad \text{ethylene glycol}$$

or

$$CH_2OHCH_2OH$$

Ethylene glycol is used as an antifreeze and engine coolant.

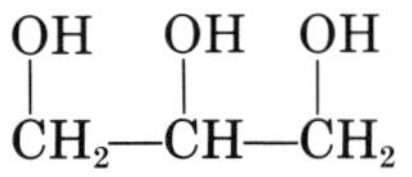

$$\begin{array}{ccc} OH & OH & OH \\ | & | & | \\ CH_2 & CH & CH_2 \end{array} \quad \text{glycerol}$$

or   or

$$CH_2OHCHOHCH_2OH \quad \text{glycerin}$$

Glycerol is used in the manufacture of plastics, drugs, cosmetics, inks, food products, and nitroglycerin, an explosive.

13-9 Aldehydes and Acids

Alcohols that have the hydroxy group attached to a carbon, which is in turn bonded to only one other carbon (R—CH_2OH), are called **primary alcohols.** When a primary alcohol reacts with certain substances called oxidizing reagents, the —CH_2OH grouping can be converted to an

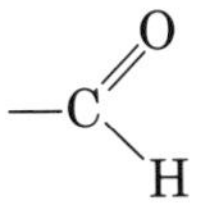

$$-C{\overset{\displaystyle O}{\underset{\displaystyle H}{\Big\backslash}}}$$ group **(aldehydo group).**

$$RCH_2OH + \text{oxidizing reagent} \rightarrow R-C{\overset{\displaystyle O}{\underset{\displaystyle H}{\Big\backslash}}}$$

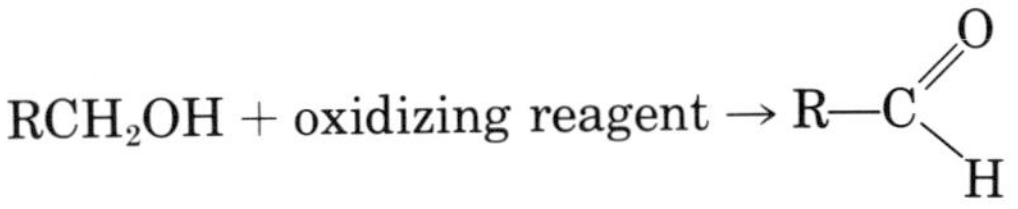

The compounds that contain an aldehydo group and correspond to for-

mula $R-C{\overset{\displaystyle O}{\underset{\displaystyle H}{\Big\backslash}}}$ or RCHO are called **aldehydes.** A few typical aldehydes

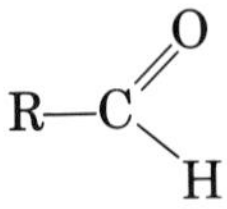

are given on the next page.

$$\begin{array}{c} H \\ \diagdown \\ C=O \\ \diagup \\ H \end{array} \quad \text{or} \quad CH_2O \quad \text{formaldehyde}$$

Formaldehyde is used for the manufacture of plastics such as Formica. Water solutions of formaldehyde, called formalin, are used as disinfectants and to preserve tissue.

$$CH_3-C\overset{\displaystyle O}{\underset{\displaystyle H}{\diagup}} \quad \text{or} \quad CH_3CHO \quad \text{acetaldehyde}$$

Acetaldehyde is used in the manufacture of plastics and for some medical purposes.

Aldehydes react with certain oxidizing reagents to convert the aldehydo group to a $-\overset{\displaystyle O}{\overset{\displaystyle \|}{C}}-OH$ called a **carboxylic group.**

$$R-C\overset{\displaystyle O}{\underset{\displaystyle H}{\diagup}} + \text{oxidizing reagent} \rightarrow R-C\overset{\displaystyle O}{\underset{\displaystyle OH}{\diagup}}$$

The compounds which include a carboxylic group and correspond to the

formula $R-C\overset{\displaystyle O}{\underset{\displaystyle OH}{\diagup}}$ or RCOOH are called the **carboxylic acids.** Some carboxylic acids are:

$$HC\overset{\displaystyle O}{\underset{\displaystyle OH}{\diagup}} \quad \text{or} \quad HCOOH \quad \text{formic acid}$$

$$CH_3C\overset{\displaystyle O}{\underset{\displaystyle OH}{\diagup}} \quad \text{or} \quad CH_3COOH \quad \text{acetic acid}$$

Vinegar is a dilute aqueous solution of acetic acid.

$$CH_3CH_2C\overset{\displaystyle O}{\underset{\displaystyle OH}{\diagup}} \quad \text{or} \quad CH_3CH_2COOH \quad \text{propionic acid}$$

$$CH_3CH_2CH_2C\overset{\displaystyle O}{\underset{\displaystyle OH}{\diagup}} \quad \text{or} \quad CH_3CH_2CH_2COOH \quad \text{butyric acid}$$

13-10   Esters

A carboxylic acid can react with an alcohol to form a product called an **ester.**

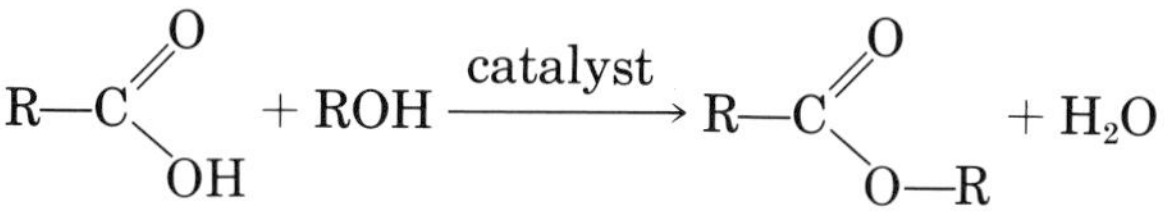

$$R-\overset{\displaystyle O}{\underset{\displaystyle OH}{C}} \;+\; ROH \;\xrightarrow{\text{catalyst}}\; R-\overset{\displaystyle O}{\underset{\displaystyle O-R}{C}} \;+\; H_2O$$

(ester)

The RO— grouping of the alcohol replaces the —OH of the acid. For example,

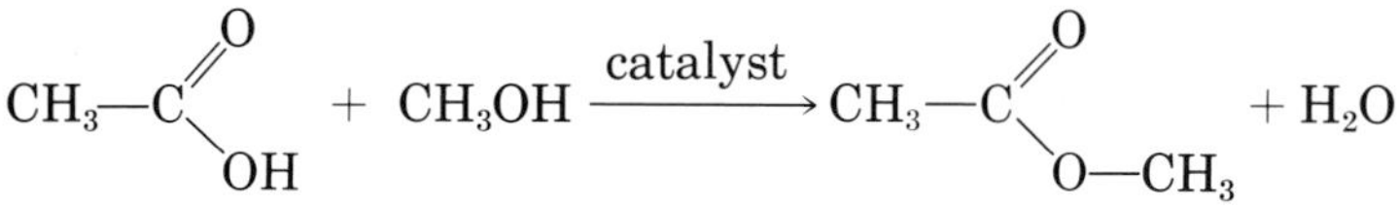

$$CH_3-\overset{\displaystyle O}{\underset{\displaystyle OH}{C}} \;+\; CH_3OH \;\xrightarrow{\text{catalyst}}\; CH_3-\overset{\displaystyle O}{\underset{\displaystyle O-CH_3}{C}} \;+\; H_2O$$

acetic acid   methyl alcohol   methyl acetate (an ester)

Many esters occur naturally and are often responsible for some of the tastes and odors of fruits. Table 13-4 lists some typical esters of this

Table 13-4   Some Esters

Name	Structure	Source or Flavor
Ethyl formate	$CH_3CH_2-O-\overset{O}{\overset{\|}{C}}-H$	Rum
Isobutyl formate	$CH_3\overset{CH_3}{\overset{\|}{C}}HCH_2-O-\overset{O}{\overset{\|}{C}}-H$	Raspberries
Ethyl acetate	$CH_3CH_2-O-\overset{O}{\overset{\|}{C}}CH_3$	Used in lacquers
n-Pentyl acetate (n-amyl acetate)	$CH_3(CH_2)_4-O-\overset{O}{\overset{\|}{C}}CH_3$	Bananas
Isopentyl acetate (isoamyl acetate)	$CH_3\overset{CH_3}{\overset{\|}{C}}HCH_2CH_2-O-\overset{O}{\overset{\|}{C}}CH_3$	Pears
n-Octyl acetate	$CH_3(CH_2)_7-O-\overset{O}{\overset{\|}{C}}CH_3$	Oranges
Ethyl butyrate	$CH_3CH_2-O-\overset{O}{\overset{\|}{C}}CH_2CH_2CH_3$	Pineapples
n-Pentyl butyrate	$CH_3(CH_2)_4-O-\overset{O}{\overset{\|}{C}}CH_2CH_2CH_3$	Apricots
"Waxes"	$CH_3(CH_2)_n-\overset{O}{\overset{\|}{C}}-O-(CH_2)_nCH_3$	$n = 23$ to 33: Carnauba wax $n = 25$ to 27: Beeswax $n = 14$ to 15: Spermaceti

type. Esters can also be formed between noncarboxylic acids and alcohols. It is possible to have phosphate esters derived from phosphoric acid,

$$HO-\overset{\overset{\displaystyle O}{\|}}{\underset{\underset{\displaystyle OH}{|}}{P}}-OH : HO-\overset{\overset{\displaystyle O}{\|}}{\underset{\underset{\displaystyle OH}{|}}{P}}-OR \quad or \quad HO-\overset{\overset{\displaystyle O}{\|}}{\underset{\underset{\displaystyle OR}{|}}{P}}-OR$$

Two examples of such phosphate esters are:

$$HO-\overset{\overset{\displaystyle O}{\|}}{\underset{\underset{\displaystyle OH}{|}}{P}}-OCH_3$$ methyl phosphate used as a gasoline additive to control the ignition process in the engine

$$HO-\overset{\overset{\displaystyle O}{\|}}{\underset{\underset{\displaystyle OCH_2CH_3}{|}}{P}}-OCH_2CH_3$$ diethyl phosphate

13-11   Amines

The structural formula of ammonia is

$$\overset{\displaystyle N}{\underset{\underset{\displaystyle H}{|}}{\diagup \diagdown}}$$
$$H \qquad H$$

Compounds in which an —H of ammonia is replaced by an alkyl group, —R, are called **amines.**

$$R-N\overset{\displaystyle H}{\underset{\displaystyle H}{\diagdown}} \qquad or \ R-NH_2 \qquad or \ RNH_2$$

The —NH_2 group is called the **amino group.** Amines with one alkyl group are called **primary amines.** Amines with two alkyl groups (which need not be alike) are called **secondary amines,** and those with three alkyl groups are called **tertiary amines.**

$$R-\underset{\underset{\displaystyle R}{|}}{N}-H \qquad\qquad R-\underset{\underset{\displaystyle R}{|}}{N}-R$$
secondary amine   tertiary amine

$$CH_3CH_2NH_2 \qquad ethylamine$$

$$CH_3-\underset{\underset{\displaystyle CH_2CH_3}{|}}{N}-H \qquad methylethylamine$$

$$CH_3-\underset{\underset{\displaystyle CH_3}{|}}{N}-CH_3 \qquad trimethylamine$$

13-12   Amides

Under certain conditions, ammonia and some amines can combine with carboxylic acids in a manner that is quite similar to the formation of an ester.

$$R-\overset{\overset{O}{\|}}{C}-\boxed{OH\quad H}-\underset{\underset{H}{|}}{N}-H \rightarrow R-\overset{\overset{O}{\|}}{C}-\underset{\underset{H}{|}}{N}-H + H_2O$$

(amide)

The product of such a reaction is called an **amide;** the linking of an amine and an acid is called an **amide linkage.** As we shall see in Section 14-4, amide linkages are involved in the formation of proteins. A few examples of amides are:

$$CH_3-\overset{\overset{O}{\|}}{C}-NH_2 \qquad \text{acetamide}$$

$$CH_3CH_2CH_2\overset{\overset{O}{\|}}{C}-NH_2 \qquad \text{butyramide}$$

$$CH_3\overset{\overset{O}{\|}}{C}-\underset{\underset{}{\overset{\overset{H}{|}}{N}}}-CH_3 \qquad \text{N-methylacetamide (the N- preceeding the methyl indicates that the methyl group is attached to the nitrogen of the amide)}$$

13-13   Amino Acids

A class of compounds that are important in biochemistry is the amino acids. The common **amino acids** are carboxylic acids with an amino group bonded to the carbon next to the carboxylic group carbon.

$$NH_2-\underset{\underset{R}{|}}{CH}-C\overset{\overset{O}{\diagup}}{\diagdown}OH$$

The —R group may be an alkyl group (i.e., —CH$_3$) or a substituted alkyl group (i.e., —CH$_2$OH). Various amino acids differ only in the structure of the —R group. The common amino acids that are found in nature have been given common names. For example,

$$NH_2CH-C\overset{\overset{O}{\diagup}}{\diagdown}OH \quad \text{glycine} \qquad NH_2CH-C\overset{\overset{O}{\diagup}}{\diagdown}OH \quad \text{alanine}$$
$$\underset{H}{|} \qquad\qquad\qquad\qquad\quad \underset{CH_3}{|}$$

The names and structures of some other amino acids are given in Table 13-5. As we shall see in Section 14-4, the amino acids are the building units of proteins found in living organisms. Since an amino acid can act

Table 13-5   Common Amino Acids

All the amino acids except proline and hydroxyproline have the general formula

$$NH_2-CH-\overset{\overset{\displaystyle O}{\|}}{C}-OH$$
$$\underset{R}{|}$$

in which R is the characteristic group for each acid. The R groups, names, and abbreviations[a] are as follows.

1. Glycine   —H   Gly
2. Alanine   —CH$_3$ Ala
3. Serine   —CH$_2$OH   Ser
4. Cysteine   —CH$_2$SH   Cys
5. Cystine   —CH$_2$—S—S—CH$_2$—   Cys—S—S—Cys[b]
6. Threonine[c]   —CH—CH$_3$ Thr
                      |
                     OH

7. Valine[c]   CH$_3$—CH—CH$_3$ Val
                        |
8. Leucine[c]   —CH$_2$—CH—CH$_3$ Leu
                            |
                           CH$_3$

9. Isoleucine[c]   —CH$\begin{smallmatrix}CH_3\\ \\CH_2—CH_3\end{smallmatrix}$ Ile

10. Methionine[c]   —CH$_2$—CH$_2$—S—CH$_3$ Met
11. Aspartic acid   —CH$_2$CO$_2$H   Asp
12. Glutamic acid   —CH$_2$—CH$_2$—CO$_2$H   Glu
13. Lysine[c]   —CH$_2$—CH$_2$—CH$_2$—CH$_2$—NH$_2$ Lys
14. Arginine   —CH$_2$—CH$_2$—CH$_2$—NHCNHNH$_2$ Arg

[a] These are the official IUPAC symbols for amino acids.
[b] Cystine involves two cysteine units joined by disulfide linkage (—S—S—).
[c] Essential amino acids needed in the diet of humans.

either as an acid or an amine, amino acids can link together through amide linkages.

13-14   **Benzene and Cyclic Compounds**

In some hydrocarbons the carbon sequence is one in which the carbons form a ring. These hydrocarbons are called **cyclic** or **ring compounds.** (See Figure 13-3.) For example, cyclohexane can be represented by the formula

$$\begin{matrix} & H_2C-CH_2 & \\ H_2C & & CH_2 \\ & H_2C-CH_2 & \end{matrix}$$

Table 13-5   (continued)

15. Phenylalanine[c] —CH₂—⟨benzene ring⟩   Phe

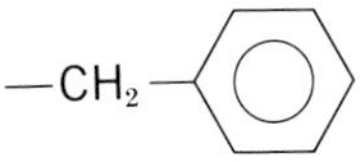

16. Tyrosine —CH₂—⟨benzene ring⟩—OH   Tyr

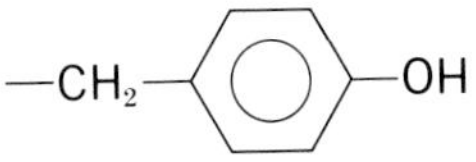

17. Tryptophan[c] —CH₂—⟨indole ring with N—H⟩   Trp

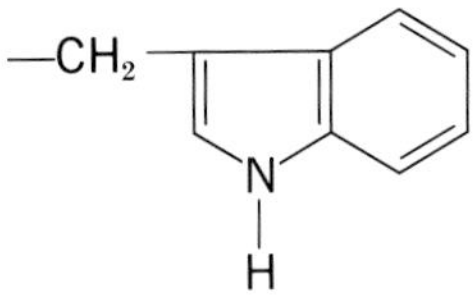

18. Histidine —CH₂—⟨imidazole ring N⟍ N—H⟩   His

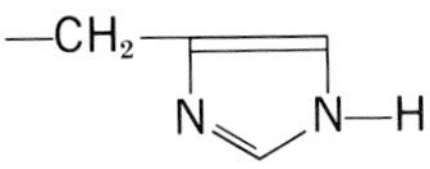

19. Proline   H₂C——CH₂
              H₂C   CHCO₂H
                 ⟍N⟋
                  |
                  H   Pro

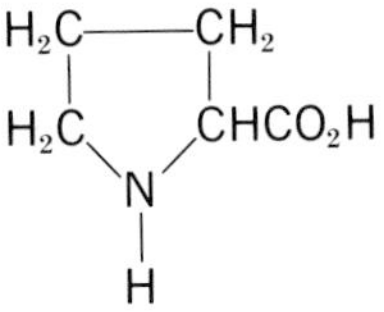

20. Hydroxyproline   HOHC——CH₂
                     H₂C   CHCO₂H
                        ⟍N⟋
                         |
                         H   Hyp

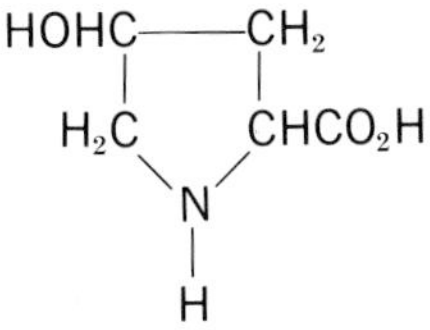

Another example is cyclopentane:

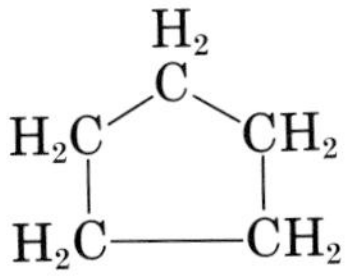

$$
\begin{array}{c}
H_2 \\
C \\
H_2C \qquad CH_2 \\
H_2C \!\!-\!\! CH_2
\end{array}
$$

Often these cyclic compounds are represented for convenience by a geometrical figure representing the carbon sequence.

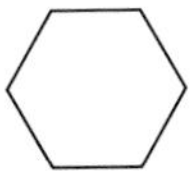

⬡ represents
$$
\begin{array}{c}
H_2C\!-\!CH_2 \\
H_2C \qquad CH_2 \\
H_2C\!-\!CH_2
\end{array}
$$
  cyclohexane

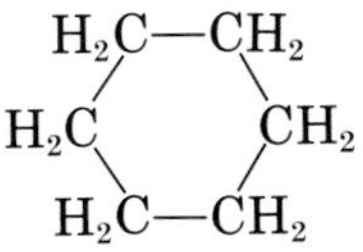

⬠ represents
$$
\begin{array}{c}
H_2 \\
C \\
H_2C \qquad CH_2 \\
H_2C\!-\!CH_2
\end{array}
$$
  cyclopentane

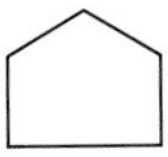

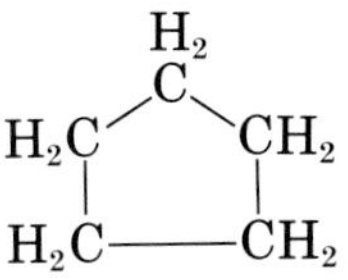

Substituted ring compounds can be represented using the figure with the group that replaced an —H connected to the figure. For instance, the alcohol, cyclohexanol ($C_6H_{11}OH$), can be represented as

A very prevalent and important ring compound is **benzene,** C_6H_6, which can be represented by the structural formulas

Benzene is an unusually stable ring compound, and many important substances are formed by substituting another group or other groups for hydrogens on the benzene ring. Notice that the structural formulas given above indicate that benzene has alternating single and double bonds around the ring. Either of the two arrangements shown could be used to represent benzene. Actually, these two formulas are different. One shows the double bonds in one possible arrangement, and the other shows the double bonds in another arrangement. However, studies have found that all the carbon-carbon bonds in benzene are the same. Thus, neither of the structures represents the actual bonding electron distribution in the benzene ring. The actual structure is something in between these two extremes. The actual structure is said to be a resonance hybrid between these two contributing structures. Then, how should we represent benzene? One possible way is to draw the structure to indicate that the carbon-carbon bonds are neither completely single nor completely double.

However, since it is not convenient to write the structure every time we want to represent benzene, it is possible to use a geometrical figure as was done with the cyclic alkanes. The symbol used to represent benzene as a resonance hybrid is which should be interpreted as representing the structure shown above. Sometimes benzene is represented by similar symbols corresponding to the resonance contributing forms.

and

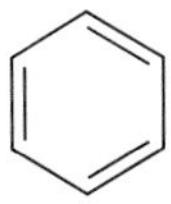

Benzene is used as a starting material to prepare many useful compounds. Over 4.5 million tons of benzene are produced annually in the United States. Compounds involving a benzene ring are sometimes called **aromatic compounds.** Substituted benzene compounds can be represented by showing the substituted group attached to the ring. For example,

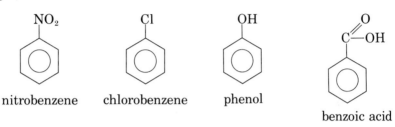

nitrobenzene   chlorobenzene   phenol

benzoic acid

When more than one substituted group appears on the ring, the positions on the ring are distinguished by numbering the carbons

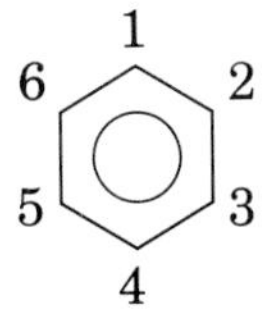

$$
\begin{array}{c}
1 \\
6 \quad \quad 2 \\
5 \quad \quad 3 \\
4
\end{array}
$$

or by using the following notation.

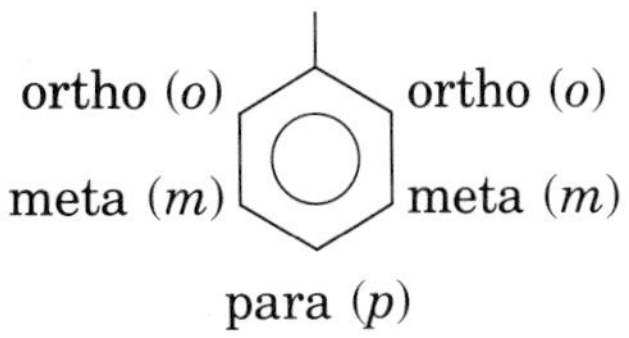

ortho (o)   ortho (o)

meta (m)   meta (m)

para (p)

To illustrate, consider the following compounds

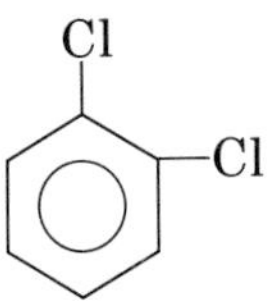

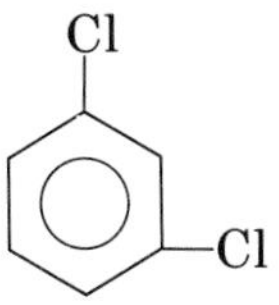

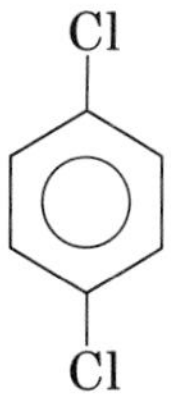

ortho-dichlorobenzene   meta-dichlorobenzene   para-dichlorobenzene
o-dichlorobenzene   m-dichlorobenzene   p-dichlorobenzene
1,2-dichlorobenzene   1,3-dichlorobenzene   1,4-dichlorobenzene

A few typical aromatic compounds are listed in Table 13-6.

Table 13-6   Some Typical Aromatic Compounds

Formula	Name	Comment
CH₃ (on benzene ring)	Toluene	Used to prepare TNT, solvent
NH₂ (on benzene ring)	Aniline	Used in dyes and drugs
COOH (on benzene ring)	Benzoic acid	Used in food preservatives
CHO (on benzene ring)	Benzaldehyde	Flavoring agent, "oil of almond"
Cl (on benzene ring, para Cl)	para-Dichlorobenzene	Moth balls
CH₃, NO₂, NO₂, NO₂ (on benzene ring)	2,4,6-Trinitrotoluene	TNT, an explosive

13-15   Polycyclic and Heterocyclic Compounds

Compounds such as cholesterol in which two or more rings mutually share carbons are called **polycyclic** (many ring) **compounds.**

Table 13-6   (*continued*)

Formula	Name	Comment
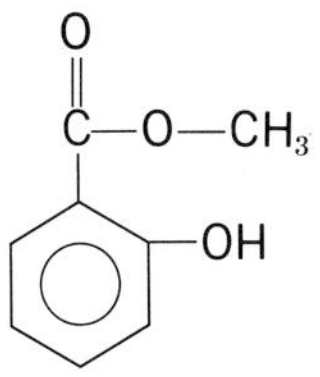	Methyl salicylate	Oil of wintergreen; used in rubbing ointments
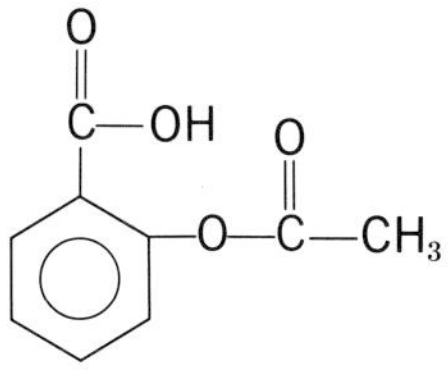	Acetylsalicylic acid	Aspirin; analgesic (pain reliever)
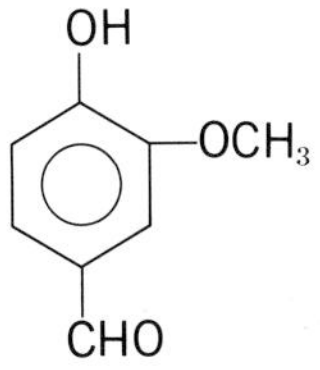	Vanillin	Vanilla flavoring
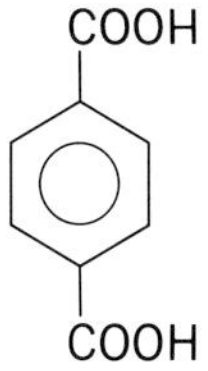	Terephthalic acid	Used in the manufacture of polyesters such as Dacron, Fortrel, and Mylar

Polycyclic compounds are common in biochemistry. Some cyclic compounds such as

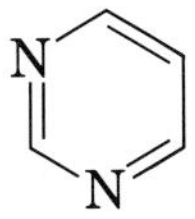

pyrimidine 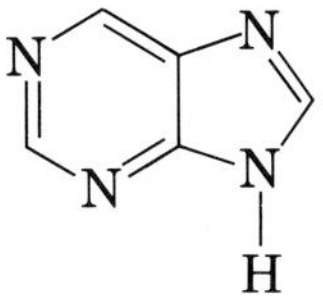   purine

have a few atoms other than carbon in the rings. Such cyclic compounds that include carbon and some other type of atom in the ring or rings are called heterocyclic compounds. Some important **heterocyclic compounds** are given below.

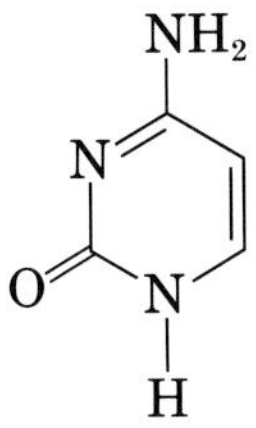

cytosine 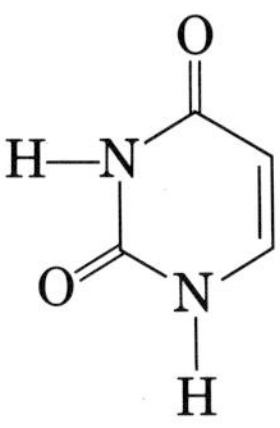   uracil 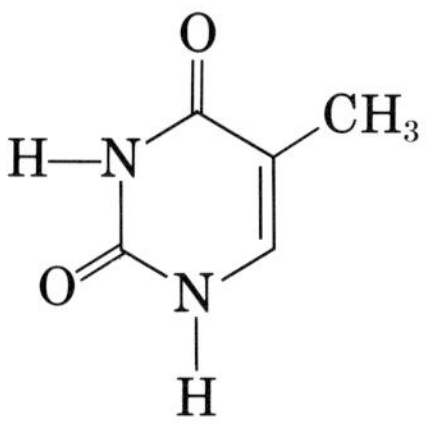   thymine 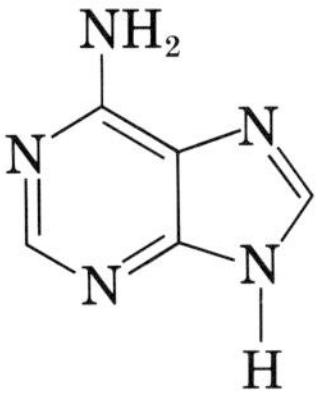   adenine 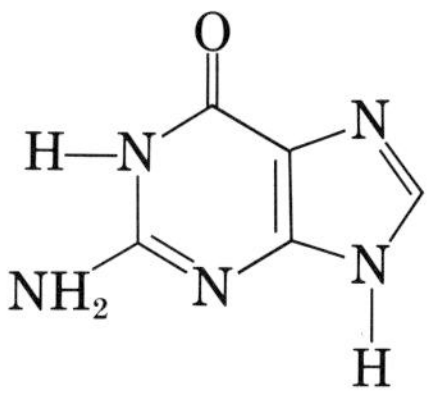   guanine

Table 13-7   Summary of Organic Chemicals

Group	General Formula	Example
Alkane	C_nH_{2n+2}	$CH_3CH_2CH_3$, propane
Alkenes	C_nH_{2n}	$CH_2{=}CH_2$, ethylene
Alcohols	ROH	CH_3CH_2OH, ethyl alcohol
Aldehydes	RCHO	CH_3CHO, acetaldehyde
Acids	RCO_2H	CH_3CO_2H, acetic acid
Esters	RCO_2R	$CH_3CO_2CH_2CH_3$, ethyl acetate
Amines	RNH_2, R_2NH, R_3N	CH_3NH_2, methylamine
Amides	RCO_2NH_2	$CH_3CO_2NH_2$, acetamide
Amino acids	NH_2CHCO_2H 	 R	$NH_2CH_2CO_2H$, glycine
Cyclic hydrocarbons	C_nH_{2n}	cyclohexane
Benzene	C_6H_6	
Aromatics	Substituted benzene	chlorobenzene
Heterocyclics	Element other than C in ring	pyridine

From *Understanding Chemistry: From Atoms to Attitudes* by T. R. Dickson, Wiley, N.Y., 1974.

These five compounds are the heterocyclic bases or heterocyclic amines, discussed in Section 14-7, which are involved in biologically important ribonucleic acids (RNA) and deoxyribonucleic acids (DNA).

Only a few types of organic compounds have been discussed in this chapter. Organic chemistry includes the study of numerous other kinds of compounds and the chemical reactions they undergo. The organic chemistry considered in previous sections is summarized in Table 13-7.

Questions and Problems

1. What is organic chemistry?
2. What are synthetic organic chemicals?

3. Describe the group of compounds known as alkanes.

4. Give the formulas for the following alkanes.
   (a) methane
   (b) butane
   (c) pentane
   (d) propane
   (e) octane

5. What information is given by the structural formula of a compound?

6. What are structural isomers? Which of the following sets of compounds are isomers?

                  CH_3
                  |
(a) $CH_3—C—CH_3$ and $CH_3CH_2CH_2CH_2CH_3$
                  |
                  CH_3

(b) $CH_3—O—CH_3$ and CH_3CH_2OH

              CH_3
              |
(c) $CH_3CHCH_2CH_3$ and $CH_3CH_2CH_2CH_3$

7. Give the name or structural formula of each of the following alkyl groups.
   (a) $—CH_3$
   (b) ethyl group
   (c) $CH_3CH_2CH_2—$
   (d) isopropyl group

            CH_3
            |
  (e) $CH_3—CH—CH_2—$

   (f) tertiary butyl group

8. What is the distinguishing characteristic of a compound that is classed as an alkene? Give the name and structural formula of a typical alkene.

9. Describe the polymerization reaction of an alkene using the words monomer and polymer.

10. Give two examples of common plastics used in commercial products.

11. The compound styrene, $CH_2{=}C\overset{\displaystyle H}{\underset{\displaystyle C_6H_5}{}}$, polymerizes to form polystyrene.

    Give an equation for this polymerization reaction.

12. The compound methyl methacrylate, $CH_2{=}C\overset{\displaystyle CH_3}{\underset{\displaystyle CO_2CH_3}{}}$, polymerizes to

    form polymethyl methacrylate usually called Plexiglas, Lucite, or acrylic plastic. Give an equation for this polymerization reaction.

13. The compound vinyl chloride, $CH_2{=}C$ $\overset{Cl}{\underset{H}{<}}$, polymerizes to form polyvinyl chloride, PVC. Give an equation for this polymerization reaction.

14. The average mass of a polyethylene macromolecule is about 50,000 amu. Determine the approximate number of ethylene units in such a macromolecule. (Hint: There are 28 amu per ethylene unit.)

15. Keep a list of the plastics and other synthetic organic chemicals that are used in your life for one day. (See Table 13-3.)

16. Describe the chemical nature of and give an example of each of the following classes of organic compounds.
    (a) alcohol
    (b) aldehyde
    (c) carboxylic acid
    (d) ester
    (e) amine
    (f) amide
    (g) amino acid

17. What is a cyclic hydrocarbon? Give an example.

18. What symbolic formula is used to represent benzene?

19. What is an aromatic compound? Give an example.

20. What is a heterocyclic compound?

21. Classify each of the following compounds as an alkane, alkene, alcohol, aldehyde, carboxylic acid, ester, amine, amide, amino acid, aromatic, or heterocyclic. (Some may fall into more than one class.)

(a)　$CH_3CH_2CH_2OH$

(b)　$CH_3CH_2\overset{\displaystyle O}{\overset{\|}{C}}{-}OH$

(c) 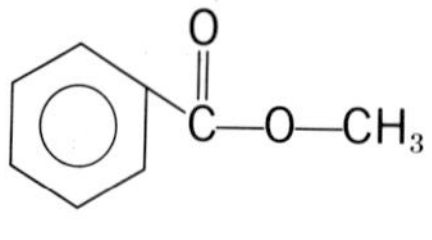　 $\overset{\displaystyle O}{\overset{\|}{C}}{-}O{-}CH_3$

(d)　$CH_3CH_2CH_2CH_3$

(e) 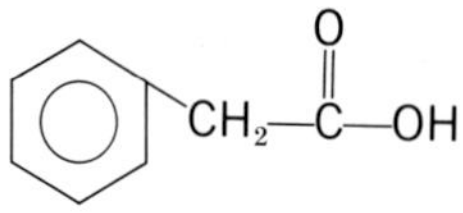　 $CH_2{-}\overset{\displaystyle O}{\overset{\|}{C}}{-}OH$

(f) 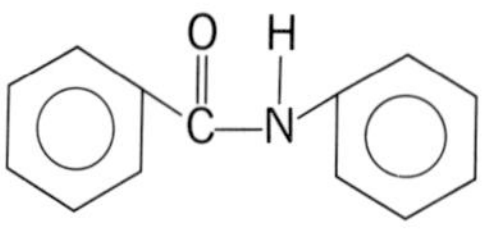　 $\overset{\displaystyle O\ H}{\overset{\|\ \ |}{C}{-}N}$

(i)　CH_3NH_2

(j)　$CH_3CH{=}CH_2$

(k) 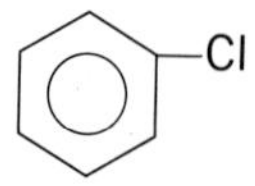　 Cl

(l)　$\overset{\displaystyle H{-}C{=}O}{\underset{}{CH_3CHCH_3}}$

(m) 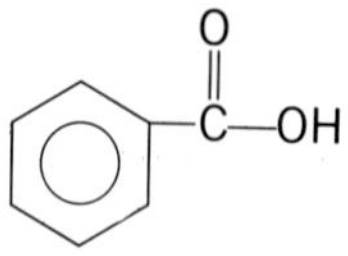　 $\overset{\displaystyle O}{\overset{\|}{C}}{-}OH$

(n)　$CH_3{-}\underset{\underset{\displaystyle CH_3}{|}}{N}{-}CH_3$

(g)

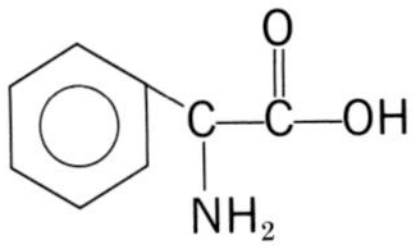

$$\underset{NH_2}{\overset{\displaystyle C6H5}{}}\;C\!-\!\overset{\displaystyle O}{\overset{\|}{C}}\!-\!OH$$

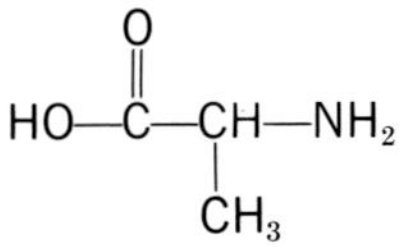

(h)

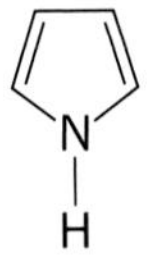

(o) HO—$\overset{\displaystyle O}{\overset{\|}{C}}$—CH—NH$_2$
                        |
                       CH$_3$

(p) 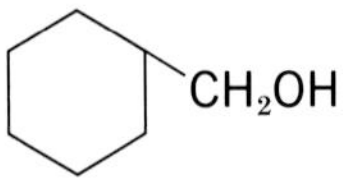 CH$_2$OH

14

Chapter 14: The Chemistry of Life: Biochemistry

Objectives

The student should be able to:

1. Describe the structure of a common carbohydrate such as glucose or ribose.
2. Recognize the cyclic form of glucose or ribose from the formula.
3. Describe the structure of maltose, sucrose, starch, or cellulose in terms of glucose.
4. Describe the structure of simple lipids.
5. Describe how amino acids link to form peptides.
6. Describe the primary, secondary, and tertiary structures of proteins.
7. Describe the lock-and-key theory of enzymes.
8. Describe the primary structure of DNA and RNA and the double-helix structure of DNA.
9. Describe the genetic code.
10. Describe enzyme synthesis that takes place in cells under control of cellular DNA.

Terms to Know

Biochemistry
Carbohydrates
Monosaccharide
Disaccharide
Polysaccharide
Starch
Cellulose
Lipids
Fats
Oils
Steroids
Peptides
Peptide bonds
Proteins or polypeptides
Enzymes
Genes
DNA
RNA
Double-helix theory
Genetic code
Genetic defects

14

The Chemistry of Life: Biochemistry

14-1 Biochemistry

Certain organic and inorganic compounds are constituents of animal and plant life. Many of these compounds are fundamental to the processes of life. The study of compounds involved in biological processes and the changes they undergo during life processes is called **biochemistry.** The field of biochemistry is one of the frontiers of science. Many scientifically exciting discoveries involving life processes are being made by biochemists. Such discoveries range from the establishment of a theory to explain the chemistry of heredity to the development of drugs and vaccines to cure disease.

Biochemists study the chemical aspects of living organisms. Living organisms are a complex mixture of various kinds of chemical compounds that act in concert to maintain the life of the organism. Considering the great diversity of plants and animals, it is surprising that many of the same chemical compounds and chemical processes are shared by all life forms. In this chapter, we will study some biochemically important substances and consider some of the life processes of humans.

Many chemical elements are involved in the biochemical compounds. However, since most of the biochemical compounds are organic, only a few elements make up most of these compounds. Combined hydrogen, oxygen, carbon and nitrogen make up about 99% of the atoms of living organisms. Calcium, chlorine, magnesium, phosphorus, potassium, sodium, and sulfur make up most of the remaining 1%. Many other elements are present in small amounts as trace elements.

14-2   Carbohydrates

Carbohydrates or sugars are produced by green plants during photosynthesis. Carbohydrates contain only carbon, hydrogen, and oxygen. They occur in a variety of sizes ranging from simple sugars called **monosaccharides** to polymer molecules called **polysaccharides.** Most of the common monosaccharides are polyhydroxy aldehydes. Among the most important monosaccharides are **ribose** and **glucose** (sometimes called dextrose).

These structures represent the open-chain forms of these sugars. For reference, the carbons are numbered from top to bottom. For example, the carbons in glucose are numbered

This monosaccharide very often occurs as a ring compound. The ring is usually formed by intramolecular reaction between the aldehydo group and the —OH group on the next-to-last carbon. The two ring forms of glucose can be pictured as

The—OH on carbon 1 is above the ring in the β form and below in the α form. These cyclic forms can be more conveniently written as a heterocyclic ring leaving the ring carbons out. Thus, the two forms of glucose can be represented as

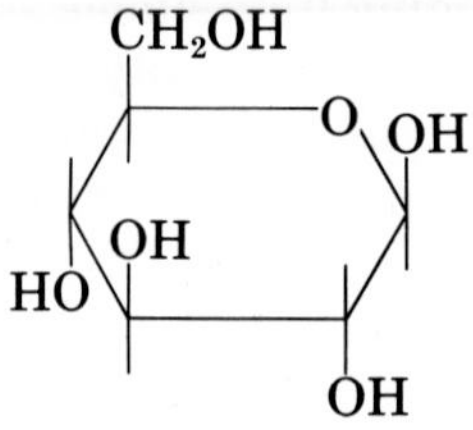

CH₂OH ... placeholder

β-D-glucose 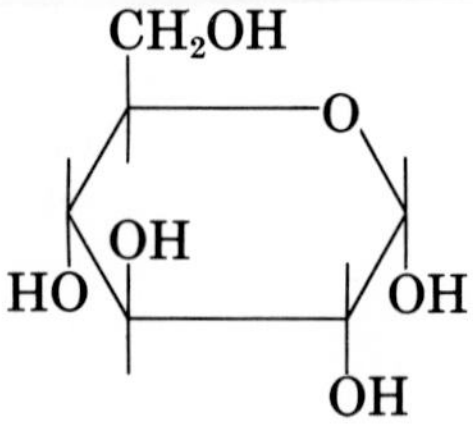 α-D-glucose

[β(beta) form C-1 —OH above ring] [α(alpha) form C-1 —OH below ring]

The biochemically important cyclic form of ribose involves a five atom ring and can be represented as

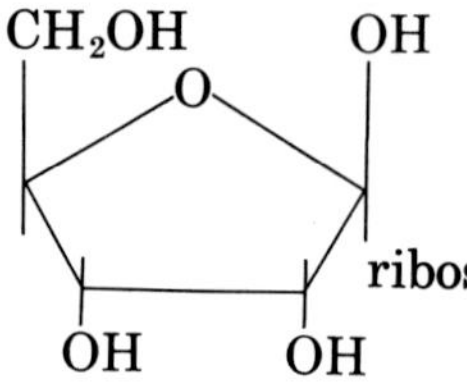

ribose

A similar cyclic compound of biochemical importance is one in which the —OH group on carbon number two has been replaced by a hydrogen. In other words, the oxygen of the —OH group can be considered to have been removed. This compound is called **deoxyribose** and can be represented as

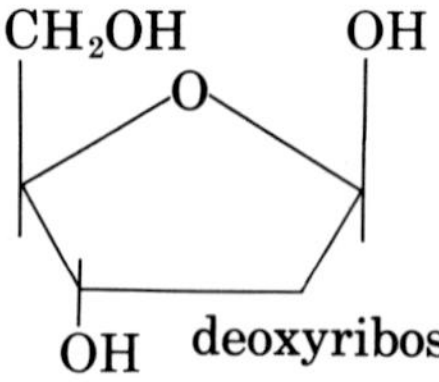

deoxyribose

As we shall see in Section 14-7, ribose and deoxyribose are involved in the structure of DNA and RNA.

Glucose in the combined form or as the free monosaccharide is undoubtedly the most abundant organic compound in nature. It is an important food source (a quick energy source) that is found in plants, fruits, and vegetables and is present in the bloodstream of certain animals. Two glucose molecules can link together through certain —OH groups to form the compound **maltose, a disaccharide** (two monosaccharides joined).

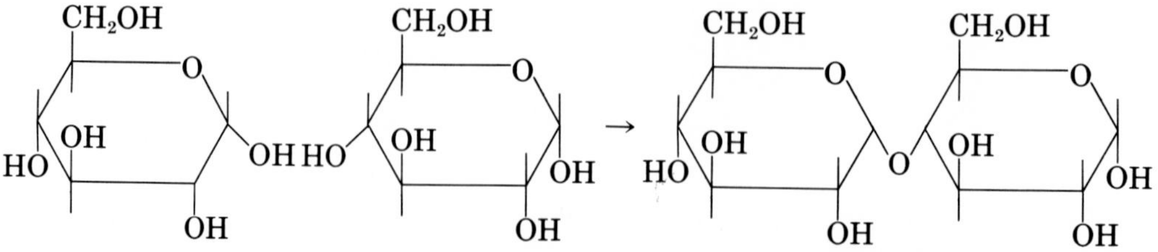

α-glucose α-glucose α-maltose

The two α-glucose units are linked through carbon 1 and carbon 4 by what is called an α linkage. Another disaccharide of interest is **sucrose** (table sugar), which involves α-glucose linked to the β form of the five-membered ring of the monosaccharide called fructose.

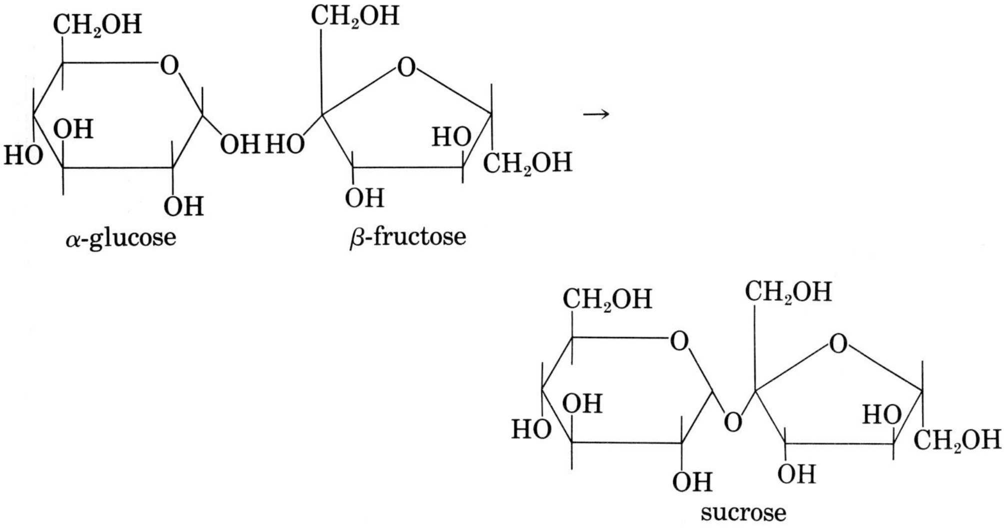

CH$_2$OH   CH$_2$OH

OH

HO   OH HO   HO

      CH$_2$OH   →

OH   OH

α-glucose β-fructose

CH$_2$OH   CH$_2$OH

OH

HO   O   HO

     CH$_2$OH

OH   OH

sucrose

Table 14-1 lists some common carbohydrates and their sources.

Many glucose units can bond together to form polymeric molecules known as **polysaccharides.** These polysaccharides are found in plants and animals. The polysaccharides involve long chains of hundreds and thousands of glucose units. **Starch** is a polysaccharide that is found in the seeds and roots of many plants. (See Figure 14-1.) It can be digested and is used as a food. Actually, there are two forms of starch that occur in plants. **Amylose** consists of long chains of α-glucose units. **Amylopectin** consists of branched chains of α-glucose units. **Glycogen** is a polysaccharide that is found in animals. Carbohydrates are stored in the body as glycogen. The structure of glycogen is similar to that of amylopectin. Another important polysaccharide is **cellulose,** which makes up the structural material of many plants. (See Figure 14-2.) Cellulose con-

Table 14-1 Common Carbohydrates

	Source
Disaccharide	
Surcose	Sugar cane, sugar beets, fruits, honey
Lactose	Milk sugar
Maltose	Hydrolyzed starch
Cellobiose	Hydrolyzed cellulose
Monosaccharide	
Glucose	Hydrolyzed sucrose, lactose, maltose, and cellobiose
Fructose	Hydrolyzed sucrose (fruits and honey)
Galactose	Hydrolyzed lactose

From *Understanding Chemistry: From Atoms to Attitudes* by T. R. Dickson, Wiley, N.Y., 1974.

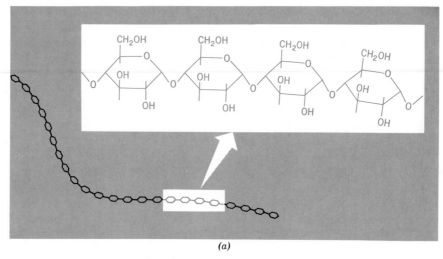

(a)

A portion of an amylose molecule.

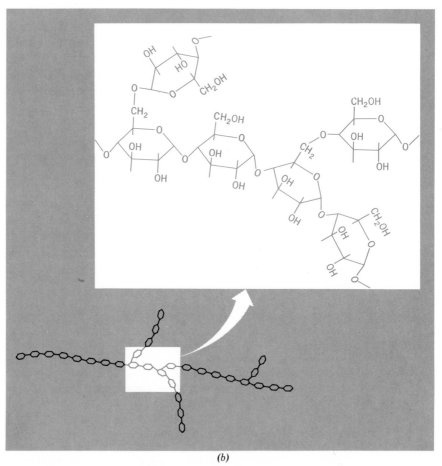

(b)

A portion of an amylopectin molecule.

Figure 14-1 Amylose and amylopectin starch. From *Understanding Chemistry: From Atoms to Attitudes* by T. R. Dickson, Wiley, N.Y., 1974.

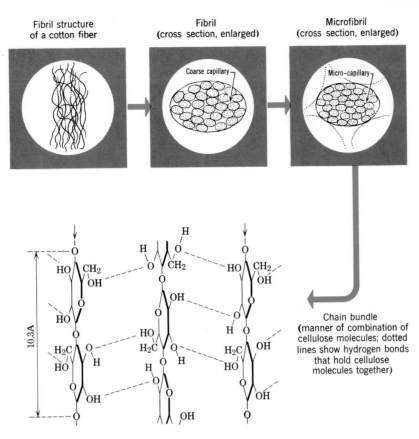

Figure 14-2   Details of the cotton fiber. Cotton is composed of intertwined cellulose molecules each containing thousands of glucose units. (Adapted from illustrations appearing in H. R. Mauersberger, editor, *Matthews' Textile Fibers,* sixth edition, Wiley, New York, 1954, pages 73 and 77.) From *Elements of General and Biological Chemistry,* 3rd ed., by John Holum, Wiley, New York, 1972.

sists of long chains of β-glucose units. It cannot be digested by humans, but it can be broken down by certain microorganisms. Cellulose is obtained in large amounts from trees and the cotton plant. It is used in the manufacture of paper, cotton cloth, cellophane, cellulose nitrate (used as an explosive), cellulose acetate, and rayon (synthetic fibers used to make textiles). Over 50 million tons of cellulose products are used in the United States each year. The difference between starch and cellulose is that starch consists of α-glucose molecules bonded to form polymer molecules, and cellulose consists of β-glucose molecules bonded to form polymer molecules.

14-3   Lipids

When the cells of living organisms are crushed and mixed with solvents certain chemical components can be dissolved. Some cellular components are water soluble while others are sparingly soluble in water.

Other components are quite soluble in nonpolar organic solvents such as carbon tetrachloride. Those biochemical compounds that can be extracted from crushed cells by organic solvents are called **lipids.** Lipids represent a wide variety of compounds as shown in Table 14-2. Only the simple lipids and steroids will be considered here.

Simple lipids are esters of glycerol and long chain carboxylic acids. These esters can be represented as

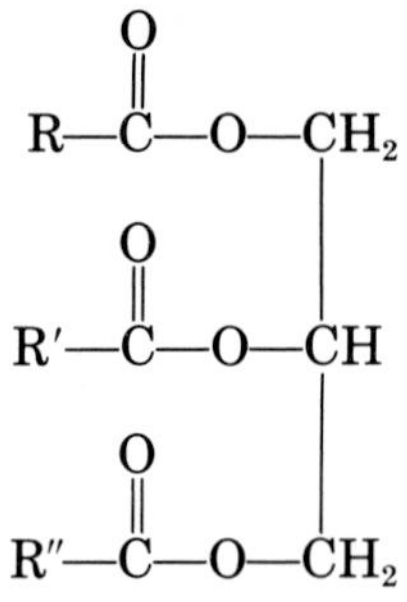

$$
\begin{array}{l}
\overset{\displaystyle O}{\overset{\|}{\text{R}-\text{C}}}-\text{O}-\text{CH}_2 \\[2em]
\overset{\displaystyle O}{\overset{\|}{\text{R}'-\text{C}}}-\text{O}-\text{CH} \\[2em]
\overset{\displaystyle O}{\overset{\|}{\text{R}''-\text{C}}}-\text{O}-\text{CH}_2
\end{array}
$$

Such esters of glycerol are called **fats** or **oils.** The difference between fats and oils is that fats are solid while oils are liquid at room temperature. Fats are found mainly in animals and oils in plants. Fats and oils are a food source. Certain fats are stored in our body in adipose tissue, which can serve as protective covering of certain parts of the body. The fat is also a reserve source of energy for the body. However, too much reserve fat is not advantageous.

Glycerol esters can be enzymatically broken down into glycerol and the constituent acids that are called **fatty acids.**

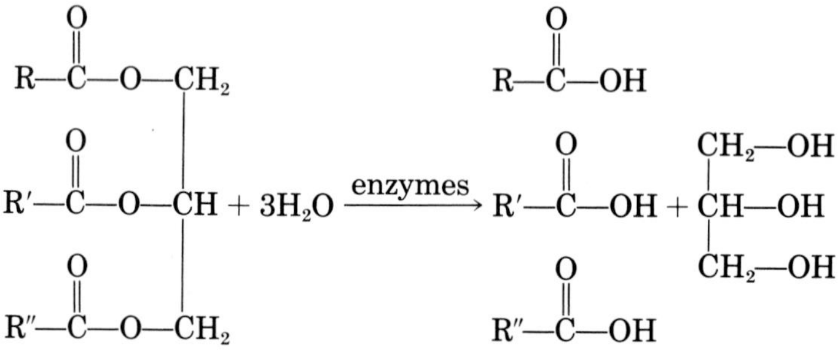

$$
\begin{array}{l}
\overset{\displaystyle O}{\overset{\|}{\text{R}-\text{C}}}-\text{O}-\text{CH}_2 \\[2em]
\overset{\displaystyle O}{\overset{\|}{\text{R}'-\text{C}}}-\text{O}-\text{CH} + 3\text{H}_2\text{O} \xrightarrow{\text{enzymes}}
\end{array}
\quad
\begin{array}{l}
\overset{\displaystyle O}{\overset{\|}{\text{R}-\text{C}}}-\text{OH} \\[2em]
\overset{\displaystyle O}{\overset{\|}{\text{R}'-\text{C}}}-\text{OH} + \\[2em]
\overset{\displaystyle O}{\overset{\|}{\text{R}''-\text{C}}}-\text{OH} \\
\text{fatty acids}
\end{array}
\quad
\begin{array}{l}
\text{CH}_2-\text{OH} \\[1em]
\text{CH}-\text{OH} \\[1em]
\text{CH}_2-\text{OH} \\
\text{glycerol}
\end{array}
$$

Table 14-3 lists some typical fatty acids that are found in simple lipids. Fats and oils include such common products as butter, lard, olive oil, coconut oil, peanut oil, corn oil, and safflower oil. Table 14-4 lists the composition of some common glycerol esters. Note that some of the vegetable oils contain high percentages of unsaturated (double bond containing) fatty acids. These oils are sometimes called **polyunsaturated oils.**

Steroids comprise another important class of lipids. Steroids have quite different chemical structures than simple lipids. Most steroids have the basic polycyclic ring structure shown in Figure 14-3. A wide variety of steroids are found in the body and most have very important

Table 14-2   Types of Lipids

Type	Example
Simple lipids	
Fatty acid esters of glycerol	Vegetable oils and animal fats
Waxes	Fruit and vegetable waxes
Steroids	Cholesterol, sex hormones
Phosopholipids	
Lecithins	
Cephalins	Lipids found in nerve tissues
Phosphatidylserines	
Sphingolipids	
Sphingomyelins	Lipids found in brain tissue
Cerebrosides	

physiological activities. Let us consider the structures and activities of some steroids.

Cholesterol is the most prevalent steroid in the body.

Table 14-3   Typical Fatty Acids

Name	Formula
Myristic acid	$CH_3(CH_2)_{12}\overset{O}{\underset{\parallel}{C}}{-}OH$
Palmitic acid	$CH_3(CH_2)_{14}\overset{O}{\underset{\parallel}{C}}{-}OH$
Stearic acid	$CH_3(CH_2)_{16}\overset{O}{\underset{\parallel}{C}}{-}OH$
Oleic acid	$CH_3(CH_2)_7CH{=}CH(CH_2)_7C\overset{O}{\diagdown}OH$
Linoleic acid	$CH_3(CH_2)_4CH{=}CHCH_2CH{=}CH(CH_2)_7C\overset{O}{\diagdown}OH$
Linolenic acid	$CH_3(CH_2)_4CH{=}CHCH_2CH{=}CHCH_2CH{=}CH(CH_2)_3C\overset{O}{\diagdown}OH$

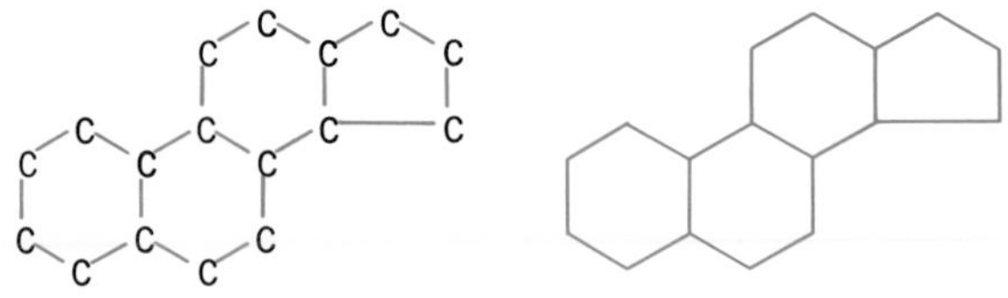

Figure 14-3   The common carbon skeletal structure of steroids.

It is found in all tissues and in the blood stream of humans and mainly in the brain, spinal cord, and nerve tissue. The physiological activity of cholesterol is unknown, but it may be an intermediate used by the body to form other steroids.

The male and female sex hormones are steroids. A variety of hormones are involved in the development and activities of human sex glands. It is known that these hormones chemically stimulate sexual processes but specifics of such functions are not known. The structure of the male hormones testosterone and androsterone produced in the testes are shown below.

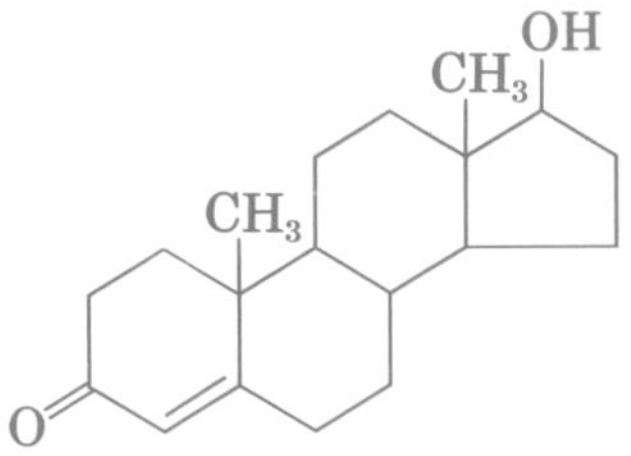

testosterone 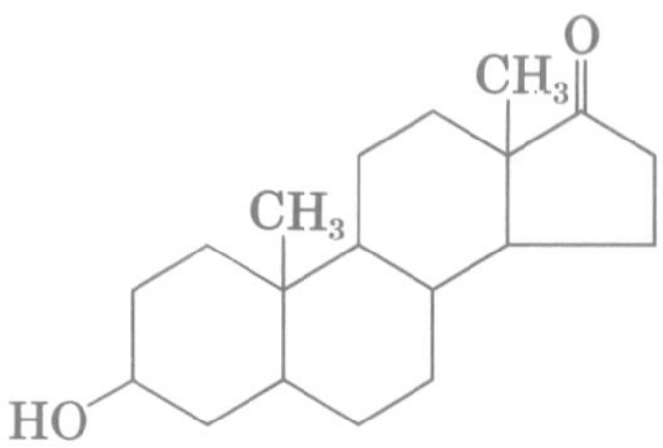   androsterone

Table 14-4   Composition of the Fatty Acids Obtained by Analysis of Common Fats and Oils

Fat or Oil	Average Composition of Fatty Acids, in Percent					
	Myristic Acid	Palmitic Acid	Stearic Acid	Oleic Acid	Linoleic Acid	Other
Animal Fats						
Butter	8–15	25–29	9–12	18–33	2–4	3–4 Butyric
Lard	1–2	25–30	12–18	48–60	6–12	1–3 Palmitoleic
Beef tallow	2–5	24–34	15–30	35–45	1–3	1–3 Palmitoleic
Vegetable Oils						
Olive	0–1	5–15	1–4	67–84	8–12	0–1 Palmitoleic
Peanut	—	7–12	2–6	30–60	20–38	0–1 Palmitoleic
Corn	1–2	7–11	3–4	25–35	50–60	0–2 Palmitoleic
Cottonseed	1–2	18–25	1–2	17–38	45–55	0–2 Palmitoleic
Soybean	1–2	6–10	2–4	20–30	50–58	4–8 Linolenic
Linseed	—	4–7	2–4	14–30	14–25	25–58 Linolenic
Safflower	—	1–5	1–5	14–21	73–78	
Marine Oils						
Whale	5–10	10–20	2–5	33–40		15–30 Other unsaturated
Fish	6–8	10–25	1–3			70 Other unsaturated

The female hormones are involved in control of the menstrual cycle and the maintenance of the fetus during pregnancy. The structures of two important female sex hormones are shown below.

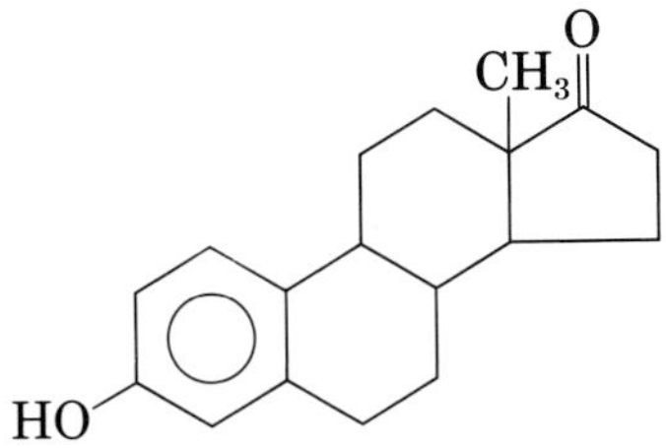

estrone 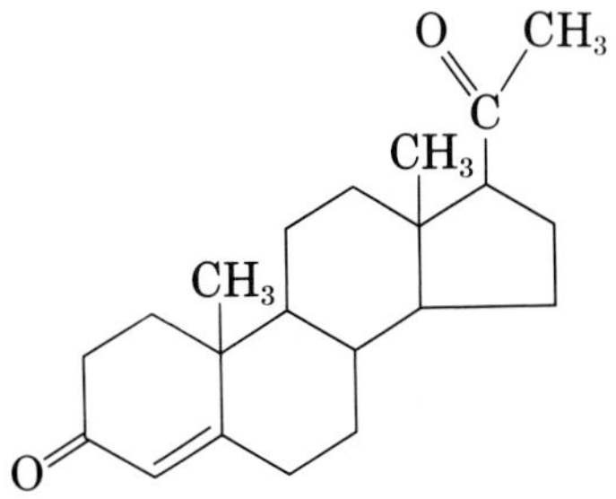   progesterone

14-4 Proteins

Proteins are polymeric materials found in living organisms. They serve as structural materials in the body and are fundamental to many of the processes of life. Proteins are polymers of amino acids and are produced by plants and animals. Proteins are synthesized from amino acids in the cells of our bodies. Proteins from animals and some plants are an important food since they provide amino acids that are essential to the body in the production of needed proteins. When proteins are digested, they are broken down by digestive enzymes into the constituent amino acids. The amino acids then become available to the cells for protein synthesis.

Table 13-5 lists some important amino acids. Recall that amino acids are both amines and carboxylic acids (See Section 13-13). It is possible for the carboxyl group of one amino acid to react with the amino group of another to form an amide linkage between the two amino acid units. For example,

$$
\underset{\text{glycine}}{NH_2CH_2\overset{\displaystyle O}{\overset{\|}{C}}\!-\!OH} + \underset{\underset{\text{alanine}}{\overset{\displaystyle |}{CH_3}}}{H\!-\!NHCH\overset{\displaystyle O}{\overset{\|}{C}}OH} \rightarrow \underset{\underset{\text{glycylalanine}}{\overset{\displaystyle |}{CH_3}}}{NH_2CH_2\overset{\displaystyle O}{\overset{\|}{C}}\!-\!NHCH\overset{\displaystyle O}{\overset{\|}{C}}OH} + H_2O
$$

Compounds involving two or more amino acids linked by amide linkages are called **peptides.** Glycylalanine shown above is a dipeptide. The amide linkages in peptides $-\overset{\displaystyle O}{\overset{\|}{C}}-\underset{|}{N}-$ are sometimes called **peptide bonds or peptide linkages.** Table 13-5 lists the official abbreviations that can be used to represent the amino acids. These abbreviations can be used to represent peptides formed from the amino acids. Thus, glycylalanine can be represented as:

Gly-Ala

Table 14-5   Functions of Protein

Structural proteins (insoluble in water)
  Collagens—found in connective tissue
  Elastins—found in tendons and arteries
  Myosins—found in muscle tissue
  Keratins—found in hair and nails
Globular proteins (can be dispersed in water solutions)
  Albumins—found in blood
  Globulins—involved in oxygen transport in the body (hemoglobin) and in
    body defense against disease (gamma globulin)
Conjugated proteins (complexes of proteins linked to other molecules)
  Nucleoproteins—protein–nucleic acid complexes
  Lipoproteins—protein–lipid complexes
  Phosphoproteins—protein–phosphorus compound complexes
  Chromoproteins—protein–pigment complexes (i.e., hemoglobin)
Enzymes (many enzymes are conjugated with coenzymes)
Hormones (not all hormones are proteins)
Antibodies involved in body defense against disease

From *Understanding Chemistry: From Atoms to Attitudes* by T. R. Dickson, Wiley, N.Y., 1974.

Table 14-6   Amino Acid Composition of Proteins[a]

| | Molar Mass of Protein | | |
Amino Acid	Human Insulin 6000	Horse Hemoglobin 68,000	Egg Albumin 45,000
Glycine	4	48	19
Alanine	1	54	35
Valine	4	50	28
Leucine	6	75	32
Isoleucine	2	0	25
Phenylalanine	3	30	21
Tryptophan	0	5	3
Proline	1	22	14
Serine	3	35	36
Threonine	3	24	16
Tyrosine	4	11	9
Hydroxyproline	0	0	0
Aspartic acid	3	51	32
Glutamic acid	7	38	52
Lysine	1	38	20
Arginine	1	14	15
Histidine	2	36	7
Cysteine	0	4	5
Cystine	3	0	1
Methionine	0	4	16

[a] Expressed as number of amino acid residues per molecule.
  Data from G. H. Haggis, D. Michie, A. R. Muir, K. B. Roberts, and P. B. M. Walter. *Introduction to Molecular Biology,* John Wiley and Sons, New York, 1964, pp. 40–41.

When using this notation, it is assumed that the amino acid on the left contributes the —OH to the formation of the peptide bond and that the amino acid on the right contributes the —NH₂ group. These abbreviations provide a more convenient way to show which amino acids comprise a peptide. Three amino acids linked by peptide bonds constitute a **tripeptide.** Using three amino acids, such as glycine, Gly, alanine, Ala, and valine, Val, it is possible to form six different tripeptides.

Gly-Ala-Val   Gly-Val-Ala

Val-Gly-Ala   Val-Ala-Gly

Ala-Val-Gly   Ala-Gly-Val

It is apparent that as the number of amino acid units increases, the number of possible combinations which form peptides increases greatly.

Proteins are peptides involving hundreds or thousands of amino acid units. Thus, proteins are called **polypeptides.** Proteins serve as structural units of cells, skin, muscles, bone interior, and nerves, as enzymes, hormones, and in many other important functions in the body. Table 14-5 lists some of the functions of proteins. In spite of the diversity of uses, all proteins are polypeptides. Analysis of proteins reveals that all proteins are made up of about 20 different amino acids. Table 14-6 lists the amino acid unit composition of three common proteins.

The sequences in which the amino acids are linked has been determined for some proteins. This requires a tremendous amount of laboratory work and analysis. The amino acid sequence involved in a protein is called the **primary structure** of the protein. The primary structure of

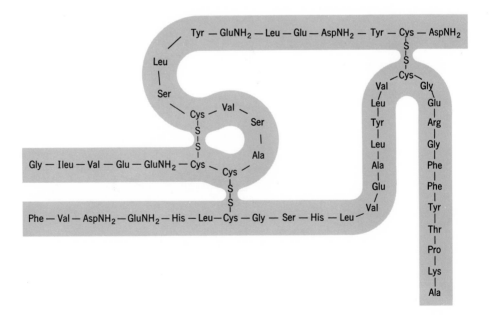

Figure 14-4   The primary structure (amino acid sequence) of beef insulin. This protein consists of 51 amino acids comprising two chains joined by disulfide linkages characteristic of the amino acid cystine. This figure shows the amino acid sequence but is not meant to convey the three-dimensional shape of the protein. Insulins from other animals have similar primary structures.

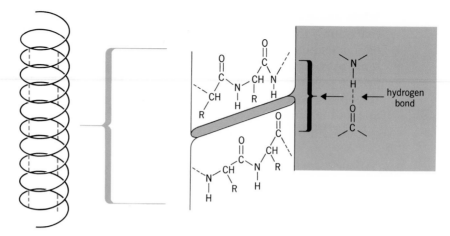

Figure 14-5   The secondary structure of a protein. Some proteins tend to coil in a form called an α-helix. This coiling produces a three-dimensional tubular aspect to the protein chains. As shown hydrogen bonds along the chain tend to hold the helix in place. Under certain drastic conditions (such as heating) the hydrogen bonds can break, and the chain takes on a random form. This is called denaturation of the protein.

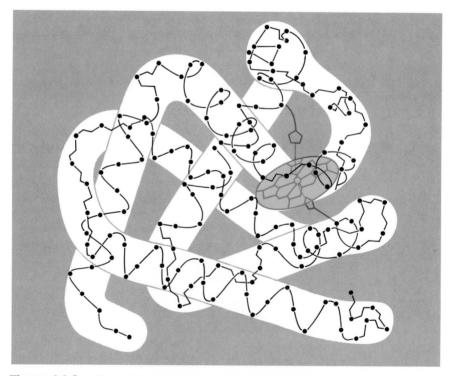

Figure 14-6   The tertiary structure of the myoglobin molecule. Myoglobin is a protein involved in the storage and transport of oxygen in muscle. The dots represent amino acid units and, the disc-shaped part is a heme unit. Approximately 70% of the amino acid sequence in the molecule is in the form of the α-helix secondary structure. Adapted from *The Structure and Function of Enzymes* by Sidney Bernhard, Benjamin, N.Y., 1968.

beef insulin is shown in Figure 14-4. It consists of 51 amino acid units composing two chains. The chains are bonded by two disulfide linkages (—S—S—) that are characteristic of the amino acid cystine (see Table 13-5 for the formula of cystine).

The primary structure of a protein gives no indication of the three-dimensional arrangement of the protein. Actually, the protein chain can take on a shape that is determined by hydrogen bonding and other forces of attraction between the amino acid groups along the chain. The configuration of the protein chain is called the **secondary structure** of the protein. A common secondary structure for proteins is the α-helix in which the protein chain is coiled in a three-dimensional helical shape. The α-helix is illustrated in Figure 14-5. Other secondary structures are known. These protein helices may actually exist in some folded or twisted form. The overall three-dimensional shape is called the **tertiary structure** of the protein. Once this structure is known, it may be possible to determine how the protein functions in the body. The tertiary structures for a few proteins have been determined; one for the protein myoglobin is illustrated in Figure 14-6. Myoglobin stores and transports oxygen in muscle tissue.

14-5   Enzymes

The numerous chemical reactions that occur in the body are referred to collectively as **metabolism.** Metabolic reactions include the reactions of food digestion and the reactions in which certain food molecules are utilized by the body for energy. Other metabolic reactions are involved in the breakdown of certain body cells and the formation of new cells. Most of the thousands of metabolic reactions require specific catalysts. These catalysts are synthesized by the body. Such biological catalysts are called **enzymes.**

Enzymes are large, polymeric molecules of varying complexity. All enzymes contain protein. Some enzymes are composed entirely of protein. Many enzymes consist of a protein portion called the **apoenzyme** and a nonprotein portion called the **coenzyme** or **cofactor.** These enzymes cannot function unless the apoenzyme and cofactor are joined. The detailed method by which enzymes catalyze chemical reactions is not known. The chemical reactions involved in life processes need specific enzymes that must be present at the correct time and place. It is thought that an enzyme functions by interacting with one of the reactants involved in a reaction. Such a reactant is called the **substrate.** A given enzyme will act as a catalyst for only one kind of reaction. An obvious question is how can an enzyme distinguish one kind of reactant from another?

A commonly accepted theory of enzyme behavior is the **lock-and-key theory** illustrated in Figure 14-7. According to this theory the enzyme has a definite three-dimensional structure that is arranged correctly so that the substrate molecule can just fit into the structure. In this way only a specific kind of substrate can fit into a given enzyme. Once the enzyme and substrate form an aggregate the substrate is exposed for the

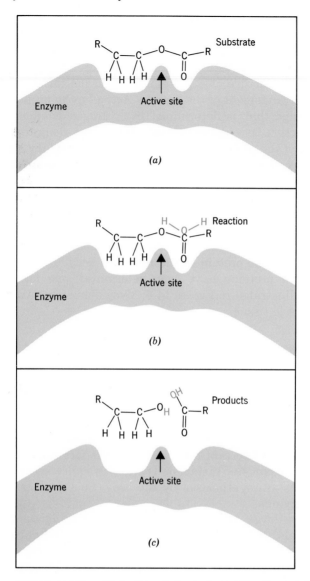

Figure 14-7   Illustration of the lock and key theory of enzymatic activity. (a) The three-dimensional shape of the enzyme in the vicinity of the active site is arranged to accommodate the substrate molecule. The active site is the portion of the enzyme that catalyzes the reaction of the substrate. (b) The substrate is held in position by the enzyme while the reaction of the substrate occurs. (c) The products or product of the substrate reaction leave the enzyme, and the enzyme is available for another substrate. From *Understanding Chemistry: From Atoms to Attitudes* by T. R. Dickson, Wiley, N.Y., 1974.

reaction. After the reaction occurs, the products move away from the enzyme leaving it available for another substrate.

Certain metabolic processes involve a long sequence of enzyme catalyzed reactions. Each enzyme must be present at the correct time for the process to occur. If any of the required enzymes is missing the process is disrupted. Since such metabolic processes are vital to life, the absence of an enzyme can lead to illness and death. Some hereditary diseases result from the inability of the body to produce certain enzymes. The absence of certain chemicals (vitamins and minerals) in the diet can lead to enzyme deficiencies. Several potent poisons, such as cyanide and mercury, inhibit the functioning of critical enzymes. A sufficient dose of cyanide ion can result in death in a matter of seconds by interfering with an enzyme involved in cell metabolism. The absence of this enzyme causes death due to lack of the ability to use oxygen.

14-6   Body Cells

Plants and animals consist of a large number of microscopic cells held together by structural compounds. Cells differ in size and shape depending on their function. However, cells have several subcellular components in common. A typical animal **cell** is illustrated in Figure 14-8. Important bodily processes occur in such cells. The cell is surrounded by

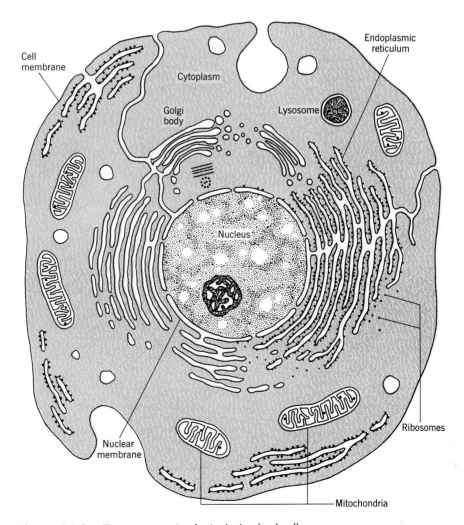

Figure 14-8   The components of a typical animal cell.

a cell **membrane** that allows passage of food molecules and chemical wastes in and out of the cell. The **cytoplasm** of the cell contains various specific cellular components. The cytoplasm consists of cellular fluids containing such substances as proteins and enzymes. The **nucleus** of the cell, surrounded by a nuclear membrane, contains the **chromosomes** that control the hereditary characteristics of the cell and are involved in cell division and protein synthesis. The **mitochondria** are membrane contained components of the cell that are involved in energy production of the cell. The **endoplasmic reticulum** are areas within the cell containing the **ribosomes** that are involved with protein synthesis. The **lysomes** of the cell appear to be involved with breakdown and dissolving of the cell if the cell membrane is ruptured.

14-7   Nucleic Acids: The Chemicals of Heredity

It has been found that a polymeric substance called **deoxyribonucleic acid (DNA)** is the fundamental constituent of genes found in the chromosomes of the cell nucleus. The **genes** are hereditary units of the cell, so an understanding of the structure of DNA is needed to interpret hereditary processes. Actually, DNA belongs to a class of polymeric sub-

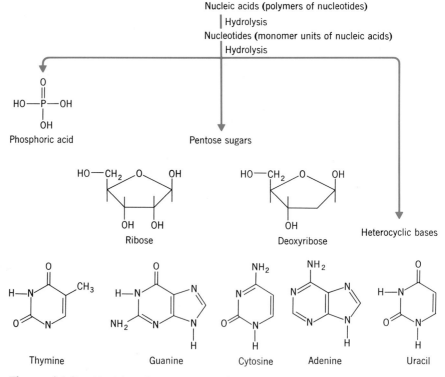

Figure 14-9   Nucleic acids form nucleotides upon hydrolysis. Upon further hydrolysis the nucleotides are broken down into phosphoric acid, a sugar (ribose or deoxyribose), and heterocyclic bases.

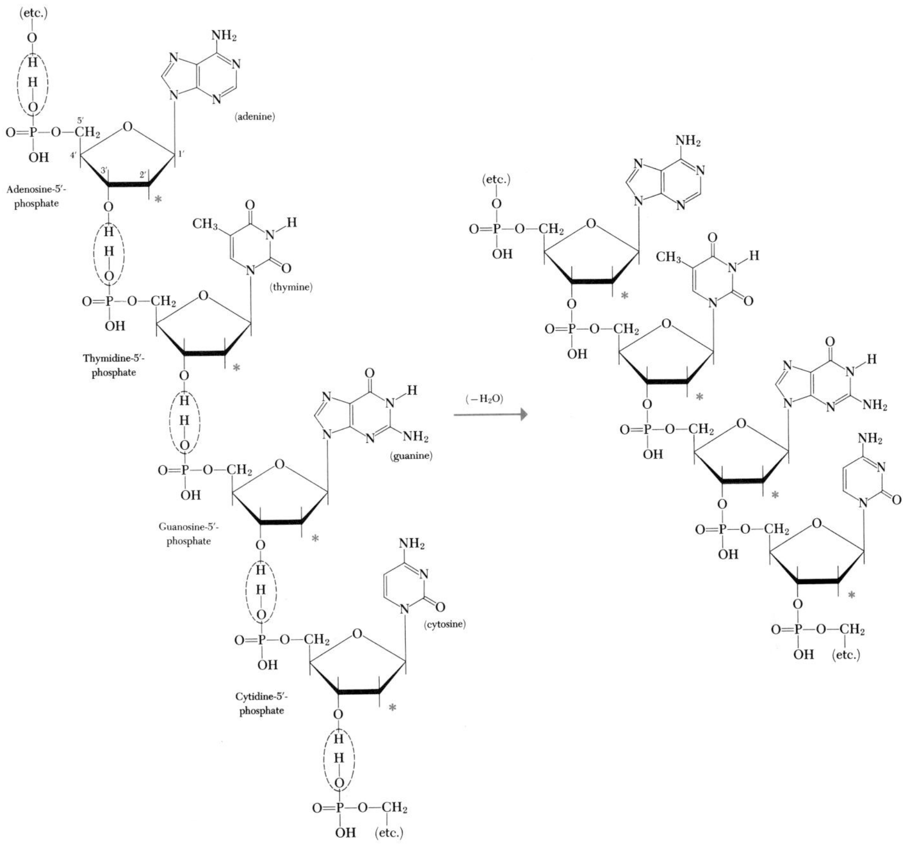

Figure 14-10   Manner of formation of a nucleic acid chain. Shown here is a segment of a DNA chain. If the sites marked by asterisks were given —OHs, the example would be for RNA (assuming uracil replaced thymine). The sequence of the heterocyclic amines is purely arbitrary in this drawing, but one each of the four amines common to DNA has been included. A molecular weight of 2.8×10^9 for the DNA of one species (*E. coli*) has been reported. Using a value of 325 as the average formula weight of each nucleotide, this DNA would be made of 8,600,000 nucleotide units. It is probable that such a DNA molecule would make up a collection of genes rather than just one. Genetic studies indicate that the average gene size is 1500 nucleotide pairs (of a double helix). From *Elements of General and Biological Chemistry*, 3rd ed., by John Holum, Wiley, New York, 1972.

stances called **nucleic acids.** Nucleic acids can be broken down into monomer units called **nucleotides.** The nucleotides can be further broken down into phosphoric acid, some heterocyclic amines or bases and either deoxyribose or ribose sugar. This is illustrated in Figure 14-9. Nucleic acids that contain deoxyribose are called **deoxyribonucleic acids (DNA)** and those that contain ribose are called **ribonucleic acids (RNA).**

The nucleotide monomer units of nucleic acids are composed of a phosphoric acid unit, a ribose or deoxyribose unit, and a unit of adenine, guanine, cytosine, and thymine in DNA or uracil instead of thymine in RNA. The formation of a typical nucleotide is illustrated below.

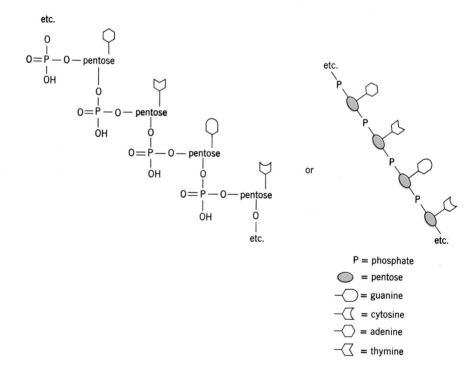

These nucleotide units are linked together to form the nucleic acid polymer as shown in Figure 14-10. Thus, the nucleic acids consist of a sequence of alternating phosphate units and sugar units. Protruding from this sequence are the heterocyclic amines. A shorthand representation of nucleic acids is given in Figure 14-11. The secondary structure of DNA is important to the visualization of how it functions in the cell. One of the most significant theories of contemporary science involves the structure of DNA as proposed by F. H. C. Crick and J. D. Watson in 1953. They proposed the **double-helix structure** of DNA in

P = phosphate

= pentose

= guanine

= cytosine

= adenine

= thymine

Figure 14-11   A shorthand representation of a nucleic acid chain that consists of an alternating sequence of phosphate units and sugar units from which the heterocyclic amines protrude.

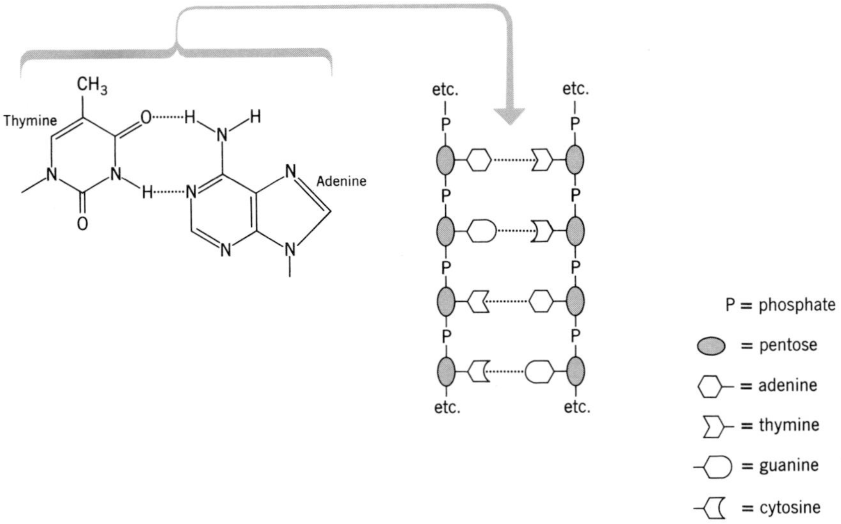

Figure 14-12   The DNA strands in the double helix form of DNA are held together by hydrogen bonding between the heterocyclic amines.

which two DNA strands intertwine in the form of two helices. The intertwining is accommodated by the hydrogen bonding between thymine and adenine units and guanine and cytosine units on opposite strands. This hydrogen bonding is illustrated in Figure 14-12, and a model of the double helix structure of DNA is shown in Figure 14-13. DNA in the genes serves as a starting point for the biological processes that are carried out in the cell. Apparently, genetic information is contained in the genes by specific sequences of heterocyclic bases on the DNA strands. In other words, the genetic information of a cell is contained as various sequences of these heterocyclic bases that are present along the DNA strands which comprise the genes of the chromosomes. These sequences constitute the **genetic code** of the genes. The term genetic code refers to hereditary information that can be transmitted to new cells and to new generations.

14-8   DNA and the Genetic Code

All cells in our bodies carry the genetic code. The vast majority of cells that structure our bodies are called **somatic** (Greek: body) cells. Certain cells located in the gonads are capable of being formed into sperm or eggs. These are called **germ** (Latin: sprout, bud) cells. Human somatic cells contain 46 chromosomes, but the germ cells contain only 23. At sexual maturity the germ cells become active enabling the female to ovulate and the male to produce sperm. Human sperm consists of a protein sheath surrounding the DNA-structured male chromosomes. The female egg carries the female chromosomes. At conception the egg is united with the sperm and the chromosomes pair up to form a new cell called a **zygote**—a fertilized egg. The zygote multiplies by cell replica-

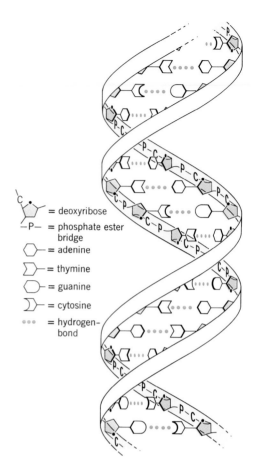

= deoxyribose

—P— = phosphate ester bridge

= adenine

= thymine

= guanine

= cytosine

••• = hydrogen-bond

Figure 14-13   Schematic representation of a DNA double helix. From *Elements of General and Biological Chemistry*, 3rd ed., by John Holum, Wiley, New York, 1972.

tion as the development of a new individual takes place. This new being carries within its cells the genetic codes donated by the mother and the father. As development proceeds, the number of cells increases rapidly, and some take on different forms and functions that results in the various specialized parts of the body. That is, some cells structure the muscles, some structure the body organs, and so on. The new individual takes on inherited characteristics of the parents. Sometimes one characteristic will dominate over the other. For instance, the brown-eye trait of the mother may dominate over the blue-eye trait of the father and the offspring will have brown eyes. All of the cell specialization and inherited characteristics are related to the genetic code contained in the cellular genes. As a result of various factors including selective inherited characteristics, conditions of growth, and the environment, the new individual does not turn out to be an exact physical replica of the mother or the father. Nevertheless, it is through the genetic code that humans propagate a physical pattern of themselves to new generations.

When cells divide (undergo mitosis) to form new cells, it is necessary for the genes to be replicated before the cell splits, so that the new cells will contain the genetic code. As shown in Figure 14-14, the double-helix theory provides an explanation of how the **replication of DNA** can take place. Furthermore, the DNA provides a pattern on which the

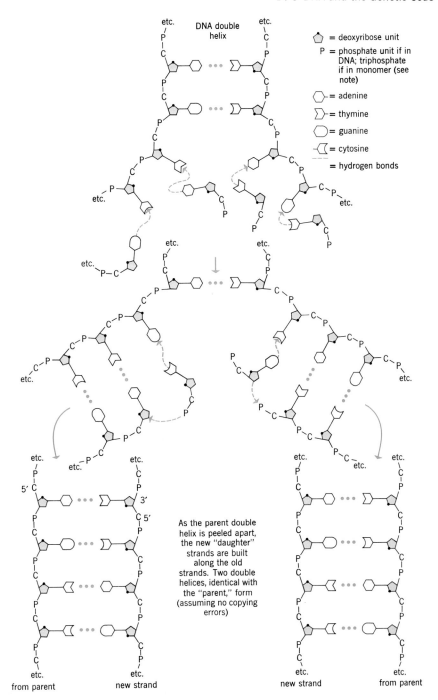

Figure 14-14   A possible mechanism for the replication of DNA. From *Elements of General and Biological Chemistry*, 3rd ed., by John Holum, Wiley, New York, 1972.

synthesis of RNA can take place. The RNA is involved in the synthesis of proteins and enzymes. **Enzyme synthesis** is a very important cellular process. The enzymes of the body are being continuously replaced, and new cells are being produced. The amino acids for these enzymes come from the breakdown of old protein and from protein-containing foods. A variety of amino acids called the amino acid "pool" is normally available in a cell. Specific enzymes are synthesized in a cell according to the genetic code of the genes. These enzymes catalyze specific reactions taking place in the cells. The genetic code carries information which produces a variety of specific enzymes resulting in the entire organism taking on various inherited characteristics.

14-9   Enzyme Synthesis

Enzyme and protein synthesis is a complicated process but the basic process is summarized in Figure 14-15. As shown in the figure, **RNA synthesis** occurs in the nucleus using the DNA as a pattern. There are three

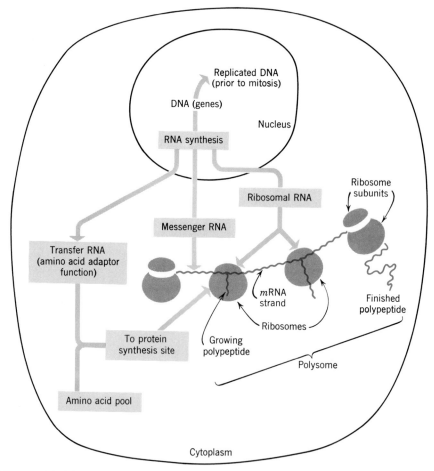

Figure 14-15   The relations of DNA to various RNAs and protein synthesis. From *Elements of General and Biological Chemistry*, 3rd ed., by John Holum, Wiley, New York, 1972.

major types of RNA produced. The **ribosomal RNA** migrates from the nucleus and becomes incorporated in the **ribosomes.** Ribosomal RNA apparently prepares the ribosomes for protein synthesis. The **messenger RNA (m-RNA)** contains the pattern (as a sequence of bases complementary to those on DNA) upon which protein synthesis is based. The m-RNA migrates from the nucleus and becomes attached to one or more ribosomes. The point of attachment of the ribosomes and m-RNA is the region at which protein synthesis occurs. A variety of **transfer RNA's (t-RNA)** migrate from the nucleus and become attached to specific amino acids from the amino-acid pool. There are at least 20 kinds of t-RNA; one for each of the amino acids. The t-RNAs with the amino acids attached migrate to the ribosomes on the m-RNA. The point of attachment of the m-RNA and the ribosome involves a certain part of the genetic code. This part of the genetic code is catalytically activated by the ribosomes. The active genetic code calls for a specific amino acid. Thus, the t-RNA with this amino acid fits into the activated portion of the m-RNA, and the amino acid it carries breaks off to become part of a growing chain. The ribosome then moves along the m-RNA to activate a portion of the genetic code adjacent to the previously used portion. The t-RNA amino-acid pair corresponding to this new code moves into the ribosome, and the amino acid becomes linked to the previous amino acid on the protein chain. The movement of the ribosome continues with a new amino acid being added to the growing protein chain each time. Finally after hundreds or thousands of amino acids have been added, the ribosome reaches the end of the m-RNA and the newly formed protein breaks off. In this manner the genetic code carried by the m-RNA from the DNA of the genes directs the synthesis of a specific protein that then functions in a certain manner. This process is very specific for the synthesis of many proteins and enzymes. In fact human cells contain enough DNA to contain codes for about 7 million different proteins. It is thought that only about 10% of the codes are active, but this represents some 700,000 proteins to regulate human growth, development, and metabolism.

Certain diseases such as galactosemia (failure to metabolize galactose) and sickle-cell anemia (the production of abnormal red blood cells) are attributed to **genetic defects** in which the genetic code is not normal. The abnormal code can cause a **genetic disease** to be passed on to the next generation depending on whether or not the offspring develops the characteristic related to the defect. Even if the characteristic does not dominate in the offspring, it is still present in the cells and can be passed on to the next generation.

Questions and Problems

1. What is biochemistry?
2. Which four elements comprise 99% of all atoms of living organisms?

3. What is a carbohydrate? Give an example.

4. Give an example of a disaccharide and indicate which monosaccharides make it up.

5. Describe the structures of starch and cellulose in terms of α- and β-glucose units.

6. What are simple lipids? What are fatty acids?

7. What are fats and oils?

8. What does the term polyunsaturated oil mean?

9. What are steroids? Give an example.

10. What are peptides? Using the abbreviations for the amino acids, give the six possible tripeptides that can be formed from the amino acids glycine (Gly), serine (Ser), and leucine (Leu).

11. Describe the structure of a protein as a polypeptide. List some of the functions of proteins in the body.

12. What are the differences between the primary, secondary and tertiary structures of proteins.

13. What are enzymes?

14. Describe the lock-and-key theory of enzyme function.

15. Describe the primary structures of RNA and DNA in terms of the chemical building units that comprise these nucleic acids.

16. Where is DNA located in cells?

17. Give a description of the double-helix theory of the structure of DNA.

18. How is the double-helix structure of DNA involved in the replication of DNA prior to cell division?

19. What is meant by the term genetic code.

20. Describe the DNA-controlled synthesis of proteins including the role of ribosomes, m-RNA, and t-RNA.

21. What is a genetic defect or a genetic disease?

15

Chapter 15: Nuclear Energy

Objectives

The student should be able to:
1 Describe the structure of a nucleus in terms of nucleons.
2 State and interpret the symbolic representation of a nuclide.
3 Describe alpha particle decay.
4 Describe beta particle decay.
5 Describe gamma ray emission.
6 State the three major types of radiation.
7 Describe the ionizing effect of radiation.
8 Describe why ionizing radiation is dangerous.
9 Describe a nuclear transmutation.
10 Describe the fission process.
11 Describe a nuclear reactor.
12 Describe the fusion process.

Terms to Know

Proton
Neutron
Atomic number
Isotopes
Nuclides
Mass number
Radioactivity
α radiation
β radiation
γ radiation
Half-life
Curie

Ion pair
Geiger-Müller tube
Radiation sickness
Fission
Uranium-235
Critical mass
Nuclear reactor
Breeder reactor
Fusion
Fission products
Plutonium-239

15

Nuclear Energy

15-1   Nuclear Chemistry and Nuclear Energy

This chapter serves as an introduction to the field of nuclear science. The structure and properties of atomic nuclei will be discussed. The use of nuclear reactors as energy sources will be considered.

The structure and properties of the nuclei of atoms is the concern of nuclear chemists and nuclear physicists. The utilization of atomic nuclei in constructive and potentially destructive manners is important to all people. The same nuclear energy that is involved in nuclear bombs may become, through the use of nuclear reactors, the main source of useful energy available to humans. Furthermore, the same nuclear radiation that can be destructive to human life is utilized as an effective medical tool.

15-2   Atomic Nuclei

Theories of nuclear structure are not as well established as the theory of atomic structure. Nuclear structure is a very active area of research by nuclear scientists. For our purposes we can consider that atomic nuclei are aggregates of the nuclear particles, **protons** and **neutrons.** The proton and the neutron are called **nucleons.** Other particles, such as the **electron,** the **alpha particle,** the **neutrino,** and the **gamma photon,** are associated with the properties of nuclei. The symbols and properties of all of these nuclear particles are summarized in Table 15-1.

Atomic nuclei are, of course, the very massive centers of atoms about which the electrons are in motion. When discussing the properties of the nuclei we will consider them without consideration of the electrons. However, keep in mind that, with the exception of few important cases, nuclei are always parts of atoms and not separate particles.

Recall from Chapter 2 that the nuclei of atoms of a given element always have the same number of protons and that this number of protons is the **atomic number** of the element. For most elements there is

Table 15-1   Nuclear Particles

Name	Mass (amu)	Charge	Symbol
Proton	1.007825	+1	^1_1H, p
Neutron	1.008665	0	^1_0n, n
Electron	0.000549	−1	$^0_{-1}\text{e}$, e
Alpha particle	4.00260	+2	^4_2He, α
Gamma particle (photon)	0	0	γ, hν
Neutrino	0	0	ν

more than one possible combination of neutrons and protons that can make up the nuclei of the atoms. However, only certain specific combinations are observed. Atoms of a given element must contain the same number of protons, but when different numbers of neutrons are present the species are called **isotopes** of that element. The nuclei of the various isotopes of the elements are called **nuclides.** Numerous nuclides occur in nature, and others can be produced synthetically. (See Section 15-7.)

Special symbolism is used to represent a nuclide. The number of protons in a nuclide is the atomic number. The **mass number** of a nuclide is defined as the sum of the number of protons and neutrons. A nuclide is represented as

$$^M_Z X$$

where X is the symbol of the element corresponding to the nuclide, M is the mass number, and Z is the atomic number. The number of neutrons in a nuclide can be determined by subtracting the atomic number from the mass number: $M - Z$. A few examples of nuclides are

$$^1_1\text{H} \qquad ^{16}_8\text{O} \qquad ^{12}_6\text{C} \qquad ^{238}_{92}\text{U}$$

These symbols are often read as the element name followed by the mass number (i.e., ^1_1H, hydrogen-one; $^{12}_6\text{C}$, carbon-twelve; $^{16}_8\text{O}$, oxygen-sixteen; $^{238}_{92}\text{U}$, uranium-two-thirty-eight). A tabulation of all of the observed nuclides has been compiled and is called the nuclide table. Table 15-2 lists the nuclides for a few elements.

15-3   Radioactivity

Most naturally occurring nuclides are stable and retain their structure indefinitely. Some nuclides, however, are not stable and are said to be

Table 15-2   Nuclides of Some Elements (naturally occurring)

Hydrogen	^1_1H	^2_1H	^3_1H
Helium	^3_2He	^4_2He	
Carbon	$^{12}_6\text{C}$	$^{13}_6\text{C}$	$^{14}_6\text{C}$
Oxygen	$^{16}_8\text{O}$	$^{17}_8\text{O}$	$^{18}_8\text{O}$
Uranium	$^{234}_{92}\text{U}$	$^{235}_{92}\text{U}$	$^{238}_{92}\text{U}$

radioactive. **Radioactivity** can be defined as spontaneous decay of a nucleus to form another nucleus and a nuclear particle. Often in this decay a more stable nucleus is formed from a less stable nucleus. The nuclear particles produced during radioactive decay are ejected from the original nucleus with large amounts of kinetic energy (high speeds). These energetic particles make radioactive substances dangerous (See Section 15-6) but sometimes, in certain controlled situations, useful.

Radioactive decay occurs only in certain ways. These are called **modes of radioactive decay.** The three most common decay modes are described below.

Alpha Particle Decay. A nucleus, called the parent, decays by emitting a high-speed helium-four nucleus called an **alpha particle** (α) and a new nucleus, called the daughter, is formed with a mass number that is four less than the original nucleus and an atomic number that is two less. The decay can be represented by the general nuclear equation

$$_Z^M X \quad \rightarrow \quad _{Z-2}^{M-4} Y \quad + \quad _2^4 He$$

$$\text{parent} \qquad \text{daughter} \qquad \text{alpha}$$
$$\text{nucleus} \qquad \text{nucleus} \qquad \text{particle}$$

X is the parent element, and Y is the newly formed element. Some examples of alpha decay are

$$_{92}^{238} U \rightarrow _{90}^{234} Th + _2^4 He$$

Uranium-238 decays to form Thorium-234.

$$_{88}^{226} Ra \rightarrow _{86}^{222} Rn + _2^4 He$$

Radium-226 decays to form Radon-222.

The alpha particles emitted from a radioactive substance are called **alpha radiation.**

One characteristic of a nuclear equation is that the sums of the mass numbers on each side of the arrow are equal and the sums of the atomic numbers are equal.

Beta Particle Decay. A nucleus decays by emitting a high-speed electron called a **beta particle** (β), and a new nucleus is formed with the same mass number as the original nucleus and an atomic number one greater. Also, a particle called an antineutrino is emitted to conserve energy.

$$_Z^M X \rightarrow _{Z+1}^M Y + _{-1}^0 e + \nu \text{ (antineutrino)}$$

Note that the mass number M does not change, but the atomic number does. Some examples of beta decay are

$$_{93}^{238} Np \rightarrow _{94}^{238} Pu + _{-1}^0 e + \nu$$

Neptunium-238 decays to form Plutonium-238

$$_{94}^{241} Pu \rightarrow _{95}^{241} Am + _{-1}^0 e + \nu$$

Plutonium-241 decays to form Amercium-241.

The beta particles emitted from a radioactive substance are called **beta radiation.**

Gamma Ray (Photon) Emission. Sometimes the daughter nucleus formed in an alpha decay or a beta decay will be in an energetically excited state. When this nucleus drops to a lower energy state, it emits a photon of electromagnetic energy called a **gamma ray, γ.** The emission of **gamma radiation** by a radioactive substance often accompanies the alpha radiation or beta radiation produced by the decay of the nuclides in the substance.

In summary, there are three major types of radiations: alpha radiation, beta radiation, and gamma radiation. Nuclides that are **alpha emitters** produce alpha radiation (α) accompanied, in some cases, by gamma radiation (γ). Nuclides that are **beta emitters** produce beta radiation (β) accompanied, in some cases, by gamma radiation (γ).

Radioactive decay is spontaneous and cannot be prevented. A sample of a radioactive substance will continue to decay by emission of radia-

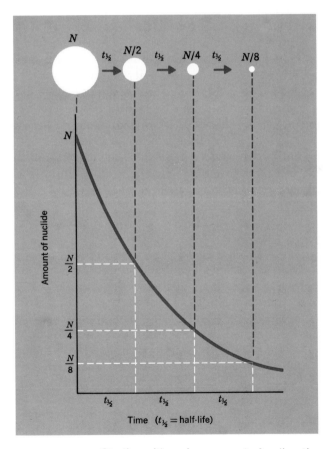

Figure 15-1   Starting with a given amount of radioactive nuclide (N), one-half the amount will remain after one half-life has passed. After a time equal to a second half-life has passed, one-fourth of the original amount remains. After the passage of another half-life, one-eighth of the original amount remains. This continues until all of the nuclides have decayed.

Table 15-3   Some Half-Lives

Nuclide		Half-Life
$^{238}_{92}$U	uranium-238	4.5×10^9 years
$^{14}_{6}$C	carbon-14	5680 years
$^{226}_{88}$Ra	radium-226	1620 years
$^{22}_{11}$Na	sodium-22	2.60 years
$^{42}_{19}$K	potassium-42	12.4 hr
$^{183}_{78}$Pt	platinum-183	6 min
$^{17}_{9}$F	fluorine-17	66 sec
$^{223}_{90}$Th	thorium-223	0.9 sec

tion until all of the sample has decayed. Not all nuclides decay at the same rate. One of the characteristics of a given radioactive nuclide is the rate at which it decays. A common way to express the rate of decay of a radioactive element is in terms of its half-life. **Half-life** is defined as the time required for the decay of one-half of a sample of a radioactive substance. Half-life is illustrated in Figure 15-1. The half-lives of nuclides vary widely, ranging from fractions of seconds to billions of years. A few typical half-lives are given in Table 15-3.

When working with a radioactive sample, it is important to know how radioactive it is (how "hot" the sample is). The curie is one unit that is used to express intensity of radioactivity. A **curie** is a unit of activity equal to 37 billion disintegrations per second. The actual amount in grams of a radioactive material that will have an activity of one curie can be determined by knowing the rate of decay of a radioactive element. For example, 1 g of radium-226 is equivalent to 1 curie. A curie is a large amount of radioactivity. Normally, for safety, **millicurie** (10^{-3} curie) and **microcurie** (10^{-6} curie) amounts of radioactive substances are used in laboratory work.

15-4   Radioactive Elements

About 330 different nuclides are found in nature. Some elements have only one naturally occurring nuclide (i.e., fluorine, $^{19}_{9}$F) while other elements have numerous natural nuclides (i.e., tin $^{112}_{50}$Sn, $^{114}_{50}$Sn, $^{115}_{50}$Sn, $^{116}_{50}$Sn, $^{117}_{50}$Sn, $^{118}_{50}$Sn, $^{119}_{50}$Sn, $^{120}_{50}$Sn, $^{122}_{50}$Sn, $^{124}_{50}$Sn). With some elements, all of their nuclides are radioactive. All isotopes of the 22 elements of atomic weight greater than bismuth (atomic number 83) are radioactive. A few nuclides of the other elements are naturally radioactive (i.e., $^{204}_{82}$Pb and $^{40}_{19}$K).

Radioactive nuclides are unstable and undergo decay to form stable nuclides or other radioactive nuclides. Studies of the naturally occurring elements have revealed that radioactive nuclides of atomic number greater than 82 belong to one or another of several **decay chains** or **series.** In other words, these elements decay by alpha or beta emission according to definite patterns. The radioactive decay series of uranium-238 is given in Figure 15-2. In this series, uranium-238 decays into thorium-234, which decays into another nuclide, and so on, until stable lead-206 is formed. Other nuclides belong to other series.

Knowledge of radioactive decay series provides a way to estimate the age of the earth. Assuming that uranium-238 was formed when the earth was formed, then a sample of undisturbed rock containing uranium would contain a certain amount of lead-206; the end product of the uranium-238 decay chain. The amount of lead-206 would depend on how long the uranium had been decaying. Since the half-life of uranium-238 is 4.5×10^9 years, a measurement of the amounts of uranium-238 and lead-206 indicates the age of the rock. In one half-life, 1 g of $^{238}_{92}U$ would decay to about 0.4 g of $^{206}_{82}Pb$, and 0.5 g of $^{238}_{92}U$ would remain. Using this method of rock dating and other similar methods, the age of the earth is estimated to be around 4.5 billion years.

15-5 Radiation Detection

Radioactivity is characterized by emission of radiation during the decay process. This radiation is used to detect and identify radioactive sub-

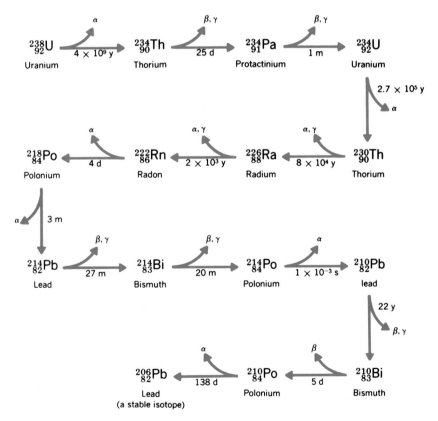

Figure 15-2 Uranium-238 radioactive disintegration series. The number beneath each arrow signifies the half-life of the preceding isotope, y = years, d = days, m = minutes, s = seconds. The small arrow that curves away from each main arrow indicates the kind (or kinds) of radiation emitted by the preceding isotope. From *Elements of General and Biological Chemistry,* 3rd ed., by John Holum, Wiley, New York, 1972.

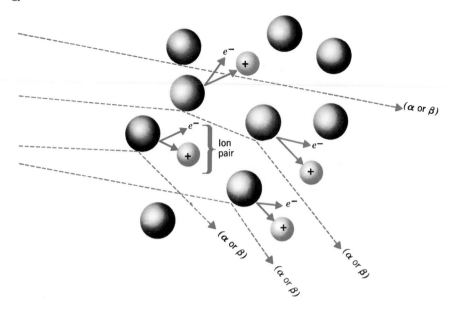

Figure 15-3   When ionizing radiation passes through matter, some of the atoms and molecules are ionized producing an ion pair involving an electron and a positive ion.

stances, and it can be very dangerous to human beings. In this section we shall consider how radiation can be detected.

The particles given off during alpha or beta decay are very high-speed, high-energy particles. These particles are ejected from the decaying nucleus and travel outward into the material surrounding the radioactive substance. The alpha and beta particles lose their energy by interacting with matter. This interaction results in an alpha or beta particle causing an electron to be lost by an atom or molecule as illustrated in Figure 15-3. In other words, these particles lose their energies by causing atoms and molecules to ionize. The electron and positive ion produced by this interaction are called an **ion pair.** As an alpha or beta particle passes through matter, it will cause the formation of many ion pairs. The more massive alpha particles ($_2^4$He) are less penetrating than beta particles ($_{-1}^0$e), but they produce more ion pairs. The typical alpha particle could travel about 6 cm in the air and produce about 40,000 ion pairs while a typical beta particle would travel 1000 cm in the air and produce about 2000 ion pairs. Gamma radiation consists of gamma photons (γ) that are photons of electromagnetic radiation similar to X rays. Gamma particles also interact with matter to form ions, and they are very penetrating. Because radiation causes the formation of ion pairs in the matter through which it passes, it is called **ionizing radiation.**

The detection of radiation is based on its ability to form ions. A typical device used to detect radiation is the **Geiger-Müller tube,** which is illustrated in Figure 15-4. The tube consists of a metal cylinder with a thin plastic window on the end. The tube is filled with a special gas and has a thin metal rod in the center. Using an external source of power, a voltage difference is maintained between the rod and the cylinder. When a particle of ionizing radiation enters the tube, some ion pairs are formed. The electrons formed in the ionization are strongly attracted to the center rod. They move toward the rod with great speed, and in the process they act essentially like beta particles in that they produce more ion pairs. This produces a large number of electrons that quickly flow toward the center rod. This flow of electrons is equivalent to a small current passing through the tube. This flow of current is carried to an electronic device, called a counter, that is connected to the tube. The counter recognizes the current flow, which it registers as an ionizing event caused by a particle of radiation. Each time a particle enters the tube and causes the electron "avalanche," the counter will tally the event. In this way the Geiger-Müller tube and the counter serve as a means of measuring the activity of a radioactive substance. The counter is used to determine the number of disintegrations per second or the number of disintegrations per minute that are associated with a radioactive sample. This is one way in which the half-life of a radioactive nuclide can be determined. The **activity** of a sample expressed in disintegrations per second is directly related to the amount of the nuclide. So, by measuring the activity of a sample over a period of time and plotting the activity versus the time, it is possible to determine the half-life of the nuclide. Generally, a counter is a convenient way to determine the activity of any radioactive substance.

15-6   Radiation Danger

The ionizing effect of radiation is the factor that makes radiation dangerous to life. Exposure to too much high-intensity radiation can cause **radiation sickness** and death. Alpha particles cannot penetrate skin and are not dangerous externally. However, if some alpha emitter is eaten or inhaled, it can cause severe damage inside the body. Gamma radiation, X rays, and neutrons can penetrate the body and, thus, are the most harmful type of radiation. It is important to guard against undue exposure to them. Beta particles can penetrate the skin some-

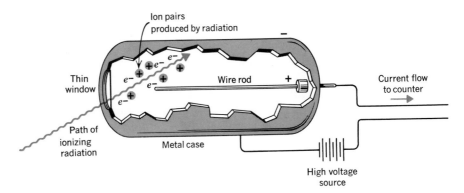

Figure 15-4   Operating principle of a Geiger-Müller tube.

what and may cause skin burns. Introduction of any source of radiation into the body can be dangerous. Ionizing radiation can cause the ionization of various compounds within the body, which will then not function correctly. It is thought that ionizing radiation interacts with water in the cells of the body to form unstable atomic aggregates called **free radicals.** These free radicals react with vital components of the cell such as the enzymes and the chromosomes thereby preventing their normal function. This leads to the malfunction or destruction of the cell. Overexposure to radiation may cause nausea, vomiting, and weakness followed by a period (days or weeks) of feeling well. This is followed by a period characterized by weakness, weight loss, fever, diarrhea, internal bleeding, and loss of hair. Finally, the victim either slowly recovers or dies.

Another danger of ionizing radiation is that it can cause **genetic damage** to body cells. Sometimes this damage leads to cancer. In the early 1900s many scientists working with newly discovered radioactive materials died of cancer. When genetic damage occurs in germ cells, cells that produce eggs or sperm, **mutations** of offspring can result. Some mutations occurred in the children of Japanese citizens exposed to radiation from nuclear bombs dropped on Japan by the United States in World War II.

Humans are constantly being exposed to ionizing radiation from naturally occurring radioactive isotopes (**background radiation**) and radioactive material that has entered the environment from nuclear power plants and fallout from nuclear weapon testing. In addition, occasional doses of radiation are obtained from medical X rays and the use of radioactive materials in medicine. Much of this exposure is unavoidable, but, since radiation can be deadly or cause cancer, it is important to keep close watch on the amounts of radioactive materials that enter the environment from human sources.

The amount of radiation absorbed by humans upon exposure to radiation can be expressed in terms of an absorbed dose unit called the **rem (roentgen equivalent man).** The technical definition of the rem is given in Table 15-4, but for our purposes we can consider the rem as a means of expressing radiation dosages to which we are exposed. Table 15-4 shows estimates of average yearly doses of radiation received by Americans (1970) according to the National Academy of Sciences. As the table shows, background radiation amounts to 102 millirems each year and the total average yearly dose is 182 millirems. The National Academy of Sciences Committee on the Biological Effects of Ionization Radiation recommends that the general population should not receive more than 170 millirems of human-made radiation each year exclusive of background and medical-radiation sources. The most likely source of additional radiation will be uranium mines, nuclear-fuel processing plants, and nuclear power plants. Radioactive material introduced from these sources will have to be closely watched (see Section 15-12). The Committee further stated that the 170 millirem value is chosen as a balance between societal needs and genetic risk, and that such a dose could lead to 6000 extra deaths by cancer each year. They further state

Table 15-4   Estimated Average Dose Rate in the United States, 1970

Roentgen equivalent man (rem) unit of radioactive dose in humans. One rem is the quantity of any radiation that, when absorbed by the human body, causes an effect equivalent to the absorption of 1 roentgen. A roentgen is a unit of radioactive exposure. One roentgen (of gamma radiation or X rays) is the amount of radiation that will produce, in 1 cubic centimeter of air, ion pairs having a total of 2.1×10^9 units of electrical charge (electrostatic units). It is estimated that the minimum lethal dose for humans (50% of those exposed would die within 30 days) is 590 rem.

Source	Average Dose Rate[a] (millirem per year)
Environmental	
Natural	102
Global fallout	4
Nuclear power	0.003
Subtotal	106
Medical	
Diagnostic X rays	72
Radiopharmaceuticals	1
Subtotal	73
Occupational	0.8
Miscellaneous	2
Subtotal	3
Total	182

[a] 1 millirem $= 10^{-3}$ rem
From *Understanding Chemistry: From Atoms to Attitudes* by T. R. Dickson, Wiley, N.Y., 1974.

that improved X ray equipment and the elimination of unnecessary X rays could greatly reduce the average yearly dose.

15-7   Nuclear Transmutations

A high-speed nuclear particle can, under certain conditions, collide with a nucleus to cause a nuclear reaction that produces a different nucleus. This process is called a **nuclear transmutation.** Some transmutations can occur naturally, but many are induced in the nuclear science laboratory. An example of a transmutation is

$$^{14}_{7}\text{N} + ^{4}_{2}\text{He} \rightarrow ^{17}_{8}\text{O} + ^{1}_{1}\text{H}$$

In this reaction, the nuclear particle $^{4}_{2}\text{He}$ is called the **projectile,** and the $^{14}_{7}\text{N}$ is called the **target nuclei.** The transmutation process involves a collision of the target nuclei with the projectiles that results in new combinations of neutrons and protons corresponding to the products.

Nuclear scientists have used nuclear transmutations as a means of preparing artificial nuclides that are sometimes called **man-made isotopes.** Nearly all of the naturally occurring nuclides have been used as

Table 15-5   Nuclear Transmutations Used to Produce Transuranium Elements

Element	Atomic Number	Reaction
Neptunium, Np	93	$^{238}_{92}U + ^{1}_{0}n \rightarrow ^{239}_{93}Np + ^{0}_{-1}e$
Plutonium, Pu	94	$^{238}_{92}U + ^{2}_{1}H \rightarrow ^{238}_{93}Np + 2^{1}_{0}n$
		$^{238}_{93}Np \rightarrow ^{238}_{94}Pu + ^{0}_{-1}e$
Americium, Am	95	$^{239}_{94}Pu + ^{1}_{0}n \rightarrow ^{240}_{95}Am + ^{0}_{-1}e$
Curium, Cm	96	$^{239}_{94}Pu + ^{4}_{2}He \rightarrow ^{242}_{96}Cm + ^{1}_{0}n$
Berkelium, Bk	97	$^{241}_{95}Am + ^{4}_{2}He \rightarrow ^{243}_{97}Bk + 2^{1}_{0}n$
Californium, Cf	98	$^{242}_{96}Cm + ^{4}_{2}He \rightarrow ^{245}_{98}Cf + ^{1}_{0}n$
Einsteinium, Es	99	$^{238}_{92}U + 15^{1}_{0}n \rightarrow ^{253}_{99}Es + 7^{0}_{-1}e$
Fermium, Fm	100	$^{238}_{92}U + 17^{1}_{0}n \rightarrow ^{255}_{100}Fm + 8^{0}_{-1}e$
Mendelevium, Md	101	$^{253}_{99}Es + ^{4}_{2}He \rightarrow ^{256}_{101}Md + ^{1}_{0}n$
Nobelium, No	102	$^{246}_{96}Cm + ^{12}_{6}C \rightarrow ^{254}_{102}No + 4^{1}_{0}n$
Lawrencium, Lr	103	$^{252}_{98}Cf + ^{10}_{5}B \rightarrow ^{257}_{103}Lr + 5^{1}_{0}n$
Kurchatovium, Kr[a]	104	$^{242}_{94}Pu + ^{22}_{10}Ne \rightarrow ^{260}_{104}Ku + 4^{1}_{0}n$

[a] Not official.

targets, and a variety of nuclear particles such as protons ($^{1}_{1}H$), deuterons ($^{2}_{1}H$), neutrons ($^{1}_{0}n$), alpha particles ($^{4}_{2}He$), and electrons ($^{0}_{-1}e$) have been used as projectiles. Often these projectiles must be given large kinetic energies before they are projected at the target nuclei. Particle accelerating devices such as cyclotrons, linear accelerators, synchrotrons, and nuclear reactors are used to produce these high-energy projectiles. Through transmutations, nuclear scientists have been able to make about 1000 different artificial nuclides. Some of the nuclides are useful in research and in medicine. For instance, carbon-14, made by the transmutation

$$^{14}_{7}N + ^{1}_{0}n \rightarrow ^{14}_{6}C + ^{1}_{1}H$$

is a radioactive (α emitter) isotope of carbon. **Carbon-14** can be incorporated in organic compounds and serve as a **radioactive tag.** The radioactive carbon can then be traced through a series of chemical reactions. Using tagged compounds provides a convenient way to study the way in which a series of reactions takes place. By use of carbon-14 tagged carbon dioxide, Melvin Calvin was able to accomplish a detailed study of photosynthesis, the process by which plants convert carbon dioxide and water into carbohydrates. The radioactive artificial nuclide **cobalt-60,** which is made by the transmutation

$$^{59}_{27}Co + ^{1}_{0}n \rightarrow ^{60}_{27}Co$$

is used as a medical tool to destroy cancerous cells in the treatment of cancer.

Among the many interesting transmutations are those used to make the so-called **man-made elements.** Uranium is the element of highest atomic number found in nature. However, by carrying out certain transmutations, nuclear scientists have been able to make elements of higher atomic number than uranium. These are called the **transuranium elements.** The transmutations have been accomplished by use of special particle accelerators and by using uranium and other previously synthe-sized transuranium elements as targets. The transmutations by which the transuranium elements have been made are given in Table 15-5. It is conceivable that more transuranium elements will be made in the future. The synthesis of element 106 was reported in September, 1974.

15-8   Nuclear Fission

A very important type of nuclear reaction is **nuclear fission,** which was discovered by Hahn and Strassman in 1938. In fission, a neutron collides with and is captured by a heavy nucleus causing the nucleus to become very unstable. The unstable nucleus then splits into two new nuclei plus a few neutrons. This fission is illustrated in Figure 15-5. The fission of uranium-235 can be represented as

$$^{235}_{92}\text{U} + {}^1_0\text{n} \rightarrow \text{fission nuclei} + \text{neutrons} + \text{energy}$$

Various fission nuclei are produced depending on how the original nucleus splits. A typical fission of $^{235}_{92}\text{U}$ is

$$^{235}_{92}\text{U} + {}^1_0\text{n} \rightarrow {}^{90}_{38}\text{Sr} + {}^{144}_{54}\text{Xe} + 2{}^1_0\text{n} + \text{energy}$$

When fission occurs, the more stable product nuclei are formed from the less stable parent nuclei, which results in a large amount of energy being produced. This energy is the most important aspect of fission since it represents **nuclear energy** that can be obtained by carrying out the fission process. The fission of a gram of uranium-235 could produce about 20 billion calories of energy. This means that 1 kg of uranium-235 can produce as much energy as the burning of about 2600 tons of coal. The energy of fission is the basis for the production of atomic energy.

Some of the products of the fission process are neutrons, which are ejected during the fission and are potentially capable of causing other nuclei to undergo further fission. Each time a fission occurs, more neutrons are produced and more fission can take place. In fact, each fission produces more neutrons than are consumed in the original fission process. Thus, if sufficient nuclei are present, it is possible to have an uncontrolled **chain or sequential reaction** of fission processes. This is illustrated in Figure 15-6.

If insufficient nuclei are present in a sample of fissionable material then enough neutrons escape without causing further fission, and the

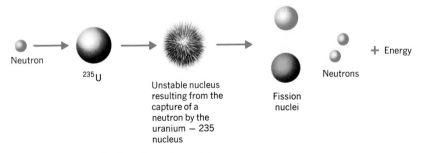

Figure 15-5   Fission of uranium-235.

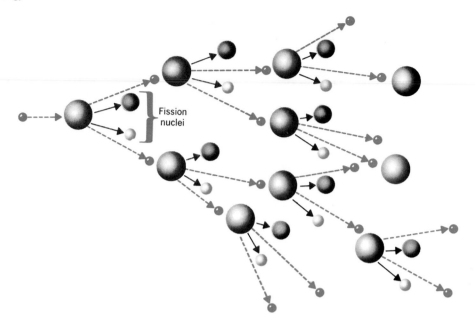

Figure 15-6   A nuclear chain reaction. The neutrons given off by the fission process can cause fission of other nuclei. Since each fission process produces about twice as many neutrons as are needed to cause fission, a nuclear chain reaction can result when a large amount of fissionable nuclei are present.

fission that does go on goes slowly enough to be under control. However, if enough fissionable material is present, a self-sustaining fission chain reaction can occur. The minimum amount of fissionable material in which a fission chain reaction can be self-sustaining is called the **critical mass** of the material. The critical mass of pure uranium-235 is less than 1 kg. A fission chain reaction can take place very quickly, and each fission process will release a large amount of energy. This accounts for the tremendous destructive power of **atomic bombs** or **fission bombs.**

15-9   Nuclear Reactors

Nuclear chain reactions can be controlled, and the energy can be utilized as a source of power. A device in which a self-sustaining fission reaction can be carried out under controlled conditions is called a **nuclear reactor.** The key to using fission in a reactor is the control of the process so that a nuclear explosion can not occur. A typical nuclear reactor is illustrated in Figure 15-7. A reactor consists of five basic components. First, the **nuclear fuel** which contains sufficient amounts of fissionable material. Typical nuclear fuel elements contain natural uranium in which the uranium-235 content has been increased from the natural level of about 0.7% to 3–4%. This uranium-235 enrichment of natural uranium is accomplished by special techniques specifically for use in nuclear reactors. There is no possibility of a critical mass of uranium-235 in a reactor and, thus, no chance of a nuclear explosion. In addition to uranium-235, the nuclides plutonium-239 and uranium-233 are also fissionable. Second, a **moderator** is used to slow down the fission neutrons. Graphite and heavy water (water with hydrogen-2 isotope rather than hydrogen-1; $^2_1H_2O$ rather than $^1_1H_2O$) are commonly used as moderators. Third, **control rods** made of cadmium or boron steel are used to control the fission. These rods are capable of absorbing neutrons. So, by moving these rods into the reactor, it is possible to carefully control the number of neutrons present. These control rods allow the correct number of neutrons to be available just to sustain continuous fission. Fourth, a **coolant** such as water or molten sodium metal is circulated about the reactor core to absorb the heat produced by fission. The heated coolant can be used to form steam which is used to generate electrical power. In fact, this is the purpose of an electrical power producing nuclear reactor. See Section 15-11. Fifth, some **shielding** material must be used to protect people from the highly radioactive core of the reactor. The shielding often consists of a thick concrete case around the reactor.

It is possible to design reactors that not only produce power but also are able to produce fissionable material. Reactors in which the fission process is used to produce more fissionable material are called **breeder reactors.** Breeder reactors function by placing certain amounts of the naturally occurring isotopes uranium-238 or thorium-232 in specially designed reactors. These isotopes are ultimately transmuted to fissionable nuclides as shown below.

$$^{238}_{92}U + {}^1_0n \text{ (from fission)} \rightarrow {}^{239}_{92}U \xrightarrow{\beta} {}^{239}_{93}Np \xrightarrow{\beta} {}^{239}_{94}Pu \text{ (fissionable)}$$

$$^{232}_{90}Th + {}^1_0n \text{ (from fission)} \rightarrow {}^{233}_{90}Th \xrightarrow{\beta} {}^{233}_{91}Pa \xrightarrow{\beta} {}^{233}_{92}U \text{ (fissionable)}$$

The breeder reactor consumes uranium-235 as fuel but produces plutonium-239 (or uranium-233) during the operation of the reactor. The

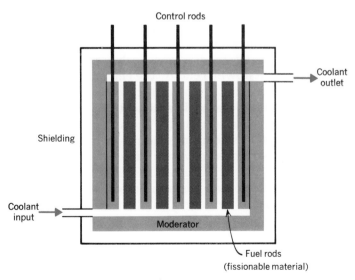

Figure 15-7   A simple nuclear reactor.

plutonium can be collected periodically and used in other reactors. A breeder reactor can produce more potential fuel than it consumes. Breeder reactors represent a vast potential source of energy as discussed in Section 15-11. However the first breeder reactor in the United States is to be built in the early 1980s.

15-10   Fusion

Fusion is another type of transmutation process that can produce energy. At high temperatures, it is possible for some lighter nuclei to fuse together to form heavier nuclei. Fusion can produce large amounts of energy. In fact, the energy produced on the sun comes from fusion. Nuclear scientists found that the temperatures produced during fission (thermonuclear temperatures) were sufficient to cause certain fusion reactions involving various hydrogen nuclides and lithium nuclides as shown below.

$$^2_1H + {}^3_1H \rightarrow {}^4_2He + {}^1_0n + \text{energy}$$
$$^2_1H + {}^2_1H \rightarrow {}^3_2He + {}^1_0n + \text{energy}$$
$$^2_1H + {}^2_1H \rightarrow {}^3_1H + {}^1_1H + \text{energy}$$
$$^1_1H + {}^7_3Li \rightarrow {}^4_2He + {}^4_2He + \text{energy}$$

The fusion process can produce larger amounts of energy for a given amount of material than the fission process. Uncontrolled **thermonuclear fusion** releases a large amount of energy and accounts for the tremendous destructive power of thermonuclear bombs **(H-bombs, hydrogen bombs,** or **fission-fusion bombs).** No way has yet been found to carry out a controlled fusion process so that the energy released can be used. However, much research is being devoted to the study of **controlled fusion,** and someday fusion reactors may be developed to provide vast amounts of low-cost energy.

15-11   Nuclear Energy

A nuclear reactor in which uranium-235 is undergoing fission can be designed to produce electrical power. Such a power reactor is illustrated in Figure 15-8. In a **power reactor** the energy of the fission is absorbed by the coolant circulating through the reactor. The hot coolant is used to heat water to generate steam. The steam turns a turbine that drives the generator that produces electricity. The steam is condensed and cooled with a cooling water system. In some power reactors the coolant in the reactor is water, which is converted directly to steam used in the turbine.

Nuclear power from fission is often predicted to be a main source of energy in the future. Currently, about 3% of the world's energy and less than 1% of U.S. energy comes from nuclear sources. The International Atomic Energy Agency predicts that by the year 2000 about 60% of the energy will come from nuclear sources and by 2010 around 76%. It is further predicted that over 50% of the electricity used in the United

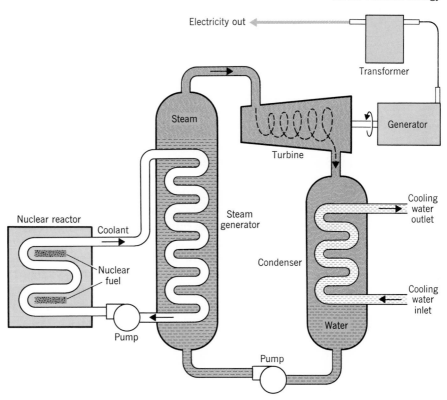

Figure 15-8   Nuclear power reactor. From *Understanding Chemistry: From Atoms to Attitudes* by T. R. Dickson, Wiley, N.Y., 1974.

States in 2000 will come from nuclear plants. However, these predictions are based on the important assumption that breeder reactors (see Section 15-9) will be functioning. More than 25 nuclear power plants are in operation in the United States, and around 117 are under construction or in the planning stage. The U.S. Atomic Energy Commission predicts that breeder reactors will be available in the 1980s. Currently all of the power reactors are designed to use uranium-235. Uranium-235 is the only common isotope that easily undergoes fission. Unfortunately, this isotope is quite rare in nature and comes from high-quality uranium ores. Predictions of the amounts of such uranium ores in the world indicate that, if wide use of uranium-235 occurs in power reactors, the world's supply will diminish around the year 2000. If this occurs, nuclear energy by fission will not be significant in the future. However, if breeder reactors are constructed that will produce fissionable uranium-233 and plutonium-239, the prospects for nuclear energy use will be increased. The isotopes uranium-238 and thorium-232 used in breeder reactors are very plentiful and would provide a vast source of nuclear fuel. In fact, it is predicted that such nuclear fuel would provide an amount of energy 10 to 100 times the amount of energy produced by all of the fossil fuels which ever existed. As was mentioned in Section 15-10, if controlled fusion reactors are ever developed, they could provide an even greater source of energy.

15-12   Pros and Cons of Nuclear Energy

There are several advantages and disadvantages to the wide-scale use of fission reactors for electrical power. Power reactors are capable of providing large amounts of energy and, thus, can replace conventional fossil fuel sources. At this time, power reactors seem to be the only feasible alternative to diminishing supplies of fossil fuels. A large-scale breeder reactor is not yet in operation. Consequently, the amounts of uranium-235 may limit nuclear fuels and make power plants uneconomical. Since the fission process is not a combustion process, power reactors do not give off large amounts of waste gases that can become air pollutants. However, some highly **radioactive gases** such as krypton-85 are given off in small amounts during fission. These gases are not the same as normal air pollutants but are an especially dangerous new kind of pollutant with which we have to contend. In fact, the use of power reactors introduces several new kinds of pollution problems. Impurities in the reactor coolant can become radioactive when exposed to the reactor core. These are known as **low-level radioactive wastes** and must be disposed of by sealing in special containers which are placed in semipermanent storage. Some low-level radioactive wastes leak into the secondary cooling waters, and, when this water is returned to the source, the radioactivity is introduced into the environment. This kind of radioactive leakage has to be continuously watched.

Figure 15-9 illustrates how radioactive substances can enter the food chain and be transmitted to humans. The inhalation of radioactive materials can induce lung cancer. Some radioactive isotopes are retained in the body because of chemical similarity to common body chemicals. Cesium-137, a β emitter of 30-year half-life, is chemically similar to potassium and can become incorporated in all of the cells of the body. Strontium-90, a β emitter of 28-year half-life, is chemically similar to calcium. Since calcium is a main constituent of bones and teeth, strontium-90 can accumulate in these areas of the body.

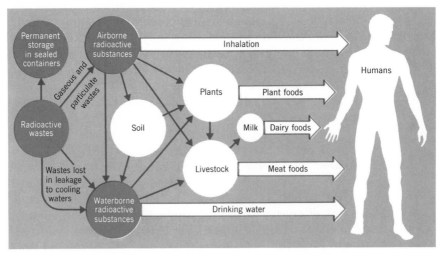

Figure 15-9   Transmission of radioactive substances to humans. From *Understanding Chemistry: From Atoms to Attitudes* by T. R. Dickson, Wiley, N.Y., 1974.

15-13   Radioactive Wastes

When fission occurs, the **fission products** produced accumulate in the reactor and have to be periodically removed. These fission products are extremely radioactive and hazardous. Such wastes are shipped in special containers to locations where they can be stored underground for safety reasons. Large amounts of such wastes are currently in storage, and if nuclear fuel usage increases, the amounts of these wastes will increase accordingly, If nuclear energy is used as predicted, it is expected that 27 billion curies of radioactive wastes will have accumulated by 2000. Transportation, handling and storage problems will be significant. In fact, these wastes will have to be stored for thousands of years in safe areas. Some scientists have suggested the construction of large pyramid structures for storage. Archeologists of the future will be quite surprised when they open these pyramids.

Another aspect of nuclear energy is the problem of **waste heat.** Currently, nuclear power plants are less efficient in energy conversion than fossil fuel plants. A conventional nuclear power plant produces 50% more waste heat than a comparable fossil fuel plant. However, more efficient nuclear plants are being developed, and breeder reactors should be as efficient as fossil fuel plants. Nevertheless, once nuclear energy begins to substantially replace fossil fuel energy, waste heat will be of greater concern.

15-14   Reactor Safety

Another problem of nuclear reactors involves **reactor safety** and **emergency-core cooling.** The core of a nuclear reactor in a power plant is maintained at a high temperature by the fission process. This temperature is kept at reasonable levels by the coolant. If the coolant flow stops, the temperature will rise. The fission process can be stopped or reduced by use of the reactor control rods (see Section 15-9). It is conceivable that the coolant flow may be cut off by accident, and, before the control rods could be used, the temperature of the core could rise so that a melting of the inner portion of the reactor takes place. The probability of such an accident is low, but the results could be disastrous. Breeder reactors are more easily damaged by coolant effects than conventional reactors. In 1955 and 1966 experimental breeder reactors being tested in the United States suffered partial core melting and had to be abandoned. A **core melting** of a reactor could result in radioactive material from the fission products and fuel rods being released into the water and air environment around the reactor. Winds or waterways could spread this material about. The major problem with this type of pollution is that it is extremely dangerous to life and cannot be seen, tasted, or smelled. It takes special instruments to detect radiation. Furthermore, some of the radioactive isotopes have long half-lives, which means that a contaminated area would have to be evacuated and would not be usable for living or agriculture for a period of time. Plutonium-239 used and produced in breeder reactors has a half-life of about 24,000 years. A nuclear accident involving plutonium could contaminate an area for an indefinite length of time. The chance of such a disaster means that the safety factors in nuclear reactors have to be carefully developed and continuously reviewed. However, the effects of earthquakes, human error of plant operators, and even the chance of sabotage are unpredictable factors. Overall, nuclear fission is potentially a very hazardous source of energy.

15-15   Plutonium

A final concern over the large-scale use of nuclear energy involves **plutonium.** A large breeder reactor requires 1000 kg of plutonium as fuel. Breeder reactors produce more fuel than they consume, which means that if such reactors are operating as expected, they will produce 80,000 kg of plutonium per year by the year 2000. At the current price of $10,000 per kg or even a lower price, plutonium sources would be attractive to criminal elements of society, and a black market might develop. Moreoever, the critical mass of plutonium-239 is about 5 kg. It is conceivable that a crude atomic bomb could be constructed from this amount of plutonium.

The desirable aspects and possible necessity contrasted with the many disadvantages of nuclear power comprise a significant environmental dilemma. We will have to make many decisions on nuclear power use in the near future.

Questions and Problems

1. Give a description of the structure of a nucleus.

2. What are isotopes? What are nuclides?

3. Define the atomic number and mass number. How can the number of neutrons in a nuclide be determined from the mass number and atomic number?

4. Give the symbol used to represent the following nuclides.
   (a) carbon-14
   (b) oxygen-16
   (c) uranium-238
   (d) radium-226
   (e) potassium-40

5. Determine the number of neutrons in the nuclei of each of the nuclides in problem 4.

6. How many protons and neutrons are in each of the following nuclides.
   (a) ^2_1H
   (b) $^{12}_6C$

(c) $^{16}_{8}O$

(d) $^{238}_{92}U$

7. The three natural isotopes of oxygen (atomic number 8) have nuclei that contain 8, 9, and 10 neutrons, respectively. Give the symbols and names for each of these oxygen isotopes.

8. Define radioactivity.

9. (a) Explain alpha particle decay in terms of what happens to the parent nucleus.
   (b) Explain beta particle decay in terms of what happens to the parent nucleus.

10. Give the nuclear equation for the following cases.
    (a) $^{49}_{20}Ca$ (β decay)
    (b) $^{238}_{92}U$ (α decay)
    (c) $^{14}_{6}C$ (β decay)
    (d) — $\rightarrow$ $^{222}_{86}Rn + ^{4}_{2}He$
    (e) — $\rightarrow$ $^{28}_{14}Si + ^{0}_{-1}e$

11. What is gamma ray emission?

12. Give a definition of the half-life of a radioactive nuclide.

13. Which of the elements are naturally radioactive and have no stable isotopes?

14. Describe what happens when radiation passes through matter.

15. Why is radiation dangerous to humans? Which type of radiation is the most dangerous?

16. How does a Geiger-Müller tube function?

17. What is radiation sickness?

18. What is a nuclear transmutation?

19. Complete the following equations representing nuclear transmutations.
    (a) $^{23}_{11}Na + ^{1}_{0}n \rightarrow$ ____ $+ \gamma$
    (b) $^{35}_{17}Cl +$ ____ $\rightarrow$ $^{35}_{16}S + ^{1}_{1}H$
    (c) $^{27}_{13}Al + ^{1}_{0}n \rightarrow$ $^{24}_{11}Na +$ ____
    (d) $^{14}_{7}N + ^{1}_{0}n \rightarrow$ $^{12}_{6}C +$ ____
    (e) $^{7}_{3}Li + ^{1}_{1}H \rightarrow$ $^{4}_{2}He +$ ____

20. What are man-made or transuranium elements?

21. Describe nuclear fission.

22. What is meant by the critical mass of a fissionable nuclide?

23. List and describe the basic components of a nuclear reactor.

24. What is a breeder reactor?

25. Describe nuclear fusion.

26. Give a description of a nuclear electric power plant.

27. Why are nuclear power plants being considered as a major source of electricity in the near future?

28. Why will large-scale use of nuclear power plants depend on the development of breeder reactors?

29. List some of the pros and cons of nuclear power plants.

Appendix I

Review of Mathematics

In chemistry it is sometimes necessary to carry out arithmetic calculations with numbers. The numbers used usually consist of sequences of decimal digits with a decimal point (273.16, 0.0821) or, if there are no digits to the right of the decimal point, the point is not included in the number (273, 760, 25). Most numbers with which we deal are called positive numbers, but some are negative numbers, since they are values that are less than zero on the number scale. Negative numbers are denoted by a negative sign (-30, -76.25). Usually the numbers used in chemistry are constants or measurements. Thus they have units associated with them.

$$273 \text{ K} \qquad 26.2 \text{ g} \qquad 25 \text{ ml} \qquad 0.521 \text{ m}$$

Addition. When adding numbers, always line up the decimal points correctly:

$$153.6 + 6.2 + 12.4$$

$$\left.\begin{array}{r} 153.6 \\ 6.2 \\ 12.4 \end{array}\right\} \text{addends}$$

$$\overline{172.2} \text{ sum}$$

When you add numbers with units, make sure that the numbers have the same units. In such cases, the sum will have the same units as the addends.

$$\begin{array}{r} 273 \text{ K} \\ 25 \text{ K} \\ \hline 298 \text{ K} \end{array} \qquad \begin{array}{r} 700 \text{ torr} \\ 20 \text{ torr} \\ 40 \text{ torr} \\ \hline 760 \text{ torr} \end{array}$$

Subtraction. When subtracting numbers, always line up the decimal points and make sure the units are the same.

$$760 \text{ torr} - 24 \text{ torr}$$

$$\begin{array}{rl} 760 \text{ torr} & \text{minuend} \\ - \ 24 \text{ torr} & \text{subtrahend} \\ \hline 736 \text{ torr} & \text{difference} \end{array}$$

Multiplication. The operation of multiplication is represented in several different ways.

$$ab \qquad a \times b \qquad a(b) \qquad (a)(b) \qquad \dfrac{a}{b}$$

When multiplying numbers

$$\left.\begin{array}{r} 24.2 \\ 2.1 \end{array}\right\} \text{factors}$$

$$\begin{array}{r} 24\ 2 \\ 484 \\ \hline 50.82 \end{array} \text{ product}$$

the number of digits to the right of the decimal point in the product equals the sum of the number of digits to the right of the decimal point in each factor.

Squaring a number means to multiply the number by itself. The square is represented by a superscript two following the number.

$$(2.2)^2 = (2.2)(2.2) = 4.84$$

Cubing a number means to multiply the number by itself three times. The cube is represented by a superscript three following the number.

$$(1.2)^3 = (1.2)(1.2)(1.2) = 1.728$$

When numbers with units are multiplied, the units become part of the product.

$$(2.2\ \ell)(1.3\ \text{atm}) = 2.86\ \ell\ \text{atm}$$

When a number with a unit is squared or cubed, the unit is also squared or cubed.

$$(2\ \text{g})^2 = 4\ \text{g}^2 \quad (\text{read as 4 square grams})$$
$$(3\ \text{cm})^3 = 27\ \text{cm}^3 \quad (\text{read as 27 cubic centimeters})$$

Likewise, when numbers with the same units are multiplied, the units are multiplied and expressed as part of the product.

$$(3.5\ \text{m})(20\ \text{m}) = 70\ \text{m}^2 \quad (\text{read as 70 square meters})$$
$$(4\ \text{cm})(2\ \text{cm})(3\ \text{cm}) = 24\ \text{cm}^3 \quad (\text{read as 24 cubic centimeters})$$

Division. The division operation is the inverse of the multiplication operation. Division involves the determination of how many times one number is contained in another. This can be denoted in several ways.

$$a\overline{\smash{)}b} \qquad a \div b \qquad a/b \qquad \frac{a}{b}$$

When dividing one number into another

$$\begin{array}{r} 11 \quad \text{quotient} \\ \text{divisor} \quad 8\,\overline{\smash{)}93} \quad \text{dividend} \\ \underline{8} \\ 13 \\ \underline{8} \\ 5 \quad \text{remainder} \end{array}$$

it is possible to carry out the calculations to obtain a quotient of any number of digits desired. The decimal point in the quotient can be found by moving the decimal point in the divisor over, so that all digits are to the left, and by moving the decimal point in the dividend the same number of positions.

$$\begin{array}{r} 24. \\ .625.\overline{)15.000.} \\ \underline{1250} \\ 2500 \\ \underline{2500} \end{array}$$

The division operation is often expressed as a fraction.

$$\frac{a}{b} \qquad \begin{array}{l} \text{numerator} \\ \text{denominator} \end{array}$$

When numbers with units are divided, the units of the result are expressed as a fraction.

$$\frac{50 \text{ g}}{25 \text{ ml}} = \frac{2 \text{ g}}{\text{ml}} \qquad \begin{array}{l} \text{(the numerical parts are} \\ \text{divided and the units are} \\ \text{expressed as a fraction)} \end{array}$$

$$\frac{75 \text{ miles}}{.50 \text{ hr}} = \frac{150 \text{ miles}}{\text{hr}}$$

Such a fraction can be read as 150 miles per hour. Sometimes the / is used as a per sign, such as 150 miles/hr. When the units in the numerator and denominator are the same, they cancel.

$$\frac{250 \text{ g}}{25 \text{ g}} = 10$$

Multiplication of Fractions. The product of two or more fractions can be determined by finding the product of all the numerators and dividing this by the product of all the denominators or by repeated divisions and multiplications.

$$\left(\frac{30}{2}\right)\left(\frac{3}{5}\right)\left(\frac{5}{6}\right)\left(\frac{2}{5}\right) = \left(\frac{900}{300}\right) = 3$$

When a series of fractions involving units is multiplied, the units of the result are determined by multiplying, dividing, and canceling the units of the factors.

$$\left(\frac{50\cancel{c}}{1 \text{ gal}}\right)\left(\frac{1 \text{ gal}}{20 \text{ miles}}\right) = \frac{2.5\cancel{c}}{1 \text{ mile}}$$

$$40 \text{ hr} \left(\frac{60 \text{ min}}{1 \text{ hr}}\right)\left(\frac{60 \text{ sec}}{1 \text{ min}}\right) = 144,000 \text{ sec}$$

Reciprocal or (Multiplicative) Inverse. If the product of two numbers or quantities is one, then each is called the reciprocal or multiplicative inverse of the other:

5 and $\frac{1}{5}$ are the inverse of one another, since

$$5 \times \frac{1}{5} = 1$$

$\dfrac{3}{4}$ and $\dfrac{4}{3}$ are reciprocals of one another, since

$$\frac{3}{4} \times \frac{4}{3} = 1$$

The reciprocal or inverse of a fraction is found by switching the positions of the numerator and denominator or "turning the fraction over." Thus the reciprocal of $\dfrac{2}{3}$ is $\dfrac{3}{2}$ and the reciprocal of

$$\left(\frac{32.0 \text{ g}}{1 \text{ mole}}\right) \text{ is } \left(\frac{1 \text{ mole}}{32.0 \text{ g}}\right)$$

When dividing a quantity by a fraction, the result can be obtained by multiplying the quantity by the reciprocal of the fraction.

$$\frac{C}{\dfrac{a}{b}} = C\left(\frac{b}{a}\right)$$

For example, $\dfrac{3}{\dfrac{1}{2}} = 3\left(\dfrac{2}{1}\right) = 6$

When dealing with numbers having units, the units are carried along with the inverting and the units of the answer are determined in the usual manner.

$$\frac{20 \text{ g}}{\left(\dfrac{5 \text{ g}}{1 \text{ ml}}\right)} = 20 \text{ g}\left(\frac{1 \text{ ml}}{5 \text{ g}}\right) = 4 \text{ ml}$$

$$\frac{64 \text{ g}}{\left(\dfrac{32 \text{ g}}{1 \text{ mole}}\right)} = 64 \text{ g}\left(\frac{1 \text{ mole}}{32 \text{ g}}\right) = 2 \text{ moles}$$

Significant Digits. When numbers are obtained from measurements, they are not exact numbers. Measurements are never exact, since the number of digits expressed in a measurement depends on the limits of the measuring instrument used. For instance, if the length of an object is found to be 152 mm using a millimeter ruler, we can say that the length is known to the nearest millimeter, which is the smallest unit of measure on the ruler. The uncertainty in the measurement is in the last digit read from the smallest unit of measure on the instrument. All of the digits in a measurement including the last digit are called significant digits. When dealing with numbers obtained from measurements, the following rules apply.

1. All nonzero digits are significant: 525.2 g (four significant digits).
2. Zeros between nonzero digits are significant: 2005 sec (four significant digits).
3. Zeros to the left of nonzero digits are never significant but are used to indicate the position of the decimal point: 0.0123 m (three significant digits).

4. When a number ends in zeros, the zeros are not significant unless otherwise specified: 3200 g (unless otherwise indicated, there are only two significant digits in this measurement). However a zero which is part of the measurement is a significant digit.

When you make a measurement you should express the proper number of digits that depends on the measuring instrument. If a measurement ever ends in zeros, make sure to note that the zeros are significant.

Rounding Off. Sometimes it is desirable to express fewer digits than the number of digits in a measurement or a calculated result. This is done by rounding off the number to the desired number of digits. The rules for rounding off numbers are as follows.

1. If the number following the digits to be retained is less than five, the digits to be retained are not altered.
2. If the number following the digits to be retained is greater than five, the last digit of those retained is increased by one.
3. If the first digit following the digits to be retained is five and has zeros or no digits following it, the last digit retained is increased by one it is odd and remains the same if it is even.

A few examples of rounding off are:
Round   127.63   to four digits – 127.6 (rule 1)
Round   251.7   to three digits – 252 (rule 2)
Round   15.999   to four digits – 16.00 (zeros significant – rule 2)
Round   0.0255   to two digits – 0.026 (rule 3)
Round   122.5   to three digits – 122 (rule 3)

Exponential Notation. In chemistry we often deal with large and small numbers. These numbers are usually written in a special manner. Before discussing this, however, we should consider how multiples of 10 are represented. Multiples of 10, the base of our number system, can be represented by expressing 10 raised to a power or exponent. For instance,

$$100 = (10)(10) = 10^2$$
$$1000 = (10)(10)(10) = 10^3$$
$$10000000000 = (10)(10)(10)(10)(10)(10)(10)(10)(10)(10) = 10^{10}$$

Fractions of 10 can be represented in a similar manner.

$$\frac{1}{100} = \frac{1}{10^2} = 0.01 = 10^{-2}$$

$$\frac{1}{100000} = \frac{1}{10^5} = 0.00001 = 10^{-5}$$

The power to be used is determined by counting the number of positions moved when the decimal point is moved to the right of the 1.

$$10000 = 10^4 \text{ (positive exponent moving left)}$$
$$0.001 = 10^{-3} \text{ (negative exponent moving right)}$$

When we write a large or small number, it is necessary to repeat many zeros.

$$60200000000000000000000 \qquad 0.00052$$

It is possible to represent such numbers in a form called exponential notation by moving the decimal point so that only one significant digit is on the left.

$$6.0200000000000000000000 \qquad 0005.2$$

Then, all significant digits are written with the decimal point in this position, and these digits are multiplied by 10 raised to an exponent corresponding to the number of positions the decimal has been moved.

$$6.02 \times 10^{23} \qquad 5.2 \times 10^{-4}$$

If the decimal is moved to the left, the power of 10 is positive and, if it is moved to the right, the power is negative.

$$523,000 = 5.23 \times 10^5$$
$$0.00721 = 7.21 \times 10^{-3}$$

Exponential notation is the best way to represent large and small numbers. Furthermore, numbers that end in zero can be written in exponential notation so that the significant digits are indicated. For example,

$$7620 = 7.620 \times 10^3$$

indicates that the zero is a significant digit. To convert a number from exponential notation to the normal form, we just move the decimal point the number of positions indicated by the power of ten.

$7.23 \times 10^6 = 7,230,000.$ (move decimal six places to the right)
$4.7 \times 10^{-2} = 0.047$ (move decimal two places to the left)

Power Rules. Sometimes exponential notation numbers are used in calculation. Consequently, it is important to know how to calculate with powers of 10.

1. The product of factors involving 10 raised to a power is 10 raised to the sum of the powers.

$$(10^a)(10^b) = 10^{a+b} \qquad (10^2)(10^3) = 10^{2+3} = 10^5$$

$$(10^3)(10^{-2}) = 10^{3+(-2)} = 10^{3-2} = 10^1$$

If one power is negative, the algebraic addition results in a subtraction.

$$(10^4)(10^3)(10^{-2})(10^1) = 10^{4+3-2+1} = 10^{8-2} = 10^6$$

Any number of factors may be involved.

2. The quotient of factors involving 10 raised to a power is 10 raised to the difference of the powers.

$$\frac{10^a}{10^b} = 10^{a-b} \qquad \frac{10^3}{10^2} = 10^{3-2} = 10^1$$

$$\frac{10^3}{10^{-2}} = 10^{3-(-2)} = 10^{3+2} = 10^5$$

A negative power in the denominator results in the sum of the powers.

3. When a factor involving 10 raised to a power is raised to a power, the result is 10 raised to the product of the powers.

$$(10^a)^b = 10^{ab} \qquad (10^2)^3 = (10^2)(10^2)(10^2) = 10^{3 \times 2} = 10^6$$
$$(10^{-3})^2 = 10^{-6}$$

4. When two factors involving the same power are divided or when two factors involving the same numerical power with opposite signs are multiplied, the result is 1.

$$\frac{10^a}{10^a} = 10^{a-a} = 10^0 = 1 \ (10^0 \text{ equals } 1)$$

$$\frac{10^2}{10^2} = 10^{2-2} = 10^0 = 1$$

$$(10^a)(10^{-a}) = 10^{a-a} = 10^0 = 1$$
$$(10^3)(10^{-3}) = 10^{3-3} = 10^0 = 1$$

The Number of Digits and Position of Decimal in a Calculated Result. When a calculation is carried out, a question arises concerning how many significant digits there should be in the answer. Since the factors involved have specific numbers of significant digits, the calculated result obtained from the factors must have a specific number of significant digits. The number of digits in a result calculated by addition or subtraction depends on the position of the decimal point. The result can have no more significant digits to the right of the decimal than the factor with the least number of digits to the right of the decimal. Often, an extra digit is carried and the result is rounded off to the correct number of digits.

$$
\begin{array}{ll}
16.00 & \text{(two digits past decimal)} \\
\underline{2.016} & \text{(three digits past decimal)} \\
18.016 \rightarrow & \text{answer is } 18.02, \text{ since the result may} \\
& \text{have only two digits past decimal}
\end{array}
$$

$$
\begin{array}{ll}
12.01 & \text{(two digits past decimal)} \\
6.048 & \text{(three digits past decimal)} \\
\underline{16.00} & \text{(two digits past decimal)} \\
34.058 \rightarrow & \text{answer is } 34.06, \text{ since the result may} \\
& \text{have two digits past decimal}
\end{array}
$$

The number of digits allowed in a result obtained when factors are multiplied and divided depends on the relative uncertainties of the factors and is not dependent on the number of digits past the decimal point. Actually, the uncertainty in each factor contributes to the uncertainty in the result. A general rule to determine the number of digits in a result obtained by multiplication and division is that the result can have no more significant digits than the factor with the least number of significant digits. Unfortunately, there are exceptions to this rule but, for our purposes, it will usually indicate the proper number of digits needed in the answer. Some examples of this rule are given below.

$$2.5 \left(\frac{27.2}{3.27} \right) = 21 \qquad$$ The answer should have two significant digits, since the least accurate factor (2.5) has only two digits.

$$10.2 \left(\frac{52.7}{262}\right)\left(\frac{1.052}{99.8}\right) = 0.0216$$

The answer should have three digits, since the least accurate factors have three digits.

Some factors involve pure numbers which should be disregarded when determining the number of digits in the answer.

$$64.0 \text{ g} \left(\frac{1 \text{ mole } O_2}{32.0 \text{ g}}\right)\left(\frac{2 \text{ moles } H_2O}{1 \text{ mole } O_2}\right) = 4.00 \text{ moles } H_2O$$

The 1 in the second factor and the 2 and 1 in the third factor are pure numbers and do not affect the number of digits in the calculated result. A pure number is an exact number and, thus, has any number of implied digits.

In a calculation involving the multiplication of several factors, the position of the decimal point in the result is not always obvious. The positioning of the decimal point can be aided by expressing the numbers involved in exponential notation form and then separating the powers of ten from the other numbers. Some examples are:

(460)(0.00021)
$(4.60 \times 10^2)(2.1 \times 10^{-4})$ Express numbers in exponential notation.
$(4.60 \times 2.1)(10^2 \times 10^{-4})$ Rearrange to collect powers of 10.
9.7×10^{-2} or 0.097   Answer.

(760)(273)(0.0821)(5.25)
$(7.60 \times 10^2)(2.73 \times 10^2)(8.21 \times 10^{-2})(5.25)$ Express numbers in exponential notation.

$(7.60 \times 2.73 \times 8.21 \times 5.25)(10^2 \times 10^2 \times 10^{-2})$ Rearrange to collect powers of 10.

894×10^2 or 8.94×10^4 Answer.

The decimal position in a calculation involving division can be determined in a similar manner.

$$\frac{862}{0.032}$$

$$\frac{8.62 \times 10^2}{3.2 \times 10^{-2}}$$ Express the numbers in exponential notation.

$$\left(\frac{8.62}{3.2}\right)\left(\frac{10^2}{10^{-2}}\right)$$ Rearrange to collect powers of 10.

$$\left(\frac{8.62}{3.2}\right)10^{2-(-2)} = \left(\frac{8.62}{3.2}\right)10^{2+2} = \left(\frac{8.62}{3.2}\right)10^4 = 2.7 \times 10^4$$

When the calculations involve several multiplications and divisions, the numerator and denominator terms can be treated separately and then combined to obtain the result.

$$30.0 \left(\frac{405}{298}\right)\left(\frac{760}{721}\right)$$

$$3.00 \times 10^1 \left(\frac{4.05 \times 10^2}{2.98 \times 10^2}\right)\left(\frac{7.60 \times 10^2}{7.21 \times 10^2}\right)$$ Express terms in exponential notation.

$$\frac{(3.00 \times 4.05 \times 7.60)(10^1 \times 10^2 \times 10^2)}{(2.98 \times 7.21)(10^2 \times 10^2)}$$ Rearrange to separate powers of 10.

$$(4.30)(10^{1+2+2-2-2}) = 4.30 \times 10^1$$

Often, calculations like those above are accomplished using a slide rule. The slide rule is very good for calculations involving factors that have numbers expressed to slide rule accuracy or less. (Slide rule accuracy refers to the number of digits that can be read on a slide rule – three significant digits for numbers starting with any of the digits two to nine and four significant digits for numbers starting with the digit 1.) Generally, the slide rule can be used to determine the digits in the answer to slide rule accuracy, but the position of the decimal point must be determined separately in the manner described above. However, the process can be simplified by rounding the numbers involved to one digit to approximate the answer. Since the actual digits in the answer are determined with the slide rule, the decimal position in the actual answer will be the same as that in the approximate answer. Some examples are:

$$30.0 \left(\frac{405}{298}\right) \left(\frac{760}{721}\right) = 430$$ These are the digits in the answer. The decimal position is determined by calculating the approximate answer.

$$3 \times 10^1 \left(\frac{4 \times 10^2}{3 \times 10^2}\right) \left(\frac{8 \times 10^2}{7 \times 10^2}\right)$$

$$\left(\frac{3 \times 4 \times 8}{3 \times 7}\right) \left(\frac{10^1 \times 10^2 \times 10^2}{10^2 \times 10^2}\right) = 4 \times 10^1 \text{ Approximate answer.}$$

Since the approximate answer is 4×10^1, the actual answer must be 4.30×10^1.

$$\left(\frac{0.500}{0.0821}\right) \left(\frac{1.237}{300}\right) = 251$$ These are the digits in the answer. The decimal position is determined by calculating the approximate answer.

$$\left(\frac{5 \times 10^{-1}}{8 \times 10^{-2}}\right) \left(\frac{1}{3 \times 10^2}\right) = \left(\frac{5}{8 \times 3}\right) \left(\frac{10^{-1}}{10^{-2} \ 10^2}\right)$$

$$\left(\frac{5}{24}\right) 10^{-1} = 0.2 \times 10^{-1} = 2 \times 10^{-1} \times 10^{-1} = 2 \times 10^{-2}$$

Since the approximate answer is 2×10^{-2}, the actual answer must be 2.51×10^{-2} or 0.0251.

$$1.873 \left(\frac{293}{273}\right) \left(\frac{760}{776}\right) (22.4) = 441$$ These are the digits in the answer. The decimal position is determined by calculating the approximate answer.

$$2 \left(\frac{3 \times 10^2}{3 \times 10^2}\right) \left(\frac{8 \times 10^2}{8 \times 10^2}\right) (2 \times 10^1) = 4 \times 10^1$$

Since the approximate answer is 4×10^1, the actual answer must be 4.41×10^1 or 44.1.

$$276 \left(\frac{425}{0.00376}\right) \left(\frac{0.0237}{3250}\right) (32.5) = 739$$

These are the digits in the answer to slide rule accuracy. The decimal position is determined by calculating the approximate answer.

$$3 \times 10^2 \left(\frac{4 \times 10^2}{4 \times 10^{-3}}\right) \left(\frac{2 \times 10^{-2}}{3 \times 10^3}\right) (3 \times 10^1)$$

$$\left(\frac{3 \times 4 \times 2 \times 3}{3 \times 4}\right) \left(\frac{10^2 \times 10^2 \times 10^{-2} \times 10^1}{10^{-3} \times 10^3}\right) = 6 \times 10^3$$

Since the approximate answer is 6×10^3, the actual answer must be 7.39×10^3.

When calculating with an electronic calculator more than enough digits and the proper decimal position are obtained. Remember to round off to the proper number of significant digits. When working with large or small numbers on an 8 digit calculator it may be necessary to convert to exponential numbers and calculate with the nonexponential parts. The proper exponent can then be joined with the nonexponential result obtained from the calculator.

Appendix II

Unit Equation and Problem Solving

Whenever we make a measurement of some property associated with an object or a phenomenon, it is important to always express the measurement in the form of a number and a unit. The unit indicates the defined reference against which the measurement is compared. A measurement without a unit does not always convey the intended meaning. The inclusion of units of measurement is very helpful when the measurements are used in calculations. In chemistry it is often desirable to describe an observation so that it includes combinations of fundamental properties. These combinations will have more than one unit and are often expressed in terms of a certain magnitude of one property compared to a fixed magnitude of another. As an example, consider how we normally express the speed of motion. When we want to express speed, we do not state the distance traveled and the time involved separately but, instead, we state the magnitude of the distance traveled in a specific unit of time. That is, rather than say that a speed was 100 miles in 2 hr or 25 miles in 1/2 hr we would normally express such a speed as 50 miles per hour or 50 miles/hr or

$$\left(\frac{50 \text{ miles}}{1 \text{ hr}} \right)$$

We would express the distance traveled per unit time. This expression of one observation per unit amount of another is merely a convenient way to state a relationship between the two properties. Expressions that relate properties to one another are very important in chemistry. In fact, such expressions are quite common in everyday life. For example, when we buy apples in the market, the price is stated in terms of the cost per pound. In other words, the price is stated in terms of the amount of money per unit amount (pound).

Often a measurement can be expressed in terms of more than one unit. Distances can be measured and expressed in units of inches, feet, or yards. Obviously, a given distance would be the same no matter what unit is used to express the measurement of the distance. Usually the relationship between two different units used in the measurement of the same property is defined or can be deduced. For example, the relationship between feet and inches can be expressed as

$$1 \text{ ft} = 12 \text{ in.}$$

This relationship could be expressed in an amount per unit amount form as

$$\left(\frac{12 \text{ in.}}{1 \text{ ft}}\right) \quad \text{(read: 12 inches per one foot)}$$

This is called a conversion factor. Any relationship between two units used to measure the same property can be expressed in this form. What good does it do us to express such relationships in the form of conversion factors? As we shall see, these factors are very useful in chemistry when we have to solve numerical problems. When a quantity is multiplied by a conversion factor, it is not fundamentally changed. However, we would not perform this multiplication without a reason. Conversion factors that express a relationship between units can be used to convert a measurement expressed in terms of one unit to a form that expresses the measurement in terms of another unit. By multiplying a measurement by a conversion factor, we would not fundamentally alter the measurement but would only change the measurement from one unit to another. Of course, converting a measurement from one unit to another changes the numerical part of the measurement as well as the unit.

Example A-1   Express the measurement 2.00 ft in units of inches. This measurement in feet can be converted to units of inches by multiplying the measurement by the conversion factor expressing the relationship between inches and feet.

$$2.00 \text{ ft} \left(\frac{12 \text{ in.}}{1 \text{ ft}}\right) = 24.0 \text{ in.}$$

Notice that the unwanted units (ft) cancel and that the numerical parts of the measurement and factor are multiplied, so that the measurement can be expressed in terms of a certain number of inches.

As can be seen in the above example, it is very important to include the units in a measurement and any factor used in the calculations. By including the units, we can make sure that the unwanted units cancel and that the proper unit remains. Keep in mind that units can be treated as algebraic quantities and, thus, it is possible to cancel multiply, and divide units. The treatment of units in calculations is very important but can be confusing. When the same unit appears in the numerator and denominator of a calculation, the unit may be canceled. When the same unit appears in two or more factors involved in multiplication operations, the units can be multiplied to form square units, cubic units, etc. For example, in the calculation

$$3.00 \text{ ft} \left(\frac{12 \text{ in.}}{1 \text{ ft}}\right) 10.00 \text{ in.} = 360 \text{ in.}^2$$

the ft units cancel and the in. units are multiplied and expressed as in.^{2} (square inches). The numerical parts of the factors are manipulated in the normal arithmetic manner.

Many problems in chemistry can be solved by use of conversion factors. Any quantity can be multiplied by any number of these factors without fundamentally changing it. The relationships expressing the amount of one property per unit amount of another are also conversion factors, since they represent an equivalency between the two properties. For example, when we state a speed, such as

$$\left(\frac{50.0 \text{ mi}}{1 \text{ hr}}\right)$$

this indicates that 50.0 mi is the distance traveled in 1 hr. Thus an expression such as this can be considered to be a property conversion factor and can be used to convert from one property to another.

Example A-2   If a car is traveling at an average speed of 50.0 mi/hr, how many feet will it travel in 2.00 hours?

This problem can be solved by converting the time to number of miles using the speed as a factor. Then the distance in miles can be converted to feet using the conversion factor relating feet to miles (5280 ft/mi).

$$2.00 \text{ hr} \left(\frac{50.0 \text{ mi}}{1 \text{ hr}}\right) \left(\frac{5280 \text{ ft}}{1 \text{ mi}}\right) = 5.28 \times 10^5 \text{ ft}$$

Notice that the unwanted units cancel and the desired unit is retained. Furthermore, the numbers involved are used to calculate the numerical part of the answer.

A conversion factor is often expressed in terms of the amount of one quantity per unit amount of another. However, any conversion factor can be expressed in a form that is the reciprocal of the normal form. The reciprocal of a quantity X is $1/X$. The reciprocal, $1/X$, is sometimes called the inverse of X. The reciprocal or inverse of a factor is formed by interchanging the numerator with its units and denominator with its units. Very often the reciprocal of a factor is needed to solve a problem. For example, we can express the relationship between feet and inches as

$$\left(\frac{12 \text{ in.}}{1 \text{ ft}}\right)$$

or in the reciprocal form which is found by inverting the above factor.

$$\left(\frac{1 \text{ ft}}{12 \text{ in.}}\right)$$

The form in which we express a factor depends on how we want to use it. That is, the first factor above can be used in the conversion of a measurement from units of feet to units of inches while the second factor can be used in the conversion of a measurement from units of inches to feet.

Example A-3   Express the measurement 72.0 in. in terms of yards. This measurement can be converted to units of yards by use of the factor relating feet and inches and the factor relating yards and feet.

$$\left(\frac{12 \text{ in.}}{1 \text{ ft}}\right) \text{ and } \left(\frac{3 \text{ ft}}{1 \text{ yd}}\right)$$

We must decide whether we want the factors in this form or in the reciprocal form course, this depends on which units are to be canceled and which are to be retained. Since we want to convert the measurements in inches to units of yards, we shall want the inches and feet units to cancel. Thus we should use the unit factors in the reciprocal forms.

$$\left(\frac{1 \text{ ft}}{12 \text{ in.}}\right) \text{ and } \left(\frac{1 \text{ yd}}{3 \text{ ft}}\right)$$

Using these factors, the conversion is accomplished as follows.

$$72.0 \text{ in. } \left(\frac{1 \text{ ft}}{12 \text{ in.}}\right)\left(\frac{1 \text{ yd}}{3 \text{ ft}}\right) = 2.00 \text{ yd}$$

Whenever we solve a problem by the conversion factor approach, we must decide what factors are needed and whether to use the factors in one form or in its reciprocal form. This decision depends on which units are to be canceled and what unit we want in the final answer. When the problem is set up the desired units should result. If not, the problem is probably set up incorrectly or the proper units were not used on all factors. The problem is set up by stating the quantity to be converted and then multiplying this quantity by the necessary factors. The choice of factors depends on the problem. Usually it is convenient to contain these factors in large parentheses, so that the numbers and units can be easily established from those of the other factors.

Example A-4   How many minutes will it take a car to travel 5.00 mi if the average speed of the car is 50.0 mi/hr?

We want the answer to have units of minutes. The distance traveled is given and the speed is given in units of miles per hour. The reciprocal of the speed can be used to convert the distance to the time required in hours. The number of hours can then be converted to number of minutes by use of the unit factor relating minutes and hours (60 min/hr). The unit factors to be used are

$$\left(\frac{1 \text{ hr}}{50.0 \text{ mi}}\right) \text{ and } \left(\frac{60 \text{ min}}{1 \text{ hr}}\right)$$

The conversion sequence used to convert the distance traveled to the number of minutes required should be

$$\text{miles} \rightarrow \text{hours} \rightarrow \text{minutes}$$

The set-up for the conversion should be

$$5.00 \text{ mi } \left(\frac{1 \text{ hr}}{50.0 \text{ mi}}\right)\left(\frac{60 \text{ min}}{1 \text{ hr}}\right) = 6.00 \text{ min}$$

Notice that the units cancel to give the desired units for the answer and that the numerical parts of the factors are used to calculate the numerical part of the answer.

Problem A-1   If the average speed of a car is 55.0 mi/hr, how many feet will the car travel in 10.0 sec? (Answer: 807 ft)

Appendix III

Predicting the Shapes of Molecules

The Lewis electron dot structure of a molecule indicates the number of electrons distributed about each atom. Within a given polyatomic molecule in which the octet rule is satisfied, a specific atom occupies a somewhat central position. Let us call this atom the central atom. For example, in methane

$$H-\underset{\underset{H}{|}}{\overset{\overset{H}{|}}{C}}-H$$

the carbon is the central atom. This will have four pairs of electrons about it. If we can deduce how those electron pairs are arranged in space, we shall be able to visualize the shape of the molecule. What is the likely arrangement of four pairs of electrons about the central atom? One answer is given by the **valence shell electron pair repulsion theory** (VSEPR) which states that the electron pairs will repel and be arranged about the central atom so that they are as far apart as possible. Thus, the likely arrangement of four pairs of valence electrons about the central atom is tetrahedral. That is, the electron pairs will be arranged so that they are directed towards the apecies of a tetrahedron. (See Figure A-1). In a **tetrahedral distribution,** the angle between a given pair of electrons will be about 109.5°. A tetrahedral distribution can be represented in two dimensions as

Knowing that four pairs of electrons will have a tetrahedral distribution, we can predict the likely shape of many molecules as shown in Table A-1. We would predict that a molecule of methane has a tetrahedral shape

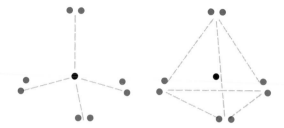

Figure A-1   Tetrahedral distribution of four electron pairs about a central atom.

The Lewis electron dot structure for water is

$$:\ddot{O}\text{—H}$$
$$|$$
$$H$$

We would predict that water would have an **angular shape**

$$\begin{array}{c} H \\ | \\ :\underset{\cdot\cdot}{O} \\ \diagdown \\ H \end{array}$$

in which the H—O—H angle is about 109.5°. The actual angle is about 105° so the predicted shape is fairly good. The Lewis electron dot structure for ammonia is

$$H\text{—}\ddot{N}\text{—}H$$
$$|$$
$$H$$

Using the tetrahedral distribution, the predicted shape for an ammonia molecule would be a **triangular pyramid**

with an H—N—H bond angle of 109.5°. The actual angle is about 107°. The electron pair repulsion theory allows for the convenient visualization of many molecules. To fit the tetrahedral case the molecule must have a central atom surrounded by four pairs of electrons with no multiple bonds. This theory is also useful for predicting the shapes of some polyatomic ions. For instance, the electron dot structure of hydronium is

$$\left[H\text{—}\ddot{O}\text{—}H \atop \phantom{H\text{—}}\underset{H}{|} \right]^{+}$$

The predicted shape would be a triangular pyramid

$$\left[\begin{array}{c} \ddot{O} \\ \diagup | \diagdown \\ H\ H\ H \end{array} \right]^{+}$$

Example A-1   Predict a likely shape for a molecule of chloroform, $CHCl_3$. The electron dot structure is

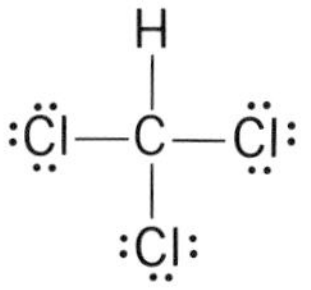

$$\ddot{C}l-C-\ddot{C}l$$

Thus, a likely shape would be tetrahedral:

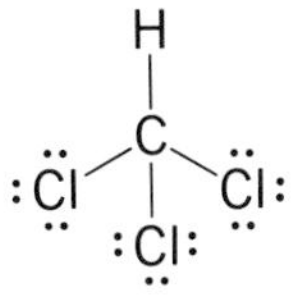

Table A-1   Possible Shapes of Species with Four Pairs of Electrons (Eight Valence Electrons) Around the Central Atom

Formula (A is central atom)	Electron dot structure		Shape
AX_4	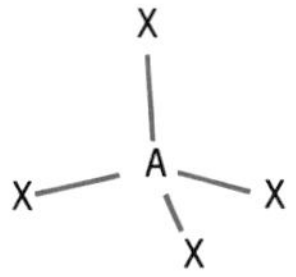	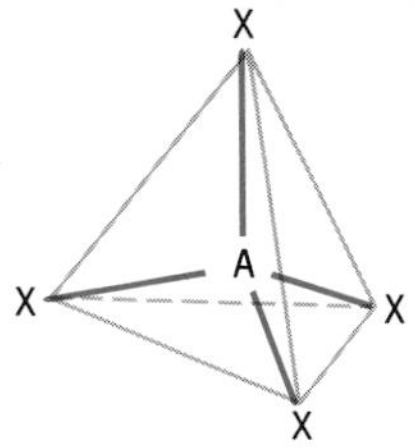	Tetrahedral
AX_3	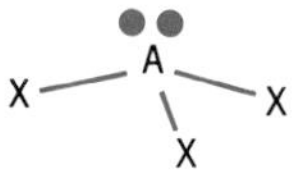	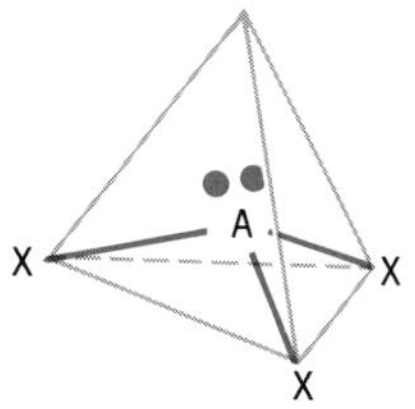	Triangular pyramid
AX_2		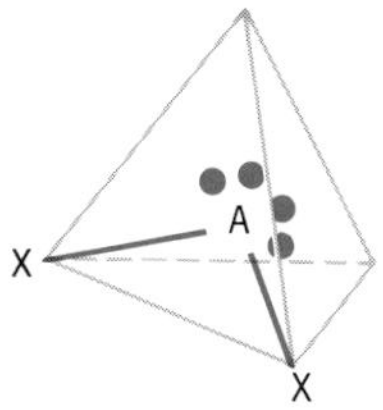 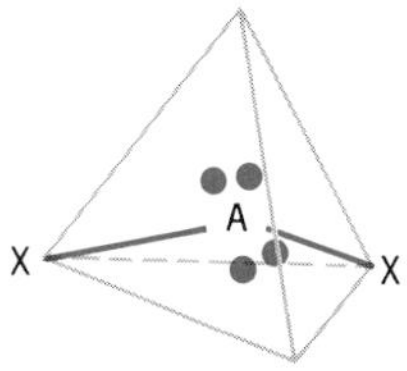	Angular

Example A-2   Predict a likely shape for a chlorate ion, ClO_3^-.
The electron dot structure is

$$\left[:\overset{..}{\underset{..}{O}}\!\!-\!\!\overset{..}{Cl}\!\!-\!\!\overset{..}{\underset{..}{O}}: \right]^-$$
$$\underset{:\overset{..}{\underset{..}{O}}:}{}$$

Thus, a likely shape would be a triangular pyramid.

$$\left[\underset{:\overset{..}{\underset{..}{O}}: \; :\overset{..}{\underset{..}{O}}: \; :\overset{..}{\underset{..}{O}}:}{\overset{\overset{..}{Cl}}{\diagup \mid \diagdown}} \right]^-$$

Problem A-1   Predict likely shapes for the following molecules and ions.
  (a) Phosphorus trichloride, PCl_3
  (b) Hydrogen sulfide, H_2S
  (c) Ammonium ion, NH_4^+
  (d) Perchlorate ion, ClO_4^-

Appendix IV

The Electronic Configurations of the Elements

Element	1s	2s	2p	3s	3p	3d	4s	4p	4d	4f	5s	5p	5d	5f
1. H	1													
2. He	2													
3. Li	2	1												
4. Be	2	2												
5. B	2	2	1											
6. C	2	2	2											
7. N	2	2	3											
8. O	2	2	4											
9. F	2	2	5											
10. Ne	2	2	6											
11. Na	2	2	6	1										
12. Mg	2	2	6	2										
13. Al	2	2	6	2	1									
14. Si	2	2	6	2	2									
15. P	2	2	6	2	3									
16. S	2	2	6	2	4									
17. Cl	2	2	6	2	5									
18. Ar	2	2	6	2	6									
19. K	2	2	6	2	6		1							
20. Ca	2	2	6	2	6		2							
21. Sc	2	2	6	2	6	1	2							
22. Ti	2	2	6	2	6	2	2							
23. V	2	2	6	2	6	3	2							
24. Cr	2	2	6	2	6	5	1							
25. Mn	2	2	6	2	6	5	2							
26. Fe	2	2	6	2	6	6	2							
27. Co	2	2	6	2	6	7	2							
28. Ni	2	2	6	2	6	8	2							
29. Cu	2	2	6	2	6	10	1							
30. Zn	2	2	6	2	6	10	2							
31. Ga	2	2	6	2	6	10	2	1						
32. Ge	2	2	6	2	6	10	2	2						
33. As	2	2	6	2	6	10	2	3						
34. Se	2	2	6	2	6	10	2	4						
35. Br	2	2	6	2	6	10	2	5						
36. Kr	2	2	6	2	6	10	2	6						
37. Rb	2	2	6	2	6	10	2	6			1			
38. Sr	2	2	6	2	6	10	2	6			2			
39. Y	2	2	6	2	6	10	2	6	1		2			
40. Zr	2	2	6	2	6	10	2	6	2		2			
41. Nb	2	2	6	2	6	10	2	6	4		1			
42. Mo	2	2	6	2	6	10	2	6	5		1			
43. Tc	2	2	6	2	6	10	2	6	6		1			
44. Ru	2	2	6	2	6	10	2	6	7		1			
45. Rh	2	2	6	2	6	10	2	6	8		1			
46. Pd	2	2	6	2	6	10	2	6	10					
47. Ag	2	2	6	2	6	10	2	6	10		1			
48. Cd	2	2	6	2	6	10	2	6	10		2			
49. In	2	2	6	2	6	10	2	6	10		2	1		
50. Sn	2	2	6	2	6	10	2	6	10		2	2		
51. Sb	2	2	6	2	6	10	2	6	10		2	3		
52. Te	2	2	6	2	6	10	2	6	10		2	4		
53. I	2	2	6	2	6	10	2	6	10		2	5		
54. Xe	2	2	6	2	6	10	2	6	10		2	6		

Element	K	L	M	4s	4p	4d	4f	5s	5p	5d	5f	6s	6p	6d	7s
55. Cs	2	8	18	2	6	10		2	6			1			
56. Ba	2	8	18	2	6	10		2	6			2			
57. La	2	8	18	2	6	10		2	6	1		2			
58. Ce	2	8	18	2	6	10	2	2	6			2			
59. Pr	2	8	18	2	6	10	3	2	6			2			
60. Nd	2	8	18	2	6	10	4	2	6			2			
61. Pm	2	8	18	2	6	10	5	2	6			2			
62. Sm	2	8	18	2	6	10	6	2	6			2			
63. Eu	2	8	18	2	6	10	7	2	6			2			
64. Gd	2	8	18	2	6	10	7	2	6	1		2			
65. Tb	2	8	18	2	6	10	9	2	6			2			
66. Dy	2	8	18	2	6	10	10	2	6			2			
67. Ho	2	8	18	2	6	10	11	2	6			2			
68. Er	2	8	18	2	6	10	12	2	6			2			
69. Tm	2	8	18	2	6	10	13	2	6			2			
70. Yb	2	8	18	2	6	10	14	2	6			2			
71. Lu	2	8	18	2	6	10	14	2	6	1		2			
72. Hf	2	8	18	2	6	10	14	2	6	2		2			
73. Ta	2	8	18	2	6	10	14	2	6	3		2			
74. W	2	8	18	2	6	10	14	2	6	4		2			
75. Re	2	8	18	2	6	10	14	2	6	5		2			
76. Os	2	8	18	2	6	10	14	2	6	6		2			
77. Ir	2	8	18	2	6	10	14	2	6	7		2			
78. Pt	2	8	18	2	6	10	14	2	6	9		1			
79. Au	2	8	18	2	6	10	14	2	6	10		1			
80. Hg	2	8	18	2	6	10	14	2	6	10		2			
81. Tl	2	8	18	2	6	10	14	2	6	10		2	1		
82. Pb	2	8	18	2	6	10	14	2	6	10		2	2		
83. Bi	2	8	18	2	6	10	14	2	6	10		2	3		
84. Po	2	8	18	2	6	10	14	2	6	10		2	4		
85. At	2	8	18	2	6	10	14	2	6	10		2	5		
86. Rn	2	8	18	2	6	10	14	2	6	10		2	6		
87. Fr	2	8	18	2	6	10	14	2	6	10		2	6		1
88. Ra	2	8	18	2	6	10	14	2	6	10		2	6		2
89. Ac	2	8	18	2	6	10	14	2	6	10		2	6	1	2
90. Th	2	8	18	2	6	10	14	2	6	10		2	6	2	2
91. Pa	2	8	18	2	6	10	14	2	6	10	2	2	6	1	2
92. U	2	8	18	2	6	10	14	2	6	10	3	2	6	1	2
93. Np	2	8	18	2	6	10	14	2	6	10	5	2	6		2
94. Pu	2	8	18	2	6	10	14	2	6	10	6	2	6		2
95. Am	2	8	18	2	6	10	14	2	6	10	7	2	6		2
96. Cm	2	8	18	2	6	10	14	2	6	10	7	2	6	1	2
97. Bk	2	8	18	2	6	10	14	2	6	10	8	2	6	1	2
98. Cf	2	8	18	2	6	10	14	2	6	10	10	2	6		2
99. Es	2	8	18	2	6	10	14	2	6	10	11	2	6		2
100. Fm	2	8	18	2	6	10	14	2	6	10	12	2	6		2
101. Md	2	8	18	2	6	10	14	2	6	10	13	2	6		2
102. No	2	8	18	2	6	10	14	2	6	10	14	2	6		2
103. Lr	2	8	18	2	6	10	14	2	6	10	14	2	6	1	2

Appendix V

Some Useful Conversion Factors

	English to Metric	Metric to English
Length		
1 inch(in.) = 2.54 centimeters(cm)	$\left(\dfrac{2.54\text{ cm}}{1\text{ in.}}\right)$	$\left(\dfrac{0.394\text{ in.}}{1\text{ cm}}\right)$
1 foot(ft) = 0.305 meters(m)	$\left(\dfrac{0.305\text{ m}}{1\text{ ft}}\right)$	$\left(\dfrac{3.28\text{ ft}}{1\text{ m}}\right)$
1 yard(yd) = 0.914 meters(m)	$\left(\dfrac{0.914\text{ m}}{1\text{ yd}}\right)$	$\left(\dfrac{1.09\text{ yd}}{1\text{ m}}\right)$
1 mile(mi) = 1.609 kilometers(km)	$\left(\dfrac{1.609\text{ km}}{1\text{ mi}}\right)$	$\left(\dfrac{0.621\text{ mi}}{1\text{ km}}\right)$
Volume		
1 pint = 0.473 liters(ℓ)	$\left(\dfrac{0.473\ \ell}{1\text{ pint}}\right)$	$\left(\dfrac{2.11\text{ pint}}{1\ \ell}\right)$
1 quart = 0.946 liters(ℓ)	$\left(\dfrac{0.946\ \ell}{1\text{ quart}}\right)$	$\left(\dfrac{1.06\text{ quart}}{1\ \ell}\right)$
1 gallon(gal) = 3.79 liters(ℓ)	$\left(\dfrac{3.79\ \ell}{1\text{ gal}}\right)$	$\left(\dfrac{0.264\text{ gal}}{1\ \ell}\right)$
Mass (Weight)		
1 pound(lb) = 454 grams(g)	$\left(\dfrac{454\text{ g}}{1\text{ lb}}\right)$	$\left(\dfrac{0.0022\text{ lb}}{1\text{ g}}\right)$
1 ounce(oz) = 28.3 grams(g)	$\left(\dfrac{28.3\text{ g}}{1\text{ oz}}\right)$	$\left(\dfrac{0.0353\text{ oz}}{1\text{ g}}\right)$
1 pound(lb) = 0.454 kilograms(kg)	$\left(\dfrac{0.454\text{ kg}}{1\text{ lb}}\right)$	$\left(\dfrac{2.20\text{ lb}}{1\text{ kg}}\right)$
Special Factors		
1 Angstrom Unit(A) = 10^{-8} centimeters	$\left(\dfrac{10^{-8}\text{ cm}}{1\text{ A}}\right)$	$\left(\dfrac{10^{8}\text{ A}}{1\text{ cm}}\right)$

1 liter(ℓ) = 10^3 cubic centimeters(cm³) $\left(\dfrac{10^3 \text{ cm}^3}{1 \ \ell}\right)$ $\left(\dfrac{1 \ \ell}{10^3 \text{ cm}^3}\right)$

1 milliliter(ml) = 1 cubic centimeter(cm³) $\left(\dfrac{1 \text{ cm}^3}{1 \text{ ml}}\right)$ $\left(\dfrac{1 \text{ ml}}{1 \text{ cm}^3}\right)$

1 atmosphere(atm) = 760 torr $\left(\dfrac{760 \text{ torr}}{1 \text{ atm}}\right)$ $\left(\dfrac{1 \text{ atm}}{760 \text{ torr}}\right)$

1 torr = 1 millimeter Hg(mmHg) $\left(\dfrac{1 \text{ mmHg}}{1 \text{ torr}}\right)$ $\left(\dfrac{1 \text{ torr}}{1 \text{ mmHg}}\right)$

Appendix VI

The Prefixes Used in the Metric System

Prefix	Symbol	Meaning	Example
tera	T	10^{12} times basic unit	Tg(teragram) $= 10^{12}$ g
giga	G	10^9 times basic unit	Gm(gigameter) $= 10^9$ m
mega	M	10^6 times basic unit	Mg(megagram) $= 10^6$ g
kilo	k	10^3 times basic unit	km(kilometer) $= 10^3$ m
hecto	h	10^2 times basic unit	hsec(hectosecond) $= 10^2$ sec
deka	da	10^1 times basic unit	dam(dekameter) $= 10^1$ m
deci	d	10^{-1} times basic unit	dg(decigram) $= 10^{-1}$ g
centi	c	10^{-2} times basic unit	cm(centimeter) $= 10^{-2}$ m
milli	m	10^{-3} times basic unit	mm(millimeter) $= 10^{-3}$ m
micro	μ	10^{-6} times basic unit	μsec(microsec) $= 10^{-6}$ sec
nano	n	10^{-9} times basic unit	ng(nanogram) $= 10^{-9}$ g
pico	p	10^{-12} times basic unit	pg(picogram) $= 10^{-12}$ g
femto	f	10^{-15} times basic unit	fsec(femtosec) $= 10^{-15}$ sec
atto	a	10^{-18} times basic unit	am(attometer) $= 10^{-18}$ m

Index

Periodic Table of the Elements

Representative Elements — s block | Representative Elements — p block | Noble gases

Key (Atomic No. Symbol Atomic Wt.): oxidation numbers · atomic number · symbol · atomic weight

Representative Elements (s block and p block), Transition Elements (d block)

Period	IA	IIA	IIIB	IVB	VB	VIB	VIIB	VIII	VIII	VIII	IB	IIB	IIIA	IVA	VA	VIA	VIIA	O (Noble gases)
1	+1 · 1 · H · 1.0079																	2 · He · 4.003
2	+1 · 3 · Li · 6.941	+2 · 4 · Be · 9.012											+3 · 5 · B · 10.81	+4 +2 · 6 · C · 12.011	+5 +3 −3 · 7 · N · 14.007	−2 · 8 · O · 15.999	−1 · 9 · F · 18.998	10 · Ne · 20.18
3	+1 · 11 · Na · 22.99	+2 · 12 · Mg · 24.30											+3 · 13 · Al · 26.98	+4 +2 · 14 · Si · 28.08	+5 +3 −3 · 15 · P · 30.97	+6 +4 −2 · 16 · S · 32.06	+7 +5 +3 +1 −1 · 17 · Cl · 35.45	18 · Ar · 39.95
4	+1 · 19 · K · 39.10	+2 · 20 · Ca · 40.08	+3 · 21 · Sc · 44.96	22 · Ti · 47.90	23 · V · 50.94	+6 +3 · 24 · Cr · 52.00	+5 +4 +2 · 25 · Mn · 54.94	+3 +2 · 26 · Fe · 55.85	+3 +2 · 27 · Co · 58.93	+3 +2 · 28 · Ni · 58.71	+2 +1 · 29 · Cu · 63.55	+2 · 30 · Zn · 65.38	+3 · 31 · Ga · 69.72	+4 +2 · 32 · Ge · 72.59	+5 +3 · 33 · As · 74.92	+6 +4 −2 · 34 · Se · 78.96	+7 +5 +3 +1 −1 · 35 · Br · 79.90	36 · Kr · 83.80
5	+1 · 37 · Rb · 85.47	+2 · 38 · Sr · 87.62	+3 · 39 · Y · 88.91	40 · Zr · 91.22	41 · Nb · 92.91	42 · Mo · 95.94	43 · Tc · 98.91	44 · Ru · 101.07	45 · Rh · 102.91	46 · Pd · 106.4	+1 · 47 · Ag · 107.87	+2 · 48 · Cd · 112.40	+3 · 49 · In · 114.82	+4 +2 · 50 · Sn · 118.69	+5 +3 · 51 · Sb · 121.75	+6 +4 −2 · 52 · Te · 127.60	+7 +5 +3 +1 −1 · 53 · I · 126.90	54 · Xe · 131.30
6	+1 · 55 · Cs · 132.91	+2 · 56 · Ba · 137.34	+3 · 57 · La · 138.91	72 · Hf · 178.49	73 · Ta · 180.95	74 · W · 183.85	75 · Re · 186.2	76 · Os · 190.2	77 · Ir · 192.22	78 · Pt · 195.09	+3 +1 · 79 · Au · 196.97	+2 +1 · 80 · Hg · 200.6	+3 +1 · 81 · Tl · 204.4	+4 +2 · 82 · Pb · 207.2	+5 +3 · 83 · Bi · 209.0	+6 +4 · 84 · Po · (210)	85 · At · (210)	86 · Rn · (222)
7	+1 · 87 · Fr · (223)	+2 · 88 · Ra · 226.0	89 · Ac · (227)	104 · Ku*	105 · Ha*													

Inner Transition Elements — f block

Lanthanum Series

58 Ce 140.12	59 Pr 140.1	60 Nd 144.24	61 Pm (147)	62 Sm 150.4	63 Eu 151.96	64 Gd 157.2	65 Tb 158.93	66 Dy 162.50	67 Ho 164.93	68 Er 167.26	69 Tm 168.93	70 Yb 173.04	71 Lu 174.97

Actinium Series

90 Th 232.0	91 Pa 231.0	92 U 238.0	93 Np 237.0	94 Pu (242)	95 Am (243)	96 Cm (247)	97 Bk (247)	98 Cf (247)	99 Es (254)	100 Fm (253)	101 Md (256)	102 No (254)	103 Lr (257)

Mass numbers of the most stable or most abundant isotopes are shown in parentheses

The elements to the right of the bold lines are called the nonmetals and the elements to the left of the bold line are called the metals.

Common oxidation numbers are given for the representative elements and some transition elements

* Kurchatovium and Hahnium are tentative names for these elements.

Atomic Weights of the Elements

(Adopted by the International Union of Pure and Applied Chemistry in 1973)

Based on $^{12}C = 12$ atomic mass units (amu)

The following values apply to elements as they exist in materials found on earth and to certain artificial elements. When used with the footnotes, they are reliable to ± 1 in the last digit, or ± 3 if that digit is in small type.